Mathe

and the Making of Modern

Ireland

Trinity College Dublin from Cromwell to the Celtic Tiger

DAVID ATTIS

Docent Press

Docent Press
Boston, Massachusetts, USA
www.docentpress.com

Docent Press publishes books in the history of mathematics and computing about interesting people and intriguing ideas. The histories are told at many levels of detail and depth that can be explored at leisure by the general reader.

Cover design by Brenda Riddell, Graphic Details.

Produced with TEX. Textbody set in Garamond with titles and captions in Bernhard Modern.

The the cover illustration of the front gate of Trinity College Dublin by James Malton is ©National Library of Ireland.

Contents

Acknowledgements   1

Telling Stories About Mathematics   4

1 The Mathematics of Conquest   27

    1.1 Puritanism and Baconian Science   31
    1.2 Mathematics in an Age of Capitalism, Colonialism, and Conquest   36
    1.3 Trinity College Dublin   38
    1.4 The Down Survey   42
    1.5 Political Arithmetic   48
    1.6 The Mathematical Principles of Natural Philosophy   57

2 Newtonianism and Natural Theology   61

    2.1 A Science Fit for Palaces and Courts   65
    2.2 The Newtonian Establishment   73
    2.3 Robinson and the Aether   76
    2.4 Materialism and Atheism   79
    2.5 Immaterialism and Religion   84
    2.6 Swift's Satire on Science   96
    2.7 We Irishmen   101

3 Enlightenment and Revolution   105

    3.1 Science, Patriotism, and Polite Culture   107
    3.2 The University and Society   115
    3.3 The Threat of French Mathematics   121
    3.4 Science and Revolution   128
    3.5 The Ideology of Ascendancy   133

4   Examining the Ascendancy   141

    4.1   The End of the Silent Sister   145
    4.2   Mathematics and Liberal Education   153
    4.3   The Age of Reform   157
    4.4   The Political Context of Lloyd's Reforms   163
    4.5   Meritocracy   166

5   Truth and Beauty   173

    5.1   The Geometry of Light   178
    5.2   Meta-Mathematics   184
    5.3   A Scientific Prophecy   189
    5.4   The Benthamites and the Coleridgians   205
    5.5   The Clerisy   209

6   Imagining Quaternions   217

    6.1   The Problem of Impossible Numbers   221
    6.2   Formalism in Mathematics   226
    6.3   The Philosophy of Pure Time   230
    6.4   Imagining Quaternions   235
    6.5   The Laws of Thought   245
    6.6   The Calculating Man   253

7   Engineering the Empire   259

    7.1   Trinity in the Service of the Empire   264
    7.2   The Science of Technology   274
    7.3   The Maxwellians   284
    7.4   Aether and Matter   290
    7.5   Theory vs. Practice   298
    7.6   Science, Industry, and Ireland   301

8   Two Revolutions   307

    8.1   The End of the Ascendancy and the Decline of Science   315
    8.2   Understanding the Atom   325
    8.3   The Hamiltonian Revival   343
    8.4   Science and the Economy   359

9   The Knowledge Economy   363

    9.1   The Rise and Fall of the Celtic Tiger   369
    9.2   Science and Economic Development   375
    9.3   Education for the New Economy   379
    9.4   Simulating Physics   388
    9.5   The Search for Symmetry   401
    9.6   The Science of Business and the Business of Science   407

Bibliography   410

Index   463

List of Figures

1	Broome Bridge, Dublin	6
1.1	The Earliest Plan of Trinity College Dublin (c. 1600)	40
1.2	Parrish Map from the Down Survey	47
1.3	Down Survey Field Book	47
1.4	Sir William Petty (1623–1687)	59
2.1	William Molyneux (1656–1698)	67
2.2	Title Page to *Christianity Not Mysterious* (1696)	80
2.3	The Flying Island of Laputa	98
3.1	The Front Gate of Trinity College Dublin	108
3.2	Dunsink Observatory near Dublin	110
3.3	Title Page to *The Ladies Diary*, (1723)	125
4.1	Title Page to *Dublin Problems* (1823)	143
4.2	Bartholomew Lloyd (1782–1837)	146
5.1	Sir William Rowan Hamilton (1805–1865)	179
5.2	Meeting of BAAS in Cork (1843)	200
6.1	Zerah Colburn (1804–1839)	218
6.2	Hamilton's Notebook on Quaternions	241
6.3	Maxwell's electromagnetic equations in quaternion form	244
6.4	George Boole (1815–1864)	247
6.5	Babbage's Difference Engine, No. 1	254
7.1	Guglielmo Marconi covering the Kingstown Regatta	260
7.2	The Atlantic Telegraph (1858)	279
7.3	The Physical Laboratory, Trinity College Dublin	283
7.4	George Francis Fitzgerald (1851–1901)	286
8.1	The General Post Office After the Easter Rising (1916)	309
8.2	Éamon de Valera (1882–1975)	312
8.3	John Joly (1857–1933)	321
8.4	E.T.S. Walton and His Particle Accelerator (1932)	338
8.5	William Rowan Hamilton Postage Stamp (1943)	345

8.6   First Meeting of the Governing Board of the School of The-
         oretical Physics at the Dublin Institute for Advanced Studies
         (November 21, 1940)   349
9.1   Trinity College Dublin in Second Life   365
9.2   IDA Ireland Advertisement featuring the Engineering Building
         at Trinity College Dublin.   372

Acknowledgements

I started the project that became this book more than twenty years ago, and over that period I have been grateful to receive support and advice from many generous people. Bruce Presley, my high school physics teacher, first introduced me to the beauty of physics, inciting a life-long passion. As an undergraduate at the University of Chicago, Bob Richards converted me to the history of science and together with Howard Stein supervised my undergraduate thesis. At Cambridge University, Simon Schaffer and Piers Bursill-Hall helped me to see new possibilities for my work. At Princeton, Norton Wise and Mike Mahoney taught me what it means to be a scholar. I first presented many of the ideas in this book to the History of Science Program Seminar at Princeton. I thank all of the participants during my years there, particularly David Brock, Jamie Cohen-Cole, Angela Creager, John Dettloff, Gerald Geison, Charles Gillispie, Ann Johnson, Jordan Kellman, Stuart McCook, Ole Molevig, and Leo Slater. Suman Seth and Eric Ash were kind enough to comment on drafts of many of the chapters. Outside of my own institution, Peter Bowler, Kathryn Olesko, and Andy Warwick have all offered me invaluable assistance. J.J. Lee graciously agreed to serve as outside reader. Daphne Wolf diligently read the entire manuscript to help me avoid any accidental transgressions on the turf of professional historians of Ireland. At Docent Press,

Mary Cronin and Scott Guthery have been incredibly supportive (and patient).

I had the privilege to spend a great deal of time in Ireland during my research, and I found that the legendary hospitality of the Irish is no myth. A great many people went out of their way to make me feel at home and assisted me in my research. They include Dick Ahlstrom, Werner Blau, Gerry Boucher, Nigel Buttimore, John Byrne, Eamonn Cahill, Michael Coey, Steven Collins, Jim Cooke, Ron Cox, Audrey Crosbie, John Donovan, Lean Doody, Ian Elliott, Orla Feely, Michael John Gorman, John Haslett, John Hegarty, Gordon Herries Davies, Roy Johnston, Kirk Junker, Patrick Kelly, Maria Lohan, P.J. Mathews, David McConnell, R.B. McDowell, James McFarland, Niall McKeith, Norman McMillan, Seamus Moran, Mary Mulvihill, Tim Murphy, Donal O'Donovan, Caroline O'Flanagan, Jane O'Mahony, Eoin O'Neill, Alicia Parsons, Mike Peardon, Patrick Prendergast, Michael Purser, James Quinn, Sinead Ryan, Brendan Scaife, Garret Scaife, Siddartha Sen, Jim Sexton, Samson Shatashvili, David Simms, David Spearman, Ken Suzuki, Brian Trench, Patrick Wayman, Trevor West, Nick Whyte, David Wilkins, and Patrick Wyse Jackson. Denis Weaire and Charles Mollan deserve special thanks both for their tremendous efforts to support the history of science in Ireland and for their unceasing generosity in supporting my own work.

The following libraries and institutions generously allowed me to use their resources: Armagh Observatory, the British Library, Cambridge University Library, Dunsink Observatory, the Library of Congress, the National Library of Ireland, the National Oceanic and Atmospheric Administration, the Public Record Office of Northern Ireland, Queen's University Belfast Library, Princeton University Library, the Royal Dublin Society, the Royal Irish Academy, Stanford University, Trinity College Dublin, the University of California at Berkeley, the University of Texas-Austin, and the University of Chicago.

I would like to thank everyone who ever gave me money to pursue my research, particularly the University of Chicago, Princeton University, the National Science Foundation, who funded both a three-year Graduate Research Fellowship and a one year Doctoral Dissertation Improvement Grant, and the Social Science Research Council, who granted me an International Dissertation Research Fellowship.

Finally I must thank those who offered essential moral and emotional support throughout this very long project: my family, Diane, Lucy, and James, for their patience, my parents, Bari and Gerry Attis, for their long-term support (especially my mother who read and corrected the entire manuscript), my mother-in-law, Catherine Kelleher, for her ongoing academic advice, and my brother-in-law, Michael Kelleher, for his help with the illustrations, and, finally, my cats, Luke and Leia, who sat patiently on my lap (and sometimes on my keyboard) as I worked.

Telling Stories About Mathematics

It's an often-told story in the history of mathematics, re-enacted every year in Dublin. It was the sixteenth of October, 1843, and Sir William Rowan Hamilton, Royal Astronomer of Ireland, was making the five-mile journey from Dunsink Observatory down to Dublin with his wife. As they walked along the Royal Canal, Sir William barely heard a word she said. Deep in thought, he remained preoccupied with a mathematical problem that had troubled him for many years, namely how to create an algebraic representation of lines in three-dimensional space. Triplets, he called them. Every morning for the past few weeks he had been greeted at the breakfast table with his sons' cries of "Well, Papa, can you *multiply* triplets?" And every morning Sir William would sigh, shake his head, and respond, "No, I can only add and subtract them."[1]

As they reached Broome Bridge, Sir William stopped suddenly. "I then and there felt the galvanic current of thought *close*," he recalled fifteen years later, "and the sparks which fell from it were *the fundamental equations between i, j, k; exactly such* as I have used them ever since."[2] Sir William took out the notebook he kept in his breast pocket and immediately jotted down the equations. He

---

[1] W.R. Hamilton to Archibald H. Hamilton (August 5, 1865) in R.P. Graves, *Life of Sir William Rowan Hamilton* (Dublin: Hodges, Figgis, 1882-89), Vol. 2, p. 434.

[2] W.R. Hamilton to P.G. Tait (October 15, 1858) in Graves 2: 435.

wrote constantly on any and every available surface; he was even known to write on the shell of his egg at breakfast. Nor could he resist the "unphilosophical impulse," as he described it, to immortalize this crucial moment by carving the fundamental equations into the stone of the bridge with his penknife. His triplets had become quaternions, four-component rather than three-component numbers, and at that moment, he explained, "I felt that it might be worth my while to expend the labour of at least ten (or it might be fifteen) years to come... I felt a *problem* to have been at that moment *solved*—an intellectual want relieved—which had *haunted* me for at least *fifteen years before*."[3] (Sir William was rather too fond of textual emphasis.)

Figure 1: Broome Bridge, Dublin, where Sir William Rowan Hamilton inscribed his quaternion equations on October 16, 1843. Illustration by Michael Kelleher

---

[3]W.R. Hamilton to P.G. Tait (October 15, 1858) in Graves 2: 436.

Even readers who have never heard this particular story before will likely find it familiar. An eccentric genius, obsessed with solving a complicated logical puzzle, barely notices the world around him. Years of intellectual struggle culminate in a moment of transcendence. Like a switch being flipped, his mind finally lands on the solution to the problem, and, once found, the answer appears self-evident. A brilliant mind sees past the shadows of everyday life and glimpses—even if only briefly or partially—a world of eternal truth. More than a decade of work and hundreds of pages of notebooks full of failed attempts are collapsed into a simple set of equations. The symbols on Broome Bridge (which has since been renamed Hamilton Bridge) appear to speak for themselves. They represented abstract truth made concrete (almost literally). In a world of uncertainty, conflict, and doubt (and Ireland in the 1840s was nothing if not this) they offered the possibility of purity, simplicity, and consensus. Even those listeners who have no idea what quaternions are or why they might be important can still walk away from the story with lessons about the nature of mathematical truth.

And yet there is an interesting tension lurking within Hamilton's story. If mathematical truth is timeless and universal, why should it matter when and where quaternions were discovered? If the equations stand on their own, why do they need a story? Hamilton's narrative emphasizes a very specific place and a very specific time—Broome Bridge in Dublin on October 16, 1843—even though these facts seem to have nothing to do with the truth of the equations. If anything, emphasizing these details could only undermine the validity of the mathematics. If the solution depends on when and where it was discovered, how could it be objective and eternal?

While mathematics as a body of knowledge aspires to universality, mathematics as a human activity has always been limited to certain places, times, and social groups. There is, in fact, a geogra-

phy of mathematics. Just as one can map the spread of the English
language, the use of bronze tools, or changing styles of dress, it is
possible to trace the spread of the calculus, imaginary numbers, or
the theory of relativity. This map of mathematical practice mirrors
maps of population, wealth, and even religion. Hamilton's creation
of quaternions, for example, depended on the existence in Dublin of
a university (Trinity College Dublin) that taught high level mathematics as well as an observatory that allowed its director plenty
of time for research. And that university in turn depended on the
wealth of the Anglican Church of Ireland and the political and cultural union between Great Britain and Ireland. Hamilton's story
(perhaps unwittingly) implies there is a connection between mathematics and Irish history.

In fact, in each period of Irish history, mathematical equations
represented not only solutions to technical problems, they also
represented a vision of what Ireland could and should be. We
don't often think of mathematicians as playing an important role
in building a nation, but it turns out that mathematics—no less
than poetry, drama, and music—has also been a medium for imagining and creating Ireland. In the seventeenth century, William
Petty's mathematics (backed by the force of Cromwell's army) was
essential to transferring the vast majority of land from Catholics
to Protestants. In the eighteenth century, Berkeley highlighted the
logical flaws at the heart of the calculus to destroy the arguments
of deists and freethinkers that threatened the established Church.
In the nineteenth century, Hamilton was convinced that his international reputation in mathematics would prove to the English
that the Irish were not intellectually inferior, thereby strengthening
Ireland's position within the Union. In the twentieth century, Éamon de Valera created the Dublin Institute for Advanced Studies
to bolster the reputation of the fledgling Republic and to keep Ireland's top mathematicians from emigrating. Today, mathematics

and data analytics have defined a vision of Ireland as a knowledge economy rooted in global networks of information.

Mathematics did not simply live in textbooks, examination halls, scholarly journals, and the odd bridge; it also appeared in sermons, political pamphlets, parliamentary debates, and government white papers. Mathematics was widely proclaimed to be the standard by which all other forms of knowledge should be judged. It exemplified truths that all rational humans had to accept regardless of nationality, religion, or political party (a very powerful idea in a country divided by nationalism, religion, and politics). And yet, precisely because mathematics represented the compelling authority of reason, it was enlisted to support a range of causes. It drew its power from its apparent objectivity and independence, and yet in exercising that power, it became embroiled in a variety of disputes. Far from being a realm of objectivity, purity, and consensus, mathematics in Ireland became a battleground.

Mathematicians and non-mathematicians alike spun a dizzying array of stories about the relevance and implications of mathematics. For some it represented access to truths that must ultimately be divine in nature, while for others it was seen as mechanical and amoral, extinguishing the spiritual for the sake of mere calculation. For some excellence in mathematics demonstrated innate intelligence that qualified students for roles in the professions, the civil service, and the political classes, while for others it was an arbitrary sorting device that recast inherited privilege as meritocracy. Some saw it as a venue where warring factions could find consensus, while others saw it as a tool of oppression by the ruling political party. Some argued that it is the ultimate source of technological progress and economic development, while others believed it was a weak attempt by useless professors to claim credit for the achievements of practical men. Over the course of four centuries, debates about the uses and the meaning of mathematics were also debates about who gets to control education, who is allowed to enter the professions, who is qualified to govern, and how society should be regulated. Despite popular stereotypes, mathematics was as much a way to engage with the world as it was way to retreat from it. It was not simply a language to describe reality; it was a tool to reshape it.

Mathematicians may well object that none of these debates have anything to do with the truth of mathematics. These people may be talking about mathematics, one might argue, but they aren't doing mathematics. And yet, these debates represent the very reasons that people pursued mathematics. The reason that we allow mathematical puzzle solvers to run our universities, create our national examinations, and regulate our economy is because they claim to be doing more than merely solving puzzles. Contrary to popular belief, mathematicians do not spend all of their time with their heads in the clouds. They seek support for their work—not only monetary support but also social, intellectual, and cultural support for the pursuit of mathematics. To find this support, mathematicians (and their allies) must tell stories, for if they were to stop talking about it, mathematics (or at least the study of mathematics) would cease to exist.

But surely, our mathematical friend might object, despite disagreements about the philosophy of mathematics or the social implications of mathematics, mathematicians at least agree on the truth of specific propositions. In every age, mathematicians have claimed that there are no controversies in mathematics: all rational people must agree on the truth of mathematical propositions because they flow necessarily from a set of indubitable assumptions.

And yet throughout history, mathematicians have in fact disagreed about what constitutes acceptable mathematical practice. Berkeley attacked Newton's use of infinitesimal quantities, questioning the logical foundations of the calculus. Others argued that imaginary numbers or negative numbers were meaningless. Hamilton and his followers forcefully lobbied for quaternions as the natural language of mathematical physics, but their arguments were sneeringly dismissed by the proponents of the vector calculus. Every mathematical practice was carefully scrutinized as to whether it matched the ideal of objective and universal truth, but even professional mathematicians sometimes differed in their conclusions.

The recognition that professional mathematicians held different opinions about what is true in mathematics is an uncomfortable one for most people. For mathematics to be true, most would agree, it is essential that its validity be independent of the beliefs of its practitioners. One need not share Hamilton's philosophy of mathematics to use quaternions (few did). Nor is it necessary to understand the social and political climate of nineteenth-century Ireland to be able to calculate with them. Most histories of mathematics, therefore, focus on consensus, shared belief, and the gradual unveiling of universal truth. This history is different. It explores the contested and contingent aspects of mathematical activity as a means to understand not the logical development of mathematical truth but the social and political development of Ireland.

The equations that Hamilton scratched into Broome Bridge soon faded. It turns out that stone is less permanent than a mathematical idea. Even a plaque placed on the bridge in the 1950s is now barely legible and covered with graffiti of a decidedly non-mathematical nature. While a group of mathematicians makes an annual pilgrimage to the bridge every year on October 16, the original meaning of Hamilton's quaternions has been forgotten by all but a few historians. Despite their impressive intellects, mathematicians are very good at forgetting. In building a new edifice of mathematical truth, they prefer to abandon the scaffolding lest it obscure the beauty of the new creation. They forget the old controversies, the outdated justifications, and the personal motivations—

the stuff of history. These are typically seen as obstacles to mathematical progress and diversions from the truth (though entertaining, perhaps, as human interest stories).

The stories that were originally used to justify Hamilton's mathematics have long since been abandoned as new stories have been adopted. Quaternions are still 'true', but they are not true in the same way or for the same reasons that Hamilton believed they were true. The mathematical equations persist, like the stone bridge, while the stories constantly shift, like the water flowing beneath. But we lose something if we dismiss these stories as simply "water under the bridge". To make progress in mathematics it is essential to abandon the old stories, but the price is that we then also forget how mathematics and mathematicians were engaged in transforming the world in which they lived. If we pull mathematics out of history, we lose sight of how mathematicians helped to build modern society, and then we can never truly understand the modern world.

Mathematics has always played an important role in history but never more so than today at the beginning of the twenty-first century. Mathematics is now built into the fabric of our everyday lives. It is not only the language through which we understand the natural universe and the social sciences, but—because of the internet—it is also the medium through which we express ourselves as consumers, as creators, as citizens, and even as 'friends' on social networks. We live in a world saturated with numbers. Scientists, engineers, pollsters, accountants, and administrators of all kinds make decisions through the medium of numbers. We have created a virtual mathematical world that is increasingly difficult to separate from the real world. Its implications are very real, and yet invisible to most.

Almost three hundred years ago, Jonathan Swift captured the way in which the abstract world of mathematics intersects with the concrete world of politics and economics. In *Gulliver's Travels*, the flying island of Laputa uses an enormous magnet to stay aloft and to move around the kingdom of Balnibarbi over which it rules. The Laputans are obsessed with pure mathematics. Servants follow them everywhere, flapping their ears and mouths to rouse them from their intense speculations. The tailor sent to outfit Gulliver in more suitable clothing takes his measurements with a quadrant, sketching a design with a ruler and compass (though the new clothes don't fit due to a mistake in the calculation).

But despite having their heads literally in the clouds, the mathematicians of Laputa exert a very real control over the people of Balnibarbi. As Laputa travels over the kingdom below, the people on the ground send up petitions and sometimes wine and food in baskets that are lowered down. If the people rebel, the king keeps the flying island over a town, blocking out the sun and rain until they relent. In extreme circumstances the king can even drop the island directly on their heads. Swift not only parodied the absurd extremes of abstract reasoning, but he also highlighted the fact that the apparently pure and disinterested pursuit of truth always relies upon (and in turn supports) a particular political and economic structure.

The great intellectual creations of mathematics do not exist only in some separate Platonic realm of forms. They are also part of the very fabric of the social and political world in which we live. And yet while most people have strong opinions about the political implications of religion, literature, or music, few feel qualified to comment on mathematics. Stories like Hamilton's have convinced us that only a special class of genius has access to mathematical truth. The algorithms that structure our lives today are not subject to the kinds of criticism and analysis to which popular writers, religious institutions, or political parties have always been. We may not all be mathematicians, but we all tell stories about mathematics. And once we recognize the way that mathematics shapes our lives, we can begin to see that the stories we tell about mathematics have real power.

The Story in Brief

The Mathematics of Conquest

In the history of science, the sixteenth and seventeenth centuries are known for the scientific revolution—the transformation of astronomy, mathematics, mechanics, and chemistry evident in the works of Copernicus, Kepler, Galileo, and Boyle, culminating in the work of Isaac Newton. In Irish history, the sixteenth and seventeenth centuries are the period of the English conquest. This was the moment when power in Ireland shifted from Catholics of Gaelic and Anglo-Norman descent to Protestants of English and Scottish descent. The scientific story is often seen as one of progress and enlightenment, the quest to find the true theory of nature and the correct method for scientific investigation. The Irish story, on the other hand, is one of sectarian hatred and unabashed colonialism. The life and work of William Petty, however, illustrate the intersection of these two stories. As both a founding member of the Royal Society of London and a member of Cromwell's army, professor of anatomy at Oxford and organizer of the Down Survey, originator of political economy and one of Ireland's wealthiest planters, Petty played key roles in both the development of modern science and the conquest of Ireland.

Far from the disinterested pursuit of truth, science during the early modern period was more commonly a means to exercise power. Mathematics grew in popularity in large part because of its role in navigation, surveying, and military engineering, and the men who brought the new science to Ireland were all members of Cromwell's army. The first professor of mathematics at Trinity College Dublin, in fact, was a surveyor employed to train Cromwell's troops in order to expedite the expropriation of Catholic lands. Petty and many of his fellow officers in Ireland saw science and mathematics as important tools for the extension of English power and English civilization. They valued them not simply because they would help them understand nature and nature's laws but also because they could use them to control nature and to impose law on Ireland.

Newtonianism and Natural Theology

As the children of the Cromwellian adventurers became landed Irish gentlemen, their tastes shifted from practical science to the pursuit of rigorous knowledge. Nobles, bishops, and administration officials actively debated questions of natural philosophy and mathematics along with theology, politics, and literature. In fact, all were intimately related. The Civil War and the Glorious Revolution had not only disrupted political, social, and religious life, they made it clear that the old truths of Aristotelian philosophy, respect for the established Church, and the divine right of kings no longer had the power to contain the disruptive forces of new ideas. The search for social order became a search for methods for generating consensus, and because mathematics was widely acknowledged as the model for proper reasoning, mathematics became the most important venue in the battle for the hearts and minds (and souls) of the English and Irish ruling class.

The magnitude of Newton's achievement obscures the fact that Newton's theories were actually quite controversial when they were first introduced. In fact, Newton's work was full of inconsistencies, problematic assumptions, and dangerous implications. The two figures who most clearly demonstrated this were both Irish—John Toland and George Berkeley. Toland was the first and the most dangerous of the free thinkers—a subversive group of deists who questioned the validity of established religion. Toland argued that

Newton's science and Locke's philosophy had unmasked revealed religion as a hoax and that the universal laws of science demand a world where all men are equal before the law. The doctrinal distinctions that Toland attacked were the very foundation of the penal laws that supported the Anglican establishment. His book *Christianity Not Mysterious* was considered so dangerous in Ireland that it was ordered to be burned by the common hangman.

While Jonathan Swift satirized the new science in literature, defenders of the faith like Berkeley rushed in to build a new philosophical bulwark to protect the establishment. Rather than defend Newtonian science, Berkeley attempted to purge it of any ideas that he believed had dangerous implications. He attacked the notion of infinitesimal quantities (the foundation of Newton's calculus), the belief that unobservable atoms in motion are the cause of our perceptions, and even the very idea that matter exists. Berkeley's philosophy would become the touchstone for a long tradition of Irish idealism from William Rowan Hamilton to George Francis Fitzgerald to William Butler Yeats.

Enlightenment and Revolution

The 1780s witnessed the social, political, and economic resurgence of Dublin. With increasing commerce, a social life second only to London, and a newfound sense of independence from the British parliament, Dublin's elite hoped to finally take their place among the enlightened capitals of Europe. Along with Dublin's great Georgian buildings, they also erected the observatories at Dunsink and Armagh and founded the Royal Irish Academy. The patriots believed that science would lead to enlightenment, prosperity, and international respect. Trinity College Dublin, however, remained essentially an ecclesiastical institution, inward looking and tradition bound. The Newtonian natural philosophy that had stirred so much controversy on its introduction was now sanctioned by nearly a century of works designed to show its harmony with Anglican theology. The fellows saw their role as the preservation of this 'Holy Alliance' between Newtonianism and Anglicanism rather than the extension of the bounds of scientific knowledge. Trinity attempted to adapt to the new cosmopolitan and improving spirit by increasing the amount of science in the curriculum and by appointing John Brinkley as professor of astronomy and the first Royal Astronomer of Ireland. Brinkley, an Englishmen and one of the first people in the British Isles to master the sophisticated French calculus, introduced the latest mathematical techniques to Trinity and made Dunsink Observatory world famous

There was a danger lurking in Brinkley's work, however. While he saw the new French science as a further support for Anglican theology, the French themselves associated their science with arguments for revolution and democracy. In the 1790s Irish radicals began to echo the claims of French revolutionaries, proclaiming that science would lead to the end of oppressive political and religious establishments. The rebellion of 1798 put an end to the enlightened dreams of the patriot movement. Trinity's model of scholarship in the service of the Anglican Church survived, while the patriot's hope that science would lead to industrial and cultural revival foundered in an atmosphere of renewed sectarianism.

Examining the Ascendancy

In 1801, the Act of Union dissolved the Irish parliament, ending Ireland's brief period of relative independence from Great Britain. While the union protected Irish Protestants from the growing demands of Irish Catholics, though, it exposed them to the demands of newly elected reformers and radicals in the British Parliament. The Church of Ireland and Trinity College Dublin stood out as the most egregious examples of minority privilege in the British Isles, and attempts were made to disestablish the Church and to take Trinity's governance away from the clergy. At the same time, new secular institutions such as the University of London and the Queen's Colleges in Ireland threatened the Anglican universities' traditional monopoly on higher education.

Trinity implemented a series of internal reforms in the 1820s and 1830s in an attempt to avoid parliamentary interference and to compete with the new trends in education. Professors were encouraged to undertake original research, written examinations were instituted, and students were offered honors courses in mathematics. Now the best students would study the latest mathematics, and fellows would be chosen for their performance in an examination dominated by mathematics. By proving their scientific mettle, the administration at Trinity hoped to convince reformers that Trinity stood for merit and progress rather than privilege and backwardness. The fellows of Trinity College promoted mathematics not as training for scientists (the word itself was not coined until 1833) but rather as the foundation of a liberal education.

Irish Protestants defended their ancient privileges by demonstrating their intellectual superiority. When the civil service opened itself to public competition in mid-century, for example, Trinity students outperformed even those from Oxford and Cambridge, proving, they believed, that they had earned the right to govern. In an age of Catholic emancipation, parliamentary reform, and popular education, Trinity felt that rigorous scientific standards for both students and professors offered the best strategy for strengthening the precarious position of the Irish Protestant university and the Protestant Ascendancy it existed to educate. Trinity's reforms worked. Not only did they successfully defend their intellectual aristocracy, they also launched a generation of brilliant mathematicians. James MacCullagh, Humphrey Lloyd, and William Rowan Hamilton were the first to fill the new research-oriented professorships and to train students for the new honors examinations.

Their work in mathematics and mathematical physics would bring Trinity an international reputation for science.

Truth and Beauty

The first major triumph of Trinity mathematics was the discovery of conical refraction in 1832. William Rowan Hamilton found through a complicated calculation that a ray of light could be refracted into cone of light within certain kinds of crystals, and his colleague Humphrey Lloyd confirmed the effect experimentally. The phenomenon itself was of no practical importance, but Hamilton's ability to predict a completely unexpected experimental result purely on the basis of mathematical theory was hailed as a triumph (and led to his knighthood). Hamilton's achievement not only supported the still-unproven wave theory of light, it also appeared to validate a philosophy of science in which mathematical hypotheses lead to testable empirical predictions. While such a hypothetico-deductive approach to science is now common, the opponents of the wave theory of light, on the other hand, railed against the use of the hypothetical ether that lay at the foundation of the wave theory. They believed that science should follow slow and careful induction from experimental facts while avoiding complicated mathematical hypotheses.

In addition to their methodological and philosophical differences, the proponents of the particle theory of light also differed from the wave theorists on social and political issues. David Brewster and Henry Brougham were Scots who made their careers outside of the university. Brougham was active in the promotion of non-denominational schools, popular education, legal reform, and Catholic Emancipation. They opposed all that the Anglican universities stood for and exerted great effort to criticize, reform, and replace them. The supporters of the wave theory, on the other hand, were all professors at Cambridge and Trinity College Dublin who saw their mathematical approach as proof of the priority of ideas over bare facts (and the priority of an intellectual elite over the masses).

Hamilton himself was less committed to the wave theory of light than to an even more abstract and idealized form of mathematical physics. He wrote to his friend Wordsworth that he always attempted to infuse "the spirit of poetry" into his science, and his work in geometrical optics and dynamics (which resulted in the famous Hamiltonian equation) have as their primary goal the search for beauty in the laws of nature. For Hamilton, the possibility of a predictive mathematical science demonstrated the harmony of the mental world and the physical world, and that harmony, he felt, is proof that both were created by the same omnipotent being. The imaginative creation of beauty in mathematical physics, he believed, is ultimately a spiritual experience, and an experience that must play a central role in an Anglican education.

Imagining Quaternions

For the rest of his career, Hamilton remained concerned about the philosophical foundations of mathematics. He desperately wanted to show that analytical mathematics (the powerful set of tools based on the calculus that had transformed physics) was just as rigorously true as Euclidean geometry. No one would deny the rigor of geometrical reasoning, but many dismissed analysis as the mechanical manipulation of meaningless symbols. Hamilton created quaternions as an analytical tool to represent lines in three-dimensional space, forging a link between geometry and analysis, and, he hoped, creating a new language for physics.

While Hamilton believed that quaternions represented the culmination of his idealist philosophy, others interpreted them in a very different way. Quaternions were the first non-commutative algebra, that is, for quaternions $A \times B$ is not equal to $B \times A$. In a sense, quaternions were the exception that proved the rule, bringing such abstract laws of algebra to the attention of mathematicians for the first time. They led mathematicians to ask themselves about the rules that any algebraic system must follow, and they demonstrated that it is possible to invent new forms of algebra with different rules. In the process they transformed algebra into the study of any self-consistent system of arbitrary rules. After quaternions, mathematics was no longer fundamentally about counting; it had become the study of formal relationships. Mathematics, therefore, could describe logic as well as physics, and George Boole, working in Cork, created the first mathematical system of logic, reducing all thought to mathematics in the hopes of finding a method to resolve the religious disputes that divided Ireland.

Calculation in the mid-nineteenth century had much broader ramifications than simply a philosophy of mathematics. Charles Babbage had shown that it was possible to design a machine that could calculate, and political economists had begun to describe human beings as mechanical calculators of self-interest. Just as Berkeley feared that Newton's calculus would weaken rigorous thinking and threaten faith in God, Hamilton feared that a belief that algebra is a mechanical process of combining meaningless symbols would destroy the spiritual experience of mathematics and degrade our very sense of humanity to animalistic responses to physical stimuli. Quaternions for Hamilton represented the power of fundamental ideas derived from the mind rather than the visible world, and he hoped they would not only be practically important to the study of science but also metaphysically significant.

Engineering the Empire

Trinity mathematics was not just about creating ideal worlds. In the 1840s, Trinity opened an engineering school, one of the first in the British Isles. Humphrey Lloyd, who helped to found the school, and George Francis Fitzgerald, who directed it at the end of the century, believed that the connection between science and industry would justify Trinity's privileged place in society. While the world might dismiss Trinity's mathematicians as irrelevant in an age of industry, Trinity would train the engineers who would run the empire. Trinity graduates worked in Canada, India, Australia, and the rest of the world, laying railroad tracks and telegraph lines, building bridges and roads. Their mathematical training at Trinity formed the fundamental tool through which they could extend British power around the world.

Fitzgerald's research, however, was not in engineering but in mathematical and experimental physics. He was one of the key figures who transformed Maxwell's groundbreaking theory of electromagnetism into a comprehensive mathematical theory. He and his colleagues not only created a new foundation for mathematical physics, but they also applied the new science to the telegraph networks that connected the British Empire and played an important role in the development of wireless telegraphy or radio. The man who first commercialized radio, however, was not an academic scientist, but a young Italian inventor, Guglielmo Marconi. Marconi (whose wife was a member of the Jameson whiskey family) performed many of his early tests in Ireland and even collaborated with Fitzgerald. But Marconi's success threatened Trinity's claim that mathematical science was essential to technological progress. The great inventions that transformed the world in the four decades from 1870 to 1910—the telephone, the internal combustion engine, the phonograph, moving pictures, radio, the electric light bulb, the steam turbine, the automobile, and the airplane—were not invented by university-trained scientists or engineers but by self-taught tinkerers and businessmen. Fitzgerald and his academic colleagues struggled to make the case that university-based research was essential for technical and industrial progress. In Ireland in particular, science failed to lead to industry, and technical education had little effect on the Irish economy. At the end of the nineteenth century, Fitzgerald's vision of a prosperous and technologically advanced Ireland collapsed, and Ireland remained an impoverished, agricultural economy well into the twentieth century.

Two Revolutions

In the first three decades of the twentieth century, a revolution occurred in physics. The new theories of relativity and quantum mechanics overthrew many of the fundamental assumptions of Maxwell's theory of electromagnetism and Newton's theory of gravitation. The ether that had served as the physical, mathematical, and philosophical basis for natural philosophy since the age of Newton was no more, leaving a generation of physicists in search of new foundations. Over this same period, a revolution occurred in Ireland. The Anglican Church of Ireland had been disestablished, land reform laws transferred most of the land in Ireland from Protestant landholders to Catholic farmers, and Ireland finally gained independence from the United Kingdom. After four hundred years, protestant dominance was decisively over (at least in the southern twenty-six counties).

The end of the Ascendancy in Ireland coincided with the end of classical physics and the (temporary) end of the mathematical tradition at Trinity College Dublin. A new generation took over the reins of power in Ireland, and a new generation took over leadership in physics. Trinity scientists who came of age in the last quarter of the nineteenth century refused to accept the new quantum mechanics just as they refused to accept the new political environment. Trinity did not disappear, but as the Ascendancy withered, so did the scientific tradition at Trinity. The best scientists left and went on to make important contributions at other universities in Ireland and throughout the Commonwealth (including

Ireland's only Nobel Prize in science for E.T.S. Walton's research at the Cavendish laboratory in Cambridge). Those who stayed (or returned, like Walton) had no time for research in the new theories. At a time when universities in England, Germany, and the United States were investing in state-of-the-art research laboratories, post-graduate research positions, and partnerships with industry and government, Trinity's small staff struggled simply to meet their weekly lecture schedules. The new University College Dublin, aligned with Catholic and nationalist interests, rapidly developed its own strong tradition in science and mathematics.

The man who did more than any other individual to shape the newly independent Ireland also played an important role in bringing the new physics to Ireland. Éamon de Valera participated in the Easter Rising, led Ireland during the War of Independence, and served as Prime Minister and later President until the age of 90. De Valera had taught mathematics before the Rising, and he never lost his passion for mathematics, especially the work of Sir William Rowan Hamilton. De Valera's personal interest in mathematical physics was the primary reason for the creation of the Dublin Institute for Advanced Studies (DIAS) in 1939, and de Valera himself recruited Austrian Nobel Laureate Erwin Schrödinger as the first director of its School of Theoretical Physics. DIAS quickly became an international center for mathematics and theoretical physics, drawing top physicists and mathematicians from around the world and helping to bring together the mathematicians from all of the Irish universities, despite their continuing political and religious differences.

Like his intellectual idol Hamilton, de Valera believed that mathematical science should be pure, independent, and above politics (even when it was being used for political purposes). They both saw mathematics as a form of high culture, unsullied by practical or commercial concerns. The elaborate mathematical theories of fundamental particle physics discussed at the DIAS seminars may have improved Ireland's intellectual reputation, but they did nothing to help the masses of the poor and unemployed. While other major economies experienced a technology-fueled post-war economic boom, Ireland was stuck with flat household incomes, high unemployment, massive emigration, and cultural stagnation. Once again, science in Ireland remained pure and elitist and failed to deliver technological and economic benefits.

The Knowledge Economy

The end of the twentieth century witnessed an economic transformation in Ireland as stunning and unprecedented in a positive sense as the Famine was in a negative sense. The Irish were among the most aggressive (and successful) at capitalizing on the global mobility of information, capital, and goods beginning in the 1980s and 1990s. The Celtic Tiger economy was built on investments from high tech multinational corporations, evolving rapidly from electronics and pharmaceutical production to business services and financial engineering as banks, hedge funds, and internet companies made Dublin their home. Ireland went from the poorest country in Europe to one of the richest in less than a generation.

Ireland's success depended on globalization and digitization, forces that not only changed the foundations of the economy but also the foundations of science. From the 1990s to the 2010s, the world became digital—not only video games, movies, and music, but also healthcare, business, engineering, physics, and astronomy were all transformed as computer processing power improved exponentially, the cost of data storage plummeted, low cost sensors became pervasive, and global networks emerged to link all of these devices together. Simulation and computation became critical tools as computers were used to model everything from weather to brain function to financial markets to supply chains. The resulting data deluge placed a new priority on the mathematical sciences. Mathematics (in the form of "big data" and business analytics) and physics (in the form of nanoscience) became the new hope for Ireland's future.

Trinity College and all of the other Irish universities have been transformed, with industrial research labs and startup companies dotting their campuses. Mathematics students now go on to software companies, consulting firms, and investment banks. As its economic importance grew, the scientific enterprise came to look more and more like the multinational corporate enterprises that supported the Celtic Tiger economy. The large global teams, expensive facilities, multimillion euro budgets, and corporate partnerships have changed the role of the scientist. The traditional roles of the mathematician as the keeper of truth, the creator of beauty, the defender of the faith, and the shaper of minds have receded in the face of the new image of the mathematician as master of the information economy. Science and mathematics are now seen as engines of economic growth, central to the new story of Ireland. Perhaps for the first time in Irish history mathematicians can be the heroes.

Chapter 1

The Mathematics of Conquest

A Blank Slate

Dr. William Petty arrived at the port of Waterford on September 11, 1652. Having abdicated prestigious positions at Oxford University and London's Gresham College, he was now Physician-General to the Parliamentary Army, "the first that ever totally subdued Ireland,"[1] as Petty proudly put it. The scene on his arrival was one of unprecedented devastation. Ireland had been at war for over ten years, since the Irish Catholics had risen up against the Protestant settlers of Ulster on October 22, 1641 in a massacre that left a permanent impression on the Irish Protestant mind. The balance of power had shifted back and forth until Oliver Cromwell arrived with his New Model Army in August of 1649 on what he believed to be a divine mission to pacify Ireland forever. At his first engagement in Drogheda (near Dublin), Cromwell's troops slaughtered even the civilians as "a righteous judgement of God upon these barbarous wretches who have imbrued their hands with so much

---

[1] William Petty, *Reflections Upon Some Persons and Things in Ireland* (London, 1660), p. 12.

innocent blood."[2] They then proceeded to lay waste to large areas
of the countryside. The statistics that Petty would later calculate indicate the extent of the carnage. Out of a total population
that he estimated to be 1,466,000 in 1641 when the rebellion first
broke out, "about 504,000 of the Irish perished, and were wasted
by the Sword, Plague, Famine, Hardship and Banishment."[3] One
Irish poet described it as "*an cogadh do chríochnaigh Éire*," the war
that finished Ireland.[4]

For Petty, however, as for other members of the English Commonwealth, Cromwell's conquest marked not an end but a beginning. The invaders described Ireland as a 'blank slate,' a '*tabula
rasa*,' a 'white paper' on which they could create a new kingdom
on Protestant and English principles.[5] Ireland was not only the
perfect place to realize the millenarian dreams of the new Puritan
regime; it also provided an ideal way to cover parliament's enormous debts. Both the Civil War and the reconquest of Ireland had
been paid for with promises of Irish land. All Catholics who could
not prove their 'constant good affection' to the parliamentary cause
had their land confiscated and were relocated to Connaught, the
westernmost of Ireland's four provinces. (Cromwell had famously
promised to drive the Catholics "To Hell or Connaught!") While
Catholics owned 60% of the land in Ireland before the rebellion, by
1660 their share had fallen to a mere 9%.[6] In an important sense,
this was the moment when power in Ireland shifted from Catholics

---

[2] Oliver Cromwell, quoted in Patrick J. Corish, "The Cromwellian Conquest, 1649-53,"
pp. 336-52 in T.W. Moody, F.X. Martin and F.J. Byrne (eds.), *A New History of Ireland,
Vol. III: Early Modern Ireland, 1534-1691* (Oxford: Clarendon, 1976), p. 340.

[3] William Petty, "The Political Anatomy of Ireland [written 1672, published 1691]," pp.
135-223 in Charles Henry Hull (ed.), *The Economic Writings of Sir William Petty* (Fairfield,
NJ: Augustus M. Kelley, 1986), pp. 149-50.

[4] Seán Ó Conaill, "Tuireamh na hÉireann," quoted in Patrick J. Corish, "The Cromwellian
Regime, 1650-60," pp. 353-86 in *A New History of Ireland*, p. 357.

[5] See T.C. Barnard, *Cromwellian Ireland* (London: Oxford University Press, 1975), pp.
14, 268. Petty himself referred to Ireland as "a white paper" in his "Treatise on Taxation
[1662]," in *Economic Writings*, p. 9.

[6] R.F. Foster, *Modern Ireland, 1600-1972* (New York: Penguin Books, 1988), pp. 115-6.

of Gaelic and Anglo-Norman descent to Protestants of English and Scottish descent. It was the occasion for what one historian has called "the most catastrophic land-confiscation and social upheaval in Irish history."[7] And Dr. Petty would personally play a key role in this process.

In the history of science, the sixteenth and seventeenth centuries are known for the scientific revolution—the transformation of astronomy, mathematics, mechanics, chemistry, and scientific method evident in the works of Copernicus, Bacon, Kepler, Galileo, and Boyle, and culminating in the work of Isaac Newton. This was also the period that saw the organization of the first scientific societies and the publication of the first scientific journals. Traditionally, it is seen as the origin of modern science. In Irish history, the sixteenth and seventeenth centuries are the period of the English conquest, with Henry VIII's resumption of direct rule in 1543, the plantations of Munster and Ulster, and Cromwell's campaign, culminating in the defeat of James II and his Catholic army by William of Orange and his Protestant army at the Battle of the Boyne in 1690. As the period in which English and Scottish Protestants first decisively gained power over the native Irish Catholics, it is often seen as the origin of modern Ireland.

On the surface these two histories appear to have little in common. The scientific story is often seen as one of progress and enlightenment, the quest to find the true theory of nature and the correct method for scientific investigation.[8] The Irish story, on the other hand, is one of sectarian hatred and unabashed colonialism. The life and work of William Petty, however, illustrate the intersection of these two stories. As both a founding member of the Royal Society of London and a member of Cromwell's army, professor of

---

[7]T.W. Moody, "Early Modern Ireland," pp. xxxix-lxiii in *A New History of Ireland*, p. xliv.

[8]See David C. Lindberg, "Conceptions of the Scientific Revolution from Bacon to Butterfield: A Preliminary Sketch," pp. 1-26 in David C. Lindberg and Robert S. Westman, *Reappraisals of the Scientific Revolution* (Cambridge: Cambridge University Press, 1991).

anatomy at Oxford and organizer of the Down Survey, creator of political economy and one of Ireland's wealthiest planters, Petty played key roles in both the development of modern science and the conquest of Ireland.

Far from the disinterested pursuit of truth, science during the early modern period was more commonly a means to exercise power. Two characteristics that marked the new science—the use of mathematics and an emphasis on the practical application of knowledge—were also intimately related to the conquest of Ireland. Mathematics, for example, grew in popularity in large part because of its role in navigation, surveying, and military engineering, all essential to the work of conquest. Meanwhile, the new Baconian or practical sciences were promoted particularly by the Puritans of the English Commonwealth as a means to regain man's dominion over nature lost at the Fall. Petty and many of his fellow officers in Ireland saw the new science and mathematics as important tools for the extension of English power and English civilization, and Ireland, they felt, offered an ideal laboratory.

The men who brought the new science to Ireland were all members of Cromwell's army, and they valued it not simply because it would help them understand nature and nature's laws but because they could use it to control nature and to impose law on Ireland. They saw this science, moreover, as evidence of their superiority over the native Irish and a justification for extending their power in Ireland. Man has an obligation to make the most of nature's resources, they believed, and if the native Irish could not improve their estates according to the most modern methods, the Puritans felt they had an obligation to do it for them. But it was not simply a matter of transplanting 'English' science to Ireland. The soldiers who promoted the new science in Ireland (Petty in particular) developed new methods and approaches tailored to the specific circumstances they found there.

In early modern Ireland, science was not the pastime of disinterested professors contemplating nature in isolated universities. It was a practical tool for the extension of Protestant rule and English commerce. When one considers the history of science in this broader light rather than simply looking at the great eponymous theories of the scientific revolution, one begins to see, not only the interactions between science and society, but also the unique and essential role that Ireland played in the development of modern science, and the role that science played in the development of modern Ireland.

1.1   Puritanism and Baconian Science

The proponents of the new science in England in the second half of the seventeenth century saw it as closely connected with Protestant and especially Puritan beliefs.[9] Historian Charles Webster explains that the Puritan Revolution was "seen as a period of promise, when God would allow science to become the means to bring about a new paradise on earth."[10] Science itself became a religious activity. God had provided the fruits of the earth for man to exploit, and many Protestants felt that the application of practical science was their religious duty. Francis Bacon, apostle of the scientific method, wrote, "For man by the Fall fell both from his state of innocence and his dominion over creation. Both of these, however, can even in this life be to some extent made good; the former by religion and

---

[9]See esp. Robert K. Merton, *Science, Technology and Society in Seventeenth Century England* [1938] (New York: Howard Fertig, 1970), Christopher Hill, *Intellectual Origins of the English Revolution* (Oxford: Clarendon, 1965), Charles Webster, *The Great Instauration: Science, Medicine and Reform, 1626-1660* (London: Duckworth, 1975), and I. Bernard Cohen (ed.), *Puritanism and the Rise of Modern Science: The Merton Thesis* (New Brunswick: Rutgers University Press, 1990). As Cohen's collection makes clear, the relationship between Puritanism and science is still heavily disputed.

[10]Webster, *The Great Instauration*, p. xvi.

faith, the latter by arts and sciences."[11] Bacon and other advocates of the new science attacked the dogmatism of Catholicism along with the closely associated scholastic philosophy. Bacon claimed Luther as his predecessor and argued that the reformation in religion should be followed by a reformation in natural philosophy. If the scholastic philosophy suited a world of university scholars and clerics, the Baconian philosophy was designed for a world of merchants and colonists. The application of science would lead to economic progress, he believed, and he found common cause with craftsmen rather than scholars. Moreover, Bacon (who served as Lord Chancellor) saw that this new practical science could play an important role in expanding the nascent English Commonwealth. His *New Atlantis* described what a state-supported institution for scientific research and exploration might look like.

By the mid-seventeenth century, a number of Puritan intellectuals had adopted Bacon's philosophy of a practical science founded on observation and experiment. Foremost among them was Samuel Hartlib, a Pole who had fled the terrors of the Thirty Year's War in Prussia, arriving in England around 1628.[12] Hartlib churned out pamphlets on everything from education and religion to animal husbandry and constantly proposed schemes that might transform his ideas into reality. Hartlib, like Bacon, was also a major proponent of the government support of science, and he saw Cromwell's Commonwealth and the conquest of Ireland as his best opportunity for the endowment of some sort of scientific organization. The Puritan regime was also sympathetic to an alliance between science, religion, and the state, and in June of 1641 the House of Commons

[11]Francis Bacon, *Novum Organum*, Trans. and Ed. by Peter Urbach and John Gibson (Chicago: Open Court, 1994), p. 292.

[12]See Hill, *Intellectual Origins of the English Revolution*, pp. 100-109, Charles Webster (ed.), *Samuel Hartlib and the Advancement of Learning* (Cambridge: Cambridge University Press, 1970), Webster, *The Great Instauration*, G.H. Turnbull, "Samuel Hartlib's Influence on the Early History of the Royal Society," *Notes and Records of the Royal Society* (1953) 10: 101-30, Mark Greengrass, Michael Leslie and Timothy Raylor (eds.), *Samuel Hartlib and Universal Reformation* (Cambridge: Cambridge University Press, 1994).

decided that all of the Irish lands confiscated from deans and chapters should go to the advancement of learning and piety. Seeing his opportunity, Hartlib published a book in October of that year timed to coincide with the beginning of Parliament's second session. The book, which he dedicated to the Parliament, described Hartlib's own utopian vision of a kingdom where the state and the religious establishment support trade, medicine, agriculture, and the mechanical arts. Six years later, Hartlib was still lobbying for a state-supported office for science and had enlisted the support of an inventor living in London, William Petty. In 1648, Petty published "The Advice of W.P. to Mr. Samuel Hartlib for the Advancement of Some Particular Parts of Learning."

Petty's scheme was in many ways simply a more detailed version of Bacon's *New Atlantis*. Without any institutional home, he explained, the practitioners of the new science were like 'so many scattered coals.' He proposed an Office of Address that would be a place "where Men may know what is already done in the Business of Learning, what is at present in Doing, and what is intended to be done...."[13] Petty also offered extensive educational reforms, recommending that all boys be taught 'some genteel Manufacture', such as making mathematical instruments, engraving, grinding lenses, or making mariner's compasses. Rather than puzzling themselves "about mere Words and chimerical Notions," he explained, "they shall not only get Honour by shewing their Abilities, but Profit likewise by the Invention of fructiferous Arts."[14] For Petty and his fellow Baconians, the scholastic philosophy was useless logic chopping. The new experimental and practical science, by contrast, offered important benefits to manufacturing and trade and therefore to the state. Practicality and profitability were to be the new watchwords in the 'business of learning.' Knowledge,

---

[13]William Petty, "The Advice of W.P. to Mr. Samuel Hartlib for the Advancement of Some Particular Parts of Learning [1648]," *Harleian Miscellany* (1745) 6: 1-13, p. 2.

[14] Petty, "The Advice of W.P.," p. 11.

Petty believed, would arise from the collection and organization of empirical facts, just as wealth grows from the accumulation of capital.

While Hartlib's attempts to convince the government to support his vision of practical science ultimately failed, he was enormously successful at organizing the informal efforts of his steadily growing network of associates. One group came to be known as the Invisible College, perhaps in contrast to the government supported 'visible' college that had failed.[15] The existence of this group is known primarily from a few obscure references by the nineteen-year-old Robert Boyle, who became involved with them around 1646. It was through the Invisible College that Hartlib became more directly involved with Irish affairs.

Robert Boyle, who would go on to become one of the founders of modern chemistry, was the youngest son of Richard Boyle, first Earl of Cork and Ireland's richest landlord. He was on a grand tour of Europe when the Irish rebellion broke out, and when his father's rents dried up, he was forced to return to London in 1644. It was in London that his sister, Lady Ranelagh, herself a strong supporter of the Parliamentary cause, introduced him to Hartlib.[16] Boyle's first scientific associate was Benjamin Worsley. Worsley had served in Strafford's government in Ireland in the 1630s, eventually becoming an army physician at the outbreak of the rebellion and finally returning to London in the mid-1640s.[17] When Worsley and Boyle met in 1646, Worsley was proposing a saltpetre works that would use chemistry to manufacture both gunpowder and fer-

---

[15] On the Invisible College, see Webster, *The Great Instauration*, pp. 57-67.

[16] See Charles Webster, "New Light on the Invisible College," *Transactions of the Royal Historical Society* (1974) 24: 19-42. On Boyle and the Hartlib Circle, see J.R. Jacob, *Robert Boyle and the English Revolution: A Study in Social and Intellectual Change* (New York: Burt Franklin & Co., 1977), pp. 16-38.

[17] See Charles Webster, "Benjamin Worsley: Engineering for Universal Reform From the Invisible College to the Navigation Act," pp. 213-35 in Greengrass et al. (eds.), *Samuel Hartlib and Universal Reformation*.

tilizer.[18] Boyle hoped to learn more about practical chemistry from Worsley, and Lady Ranelagh used her influence to extract funds from parliament for Worsley's scheme.

The Invisible College therefore had a particularly strong Irish component. Many of its members were Protestant exiles from Ireland living in London in the late 1640s like Boyle's elder brother, Lord Broghill, and his close associates Sir William Parsons and Sir John Temple.[19] Ireland appeared to be an ideal site for the pursuit of Baconian science. The practical arts of navigation, fortification, surveying, and animal husbandry provided the tools to develop Ireland, while the Puritan emphasis on hard work, the control of nature, and the errors of alternative systems of belief convinced them that conquest was a religious obligation. They saw the Irish people, like the Irish countryside itself, as wild and barbarous, in need of civilization and improvement.[20] The native Irish were portrayed as no better than animals, steeped in sinfulness and wasting their God-given resources.

These members of the Hartlib circle were the first to bring the new science to Ireland, and all of them were actively involved in the English conquest.[21] Miles Symner, who would go on to become the first professor of mathematics at Trinity College Dublin, was Chief Engineer of the army. Benjamin Worsley served as both Surgeon General and Surveyor General. And Gerard and Arnold Boate both served as physicians to the army. Science, therefore, came to Ireland in the hands of the English colonizer. But this was

---

[18]See Webster, *The Great Instauration*, pp. 377-82 and Webster, "Benjamin Worsley."

[19]Webster, "New Light on the Invisible College," p. 41.

[20]T.C. Barnard, "Gardening, Diet and 'Improvement' in Later Seventeenth-Century Ireland," *Journal of Garden History* (1990) 10: 71-85, T.C. Barnard, "The Hartlib Circle and the Cult and Culture of Improvement in Ireland," pp. 281-97 in Greengrass et al. (eds.), *Samuel Hartlib and Universal Reformation.* See also D.B. Quinn, *The Elizabethans and the Irish* (Ithaca: Cornell University Press, 1966).

[21]See Barnard, *Cromwellian Ireland*, pp. 213-48, K.T. Hoppen, "Correspondence: The Hartlib Circle and the Origins of the Dublin Philosophical Society," *Irish Historical Studies* (1976) 20: 40-48.

not simply a case of English science imported to Ireland; the Irish
context played a key role in shaping English attitudes.[22] Ireland
offered an unprecedented opportunity to demonstrate the power
and the profitability of the new scientific approach to areas like
agriculture, mining, and warfare.

1.2   Mathematics in an Age of Capitalism, Colonialism, and Conquest

While the Baconian approach was one key aspect of the new sciences and an important feature of the conquest of Ireland, mathematics also played a critical role. Navigation, surveying, artillery
ranging, and other mathematical techniques had by the sixteenth
century become crucial components of England's arsenal. John
Wallis, an English mathematician educated at Cambridge in the
1630s, recalled, "Mathematicks at that time, with us, were scarce
looked upon as Academical Studies, but rather Mechanical; as the
business of Traders, Merchants, Seamen, Carpenters, Surveyors of
Lands."[23] Those with the keenest interest in mathematics were not
university professors but an odd assortment of instrument makers
and private mathematical tutors commonly referred to by historians as 'mathematical practitioners'.

In sixteenth-century Britain, mathematics had a bad reputation. It was seen as dry and difficult, and its students as solitary
and vain. Its use in astrology and demonology, moreover, gave its
mysterious symbols and diagrams a dangerous air. Against this
backdrop, the mathematical practitioners sought to demonstrate

---

[22]See Nicholas Canny, *The Upstart Earl: A Study of the Social and Mental World of Richard Boyle, First Earl of Cork, 1566-1643* (Cambridge: Cambridge University Press, 1982), pp. 142-50, Webster, *The Great Instauration*, pp. 428-46.

[23]For an alternative view, see Mordechai Feingold, *The Mathematicians' Apprenticeship: Science, Universities and Society in England, 1560-1640* (Cambridge: Cambridge University Press, 1984).

the usefulness and accessibility of mathematics.[24] Their first target was navigation. In the second half of the sixteenth century, England finally began to take part in the great wave of exploration and international trade. During this period Englishmen searched for the Northwest Passage, colonized America, and founded England's earliest joint-stock companies. English seamen, however, used to sailing within sight of the coast of Europe, were lost when they attempted to cross the Atlantic or to reach India. The mathematical practitioners responded by adapting medieval astronomical instruments—the quadrant, astrolabe, and cross staff—that mariners could use to find their position at sea. Similar instruments (such as the theodolite) soon transformed surveying into a mathematical activity. Cartography and the military arts of fortification and artillery also utilized the new mathematical tools. The mathematical practitioners, as historian Jim Bennett explains, had "a grand vision of an integrated domain of subjects all linked in the common dependence on geometry and arithmetic, and together vital to economic, commercial, political and colonial development."[25] Mathematics, they boasted, was the ultimate tool of empire.

William Petty had taken up the study of mathematics in this context. Born in 1623 to a poor clothier in Hampshire, Petty left soon after grammar school to make his fortune. His first career as a cabin boy on an English merchant ship came to a premature end when he broke his leg after ten months and was put ashore in France near Caen. Petty used the smattering of Latin he had learned in school to convince the Jesuits there to allow him to attend their college. The French at the time, particularly the Jesuits, were more advanced than the English in terms of mathemat-

---

[24] Katherine Neal, "The Rhetoric of Utility: Avoiding Occult Associations for Mathematics Through Profitability and Pleasure," *History of Science* (1999) 37: 151-78.

[25] J.A. Bennett, "Geometry and Surveying in Early-Seventeenth-Century England," *Annals of Science* (1991) 48: 345-54, p. 345.

ics and science.[26] And at Caen Petty learned "the whole body of
common arithmetic, the practical geometry and astronomy; con-
ducing to navigation, dialling, &c.; with the knowledge of several
mathematical trades...."[27] His Jesuit education meant that at age
twenty Petty had gained "as much mathematics as any of my age
was known to have." While Petty diligently adopted their mathematics, however, he must have strongly resisted their attempts to
change his religious beliefs. For he came to Ireland in the middle
of a war between Protestantism and Catholicism, and Petty sided
firmly with the Protestants.

1.3   Trinity College Dublin

Trinity College Dublin itself was an expression of the Protestant
Reformation and an explicit component of the English conquest
of Ireland. Built on the site of a former Augustinian priory in
1591, it was intended, as historians McDowell and Webb explain,
"as a powerful auxiliary to military force, by breaking down the
two great barriers to the spread of English influence: Catholicism
and the Gaelic cultural tradition."[28] The English crown supported
it accordingly. Between 1610 and 1613, James I gave the College
20,000 acres of confiscated land.

While the government supported Trinity's Protestant education,
it did its best to extinguish all Catholic education. In the Middle
Ages, Ireland had been known as the 'island of saints and scholars'
for its role in the preservation of Christian learning after the fall
of Rome, but by the late sixteenth century, Irish Catholic schol-

---

[26]On the crucial role of Jesuits in the development of science see Peter Dear, *Discipline and Experience: The Mathematical Way in the Scientific Revolution* (Chicago: University of Chicago Press, 1995), esp. pp. 32-62.

[27]Sir William Petty's Will in George Petty Fitzmaurice, *The Life of Sir William Petty, 1623-1687* (London: John Murray, 1895), pp. 318-9.

[28]R.B. McDowell and D.A. Webb, *Trinity College Dublin, 1592-1952: An Academic History* (Cambridge: Cambridge University Press, 1982), p. 2.

ars had, for the most part, been driven out of Ireland.[29] Between 1590 and 1681, they established twenty Irish Colleges throughout Europe. Looking back, a nineteenth-century history of Irish scholarship saw the role of the Irish Colleges in the context of a great war: "[W]hile Erin's exiled warriors were cutting their way to fame, rank and glory upon the reddest battle-fields of Spain, France, Italy, Austria and the Netherlands, her scholars were bravely climbing the steep, rough hills of science and sanctity in classic halls at Antwerp and Louvain, at Lisle, Douay, Bordeaux, Rouen and St. Omer, at Salamanca and Alcala, at Coimbra and Prague, and at St. Isidore's at Rome."[30] James I ordered all Catholic priests to leave the country, and in 1653, the Commonwealth threatened them with execution if they returned.

---

[29]See Martin J. Counihan, "Ireland and the Scientific Tradition," pp. 28-43 in P. O'Sullivan (ed.), *The Creative Migrant* (Leicester University Press: Leicester, 1994) and Thomas Cahill, *How the Irish Saved Civilization* (New York: Doubleday, 1995).

[30]William P. Treacy, *Irish Scholars of the Penal Days: Glimpses of Their Labors on the Continent of Europe* (New York: 1889), p. vii. See also John Silke, "Irish Scholarship and the Renaissance, 1580-1673," *Studies in the Renaissance* (1973) 20: 169-206.

Figure 1.1: The Earliest Plan of Trinity College Dublin (c. 1600).
Credit: Original MS at Hatfield House

Trinity, far from the gaze of the staunch Anglicans of Oxford and
Cambridge, and on the front line in the battle against Catholicism,
soon became a Puritan stronghold. John Winthrop, the Puritan
governor of Massachusetts, even sent his son (later the first governor of Connecticut) to be educated at Trinity in 1622.[31] Trinity's
most famous scholar during this early period was James Ussher.[32]
One of the first students to attend the College, he went on to serve
as Professor of Theological Controversies, Vice Chancellor of the
University, Archbishop of Armagh, and Primate of Ireland. Ussher
was one of the leading scholars of the British Isles, a man of truly
European fame. He was known at the time for his works on chronology, geography, linguistics, church history, and controversial theol-

---

[31] Robert H. Murray, *Dublin University and the New World* (London: Society for Promoting Christian Knowledge, 1921).

[32] See R. Buick Knox, *James Ussher, Archbishop of Armagh* (Cardiff: University of Wales
Press, 1967) and Alan Ford, "Dependent or Independent? The Church of Ireland and Its
Colonial Context, 1536-1649," *The Seventeenth Century* (1995) 10: 163-87.

ogy, though he is remembered today primarily for his calculation that the world was created in 4004 BC (on the night before Sunday, October 23, to be precise). His enormous library contained an impressive collection of scientific and mathematical works including Copernicus's *De Revolutionibus*.[33] Ussher also patronized Hartlib and was a close friend of Richard Boyle. Henry Cromwell's officers purchased Archbishop Ussher's library (all 10,000 volumes) on his death to donate to the new College. Trinity, they hoped, would become a great center for Puritan learning. Samuel Mather, fellow of Harvard College, came to Ireland in 1654 and was made a Fellow of Trinity. Increase Mather, also a fellow of Harvard College, joined his brother briefly at Trinity, before returning to Harvard where he later became president. The Cromwellian government also appointed Trinity's first professor of mathematics in 1652. The impetus was related more to the immediate needs of the government than an appreciation for pure mathematics. The professor was needed, they explained, "forasmuch as there is a great occasion for surveying of lands in this country, and there are divers ingenious persons, soldiers and others who are desirous to be instructed and fitted for the same...."[34] The first person to hold the new professorship was Major Miles Symner who had planned fortifications during the war and sat on the committee for the satisfaction of the arrears of the army. In 1668 the Earl of Donegal endowed a chair of mathematics, and Symner (still waiting for his own arrears from the army) was eventually appointed to it.

---

[33]Elizabethanne Boran, "The Libraries of Luke Challoner and James Ussher, 1595-1608," pp. 75-115 in Helga Robinson-Hammerstein (ed.), *European Universities in the Age of Reformation and Counter Reformation* (Dublin: Four Courts, 1998), esp. pp. 80-85.

[34] Trinity College Dublin General Registry from 1626, p. 95, cited in T.C. Barnard, "Miles Symner and the New Learning in Seventeenth-Century Ireland," *Journal of the Royal Society of Antiquaries of Ireland* (1972) 102: 129-142, p. 130.

1.4   The Down Survey[35]

The original establishment of mathematics at Trinity, therefore, grew out of the very practical need for surveyors. When Petty arrived in Ireland in 1652, there were said to be 35,000 legal claims for land by soldiers waiting for their pay and investors who had 'adventured' money for the parliamentary cause. More than half of the land in Ireland stood forfeited. As they began the process of distributing the land, the parliamentary officers soon realized the difficulty of the situation. Petty's biographer explains,

> Quartermaster-General Goulding, for example, might have a debt of 232 l. 14s. 9d., which, calculated at the army rates in Connaught, was worth in the County of Sligo 465 a. 1r. 24p.; but how was Quartermaster-General Goulding to know where his particular 465 a. 1r. 24p. exactly lay, and prove his title against all comers to enter on those lands and no others, and keep them on a secure title; and how was he, on the other hand, to prove that he had not obtained a great deal more than he was entitled to by force or impudence, or by fraudulent or incompetent measures?[36]

The settlement of Ireland and the imposition of English government depended absolutely upon the creation of a rational system of landed property based on a detailed survey.

Seeing his opportunity, Petty approached the government in September of 1654 with a bold proposal for a new survey. He

---

[35]See T.A. Larcom (ed.), *A History of the Survey of Ireland Commonly Called the Down Survey by Doctor William Petty, A.D. 1655-6* (Dublin: 1851), Fitzmaurice, *Sir William Petty*, pp. 23-68, E. Strauss, *Sir William Petty: Portrait of a Genius* (London: The Bodley Head, 1954), pp. 54-90, J.H. Andrews, *Plantation Acres: An Historical Study of the Irish Land Surveyor and His Maps* (Omagh: Ulster Historical Foundation, 1985), pp. 61-73, J.H. Andrews, *Shapes of Ireland: Maps and Their Makers, 1564-1839* (Dublin: Geography Publications, 1997), pp. 118-52.

[36]Fitzmaurice, *Sir William Petty*, p. 28. Currency is in pounds, shillings and pence while land was measured in acres, roods, and perches.

offered, "To admeasure all the forfeited lands within the three provinces, according to the naturall, artificiall, and civill bounds thereoff, and whereby the said land is distinguished into wood, bog, mountaine, arable, meadow, and pasture."[37] Additionally he proposed "to add and sett out such auxiliary lines and lymits as may facillitate and ascertaine the intended subdivision without any readmeasurement." Not only would Petty measure the forfeited lands, he would also map the boundaries and classify each parcel according to its value. His final proposal, however, was even more ambitious, "To performe the same by the last of October, 1655, if the Lord give seasonable weather, and due provision bee made against tories, and that his instruments be not forced to stand still for want of bounders."[38] (The term "tories" came from the Irish *Toraidhe* for raider. It was used to describe the remnants of Irish armies from the mid-1640s and was later applied to the political supporters of the King.) Petty proposed no less than to map and classify all the forfeited lands as well as to prepare a general county and barony map of Ireland in only thirteen months. The government soon gave him the go-ahead.

Employing over a thousand hands, the survey covered an area of nearly eight and a half million acres, the greater part of twenty-nine of Ireland's thirty-four counties. Petty's survey, however, was by no means innovative in its technique. In fact, Petty ignored many of the newer scientific methods of cartography. The more mathematically adept surveyors in England scorned his instrument of choice, the circumferentor, but it was inexpensive and easy to use.[39] His innovation lay rather in the application of the division of labor, a

---

[37] *History of the Down Survey*, p. 9.

[38] *History of the Down Survey*, p. 9.

[39] In fact, the earliest surviving example of a circumferentor is a wooden Irish circumferentor of 1667 housed in the Oxford Museum of the History of Science. See Bennett, "Geometry and Surveying," p. 352. The oldest Irish signed instrument preserved in Ireland is a 1688 circumferentor. See Charles Mollan and John Upton, *The Scientific Apparatus of Nicholas Callan and Other Historic Instruments* (Dublin: St. Patrick's College Maynooth and Samton, 1994), p. 261 and J.E. Burnett and A.D. Morrison-Low (eds.) *'Vulgar and Mechanick':*

technique he probably learned from Bacon's *New Atlantis*. Petty broke down the production of instruments into distinct steps: a wire maker made the measuring chains, a watchmaker made the magnetic needles, a turner made the wooden boxes for the instruments, a pipe maker made the legs, and another worker adjusted the sights.[40] Petty also divided the work of surveying itself. To the dismay of the professional surveyors, he recruited soldiers to perform measurements in the field. Men trained in painting and drawing converted the raw measurements of distances and angles from the field books into maps on squared paper and used these maps to calculate areas. Above these, a group of supervisors remeasured the angles and compared common boundaries. Petty's goal was to complete the survey as accurately, as quickly, and as cheaply as possible. The division of labor allowed him to utilize unskilled soldiers while at the same time monitoring their work for mistakes or fraud. Petty paid the soldiers, for example, by length measured, rather than by area, to make it more difficult for them to purposefully influence the outcome. Dishonest chainmen, however, were only one of Petty's worries; the state of Ireland after ten years of war was an even greater obstacle. Throughout the countryside, dwellings had been destroyed and fields had overgrown. Bands of rebels still remained to attack the surveyors. Eight of Petty's surveyors were kidnapped by Blind Donagh O'Derrick in County Kildare, carried off into the mountains, and executed.

The scope of the project was enormous. "The quantitie of line," Petty bragged, "which was measured by the chaine and needle beinge reduced into English miles was enough to have encompassed the world neere five tymes about."[41] (Like a good fishing story,

---

The Scientific Instrument Trade in Ireland, 1650-1921 (Dublin: National Museums of Scotland and Royal Dublin Society, 1989), pp. 12-16.

[40]"A Briefe Accompt of the most materiall Passages relatinge to the Survey managed by Doctor Petty in Ireland, anno 1655 and 1656," pp. xiii-xvii in *History of the Down Survey*, p. xiv.

[41]"A Brief Accompt," p. xviii.

however, Petty's claim grew as memory faded. By 1687, he was boasting that his surveyors had measured a distance equal to eight times around the globe.[42]) The Down Survey (so-called apparently because it was put 'down' on paper) was undoubtedly an achievement. John Evelyn, one of Petty's contemporaries described it as "the most exact map that ever yet was made of any country."[43] As late as 1851, Thomas Larcom, head of the Irish Ordnance Survey, wrote, "The Down Survey stands to this day...the legal record of the title on which half the land of Ireland is held."[44] Because of its scope, its accuracy, and its timing, the Down Survey remained essential to Irish land rights for nearly two centuries.

To some, it may seem a stretch to ennoble such basic surveying techniques with the name 'mathematics.' After all, this was the age that saw the creation of the calculus. Petty's work, however, demonstrates the crucial role that such simple mathematical techniques played in seventeenth-century social and political history. Petty not only surveyed the territory, he remade it. Like the rolls of French paper on which Petty's men drew their lines, Ireland itself was a blank slate on which Petty and the parliamentary army could rewrite relations of property and political power. Traditional patterns of landholding were destroyed, ancient inhabitants transplanted, and local knowledge idealized into lines on paper. In some ways, these maps were like geometrical diagrams, representing relationships between points, lines, and angles. But the diagrams represented more than that. They represented the power of the Commonwealth government to impose its rules on Ireland. Petty later described the transformation his survey effected. His new map divided the landscape into rigid geometrical units, unlike the past in which borders shifted:

---

[42]William Petty, "A Treatise on Ireland," in *Economic Writings*, p. 615.

[43]John Evelyn, "Diary Entry for 22 March 1675," in Marquis of Lansdowne, *The Petty Papers: Some Unpublished Writings of Sir William Petty* (London: Constable & Company, 1927), p. xv.

[44]*History of the Down Survey*, p. 338.

> For as a Territory bounded by Bogs, is greater or lesser
> as the Bog is more dry and passible, or otherwise: So the
> County of a Grandee or Tierne in Ireland, became greater
> or lesser as his Forces waxed or weaned; for where there
> was a large Castle and Garrison, there the Jurisdiction
> was also large. And when these Grandees came to make
> peace, and parts one with another, the limits of their
> Land-agreements were no lines Geometrically drawn; but
> if the Rain fell one way, then the Land whereon it fell,
> did belong to A. if the other way, to B. &c.[45]

The geometrical boundary lines represented the fixity and rationality of English rule as against the ever-changeable nature of traditional Irish boundaries. English property law replaced fluid Irish
patterns of landholding just as English government replaced the
Gaelic lordships.

---

[45] Petty, "The Political Anatomy of Ireland," p. 206.

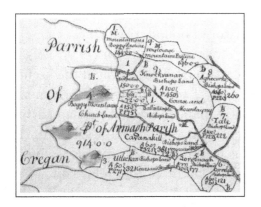

Figure 1.2: Parrish Map from the Down
Survey

Figure 1.3: Down Survey
Field Book. Credit: J.H.
Andrews , Plantation Acres:
An Historical Study of the
Irish Land Surveyor and
His Maps (Omagh: Ulster
Historical Foundation,
1985), p. 311.

There was no magical connection between lines on a map and
the exercise of power, however. Petty's men created clear corre-
spondences between rocks, hedges, bogs, and ink on a page. The
continuing presence of the army and the settlers helped to reinforce
those boundaries. In fact, when disputes over boundaries did arise
recourse was made not to the maps but to the field books, which
allowed the original lines to be redrawn on the ground itself. Petty,
however, realized the fragile nature of the survey. He feared what
would happen, "when the mearers were fled, the surveyors dead,
the marks on the land worne out, the ratts had eaten the originall
plots, and a new interest risen up."[46] (Meerers were local infor-
mants recruited to give information on property boundaries.) The
power of the map, in fact, depended on all of these factors. When
new interests did arise at the Restoration, Petty and other Protes-

[46] *History of the Down Survey*, p. 42.

tant landlords fought tooth and nail to maintain the Cromwellian land settlement (for the most part successfully). The power of the English government gave stability to the relationship between Petty's maps and the actual territorial boundaries, while the maps themselves were crucial to the exercise of that power.

With the Down Survey, Petty not only demonstrated the usefulness of practical mathematics to the Commonwealth government, he also made his fortune. He was paid handsomely (though belatedly) by a grateful government; but, like many a seventeenth-century civil servant, he made his fortune from his unique position. Almost single-handedly he ran the committee that allotted land to soldiers, and in the process he cornered the market on the debentures soldiers were given in lieu of pay. Unable or unwilling to wait for their allotted acres, they sold out to Petty at an enormous discount. Petty soon became one of Ireland's wealthiest landlords.[47] He had arrived in Ireland with less than £500 to his name, and by 1685 he was earning an income of £6,700 per year. By his death, he had amassed 50,000 acres of Irish land.

1.5   Political Arithmetic

Petty's work on the Down Survey brought him into close contact with the workings of government. He became a secretary to Henry Cromwell, Governor General of Ireland and the son of Oliver Cromwell. So from medicine to surveying, Petty now turned his attention to government and what he would later call 'Political Arithmetic'.[48] Petty's first efforts in this area were with his friend

---

[47]T.C. Barnard, "Sir William Petty, Irish Landowner," pp. 201-17 in H. Lloyd-Jones, V. Pearl and B. Worden (eds.), *History & Imagination: Essays in Honour of H.R. Trevor-Roper* (London: Duckworth, 1981).

[48]See Fitzmaurice, *Sir William Petty*, pp. 179-231, Strauss, *Sir William Petty*, pp. 181-232, Peter Buck, "Seventeenth-Century Political Arithmetic: Civil Strife and Vital Statistics," *Isis* (1977) 68: 67-84, Alessandro Roncaglia, *Petty: The Origins of Political Economy* (Armonk, NY: M.E. Sharpe, Inc., 1985), Juri Mykkanen, "'To Methodize and Regulate

John Graunt, who published in 1662 *Natural and Political Observations on the Bills of Mortality*. Graunt had collected and organized the data from London's weekly announcements of christenings and burials, resulting in a range of conclusions relating to population, death rates, the ratio of males to females, the growth of the city, and even the relative incidence of various diseases such as the plague. Petty was heavily involved in this exercise, and throughout his life a number of friends even claimed that Petty rather than Graunt had actually written the *Observations*.[49] Soon after, Petty realized that many of the same techniques could be used in other areas. He began work on his 1667 *Treatise on Taxation* and found that this subject forced him similarly to find a means to calculate population and break it down into different classifications. By 1672, Petty was working on two books, *Political Arithmetick* and the *Political Anatomy of Ireland*.

The goal of the *Political Anatomy of Ireland* was "the enriching of a Kingdom, by advancing its Trade and Publick Credit" through arguments "founded upon Mathematical Truth." The use of mathematics was key, as Petty explained, "whereupon I think depends the Political Medicine of that country; and these things too without passion or interest, faction or party; but as I think according to the Eternal Laws and Measures of Truth."[50] The problems of Ireland, he felt, could be solved mathematically. He hoped that mathematics would allow him to cut through the passionate debates surrounding Irish policy and find new techniques of government

Them': William Petty's Governmental Science of Statistics," *History of the Human Sciences* (1994) 7: 65-88, Mary Poovey, *A History of the Modern Fact: Problems of Knowledge in the Sciences of Wealth and Society* (Chicago: University of Chicago Press, 1998), pp. 120-38. For the work of Petty and Graunt and its context see Ian Hacking, *The Emergence of Probability* (Cambridge: Cambridge University Press, 1975), esp. pp. 102-10; Theodore M. Porter, *The Rise of Statistical Thinking, 1820-1900* (Princeton: Princeton University Press, 1986), esp. pp. 18-23.

[49]See *Economic Writings*, pp. xxxix-liv.

[50]William Petty, "Dedication of Second Edition of The Political Anatomy of Ireland [1719]," *Economic Writings*, p. 128.

grounded in eternal truths. Using the rhetoric of the disinterested expert, he presented himself as the doctor who could prescribe Ireland's political medicine without descending into the seediness of party politics. Ireland, Petty claimed, was a particularly appropriate environment to experiment with such new methods. He explained, "as Students in Medicine, practice their inquiries upon cheap and common Animals, and such whose actions they are best acquainted with, and where there is least confusion and perplexure of Parts; I have chosen Ireland as such a Political Animal, who is scarce Twenty years old..."[51] For Petty, Ireland as a political entity came into existence with the Act of Settlement in 1652. Not only was Ireland a blank slate, it was also a cheap and common animal, well suited for experimentation.

In the *Political Arithmetic,* Petty broadened his view from Ireland to England. The subtitle gives a sense of areas Petty felt could be reduced to mathematical terms: the Extent and Value of Lands, People, Buildings, Husbandry, Manufacture, Commerce, Fishery, Artizans, Seamen, Soldiers; Publick Revenues, Interest, Taxes, Superlucration, Registries, Banks; Valuation of Men, Increasing of Seamen, of Militia's, Harbours, Situation, Shipping, Power at Sea. Petty calculated such quantities as the total wealth of the country, the value of its land, its total exports, population, and the number of houses, all through more or less tortuous estimations and guesses. He attempted to determine whether rents were increasing or decreasing, whether the country was overpopulated or underpopulated, whether taxes were too high or too low, and perhaps most importantly, how England compared to France and the Netherlands in terms of wealth, trade, and military power. The ultimate goal was to determine accurate information about the current state of the nation in order to find a means to decide

---

[51] Petty, "The Political Anatomy of Ireland," p. 129.

between various policy recommendations in such areas as taxation or trade policy.

Petty's work on political arithmetic, however, began and ended with the problems of Ireland. An interesting example of Petty's method is his calculation of how many people were lost in the Rebellion of 1641, a number fraught with political significance.[52] Petty began with the current population of Ireland (in 1672), which he believed to about 1,100,000. (It is not clear where he found this figure.) He then estimated that since the end of the war in 1652 the population had increased by 80,000 births and 170,000 immigrants and returned exiles. This gave a population in 1652 of about 850,000. Now, in order to calculate the population in 1641, he noted that there was one third more "superfluous and spare Oxen, Sheep, Butter and Beef" (exports) in 1664 than in 1641. He concluded that the population in 1641 therefore must have been one third larger, or 1,466,000. (He actually multiplied the population figure for 1672 by $1/3$ rather than the figure for 1664, which he did not have.) Petty then subtracted the population in 1652 and found a deficit of 616,000. This represented how many more people lived in Ireland before the Rebellion than after. The proportion of British to Irish at the time, Petty continued, was 2 to 11. Therefore 112,000 English died during the Rebellion and 504,000 Irish. Of course Petty's figure included those who were exiled as well as those killed by war, plague, and famine. If these people had lived and reproduced, Petty found, the population in 1652 would have been 1,539,000 rather than 850,000, leaving 689,000 "for whose Blood some body should answer both to God and the King."[53] (Being a shrewd political operator, Petty had made the transition from Commonwealthman to monarchist with few difficulties.) Petty then went on to calculate the economic impact of the Rebellion. He began with the fact that Ireland has about eight

---

[52] Petty, "The Political Anatomy of Ireland," pp. 149-54.
[53] Petty, "The Political Anatomy of Ireland," p. 151.

million acres of profitable land. In 1641 this land was worth about
£8 million, but after the war, debentures could be purchased at 1/8
of face value (as Petty knew better than anyone), meaning the land
in Ireland was worth only £1 million. The cattle and stock, worth
£4 million in 1641 were worth only £500,000 in 1652, while the
houses lost 1/5 of their value. Petty also accounted for the value of
the people lost. While men, women, and children in England are
valued together at £70 per head, he decided to value the Irish as
slaves and Negroes are usually valued, namely £15 per head taken
together (men being worth £25 and children £5). Finally, he added
the wages of both armies and came up with a total cost for the Rebellion of £37,255,000. Twenty years of rent (from 1652 to 1673) on
the forfeited lands, he added, had not fully defrayed the cost of the
English army in Ireland. And the adventurers who invested money
in the enterprise, he found, would have made more "in open and
free market." In the end, Petty claimed that the Rebellion was a
fiasco for the English but actually a boon to the Irish. "[T]he Con-
federates gained thereby 15 Times more than they properly lost,"
while "all the several Branches of the English-Protestant Interest
lost 200 times more than they gain'd."[54] Petty not only saw the
Rebellion itself as an economic disaster for the English, he also saw
its origins as essentially economic.

> The Cause of the War was a desire of the *Romists*, to
> recover the Church-Revenue, worth about 110 M. l. *per
> Ann.* and of the Common *Irish*, to get all the *English-
> mens* Estates; and of the 10 or 12 Grandees of *Ireland*, to
> get the Empire of the whole. But upon the playing of this
> Game or Match upon so great odds, the *English* won and
> have (among, and besides other Pretences) a Gamester's
> Right at least to their Estates.[55]

---

[54] Petty, "A Treatise of Ireland," pp. 618-19.
[55] Petty, "The Political Anatomy of Ireland," p. 154.

Disregarding the political, cultural, and religious tensions, Petty preferred to see the event in purely economic and quantifiable terms.

While Petty's economic view exposed the enormous waste of the Rebellion, it also led him to a possible solution for Ireland's problems. In the *Political Anatomy of Ireland*, Petty proposed "the transmuting of one People into the other" by bringing 200,000 marriageable Irish Catholic women to England and bringing the same number of English women to Ireland.[56] In his *Treatise of Ireland*, Petty offered a more radical proposal. To effect the settlement, improvement, and union of Britain and Ireland he suggested transporting a million Irish to England and leaving the remaining 300,000 as herdsmen and dairy women.[57] While the cost of transporting so many people and the loss from abandoned property would be high (Petty estimated £4 million), the increase in the King's revenue would more than exceed the costs. By adding 1 million people to Britain's population of 7 million, Petty claimed, the King's revenue would increase by the ratio of 8 to 7 and within 20 years would cover all the expenses. This is not to mention the money saved defending and governing Ireland after it had been turned into "a Kind of Factory."[58] Essentially Petty was arguing that England's profitability would increase proportionally to its increase in population while Ireland would be worth more as a cattle factory. And he provided fifty pages of statistics to prove it.

Far from being another brutal military operation, Petty explained, this transformation would have the cooperation of all parties. "[I]t will be the Profit, Pleasure, and Security of both Nations and Religions to Agree herein,"[59] he claimed. The King could buy out Irish landowners at market prices, while Irish laborers could be

---

[56] Petty, "The Political Anatomy of Ireland," pp. 157-59.
[57] Petty, "A Treatise on Ireland," pp. 555-601.
[58] Petty, "A Treatise on Ireland," p. 560.
[59] Petty, "A Treatise on Ireland," p. 561.

offered higher wages in England. "[T]he Million of Transplantees out of Ireland," he calculated, "will after their having been Seven Years in England, become worth above 30 l. per head more than at present...."[60] Economic self-interest, he believed, would trump national pride and racial self-preservation. Perhaps the greatest benefit of the plan, according to Petty, is the way it would succeed in integrating the two nations: "The Manners, Habits, Language, and Customs of the Irish (without Prejudice to Religion) will be transmuted into English, within less than an Age, and all Old Animosities forgotten."[61]

Crucial to Petty's argument was a re-imagining of the Irish. Petty proceeded by transforming the Irish from religious, political, or racial beings into economic beings.[62] In laying out the terms of his argument, he proposed to consider the Irish "not as old Irish, or such as lived there about 510 Years ago, when the English first medled in that Matter; Nor as those that have been added since, and who went into Ireland between the first Invasion and the Change of Religion; Nor as the English who went thither between the said Change and the Year 1641, or between 1641 and 1660; Much less, into Protestants and Papists, and such who speak English, and such who despise it."[63] Rather than the common ways in which the inhabitants of Ireland identified themselves in terms of faith, language, or ancestry, he decided to consider them simply,

1. As such as live upon the King's pay

2. As owners of Lands and Freeholds

3. As Tenants and Lessees to the Lands of others

4. As Workmen and Labourers.

---

[60] Petty, "A Treatise on Ireland," pp. 576-77.

[61] Petty, "A Treatise on Ireland," p. 573.

[62] Poovey, *The Making of the Modern Fact*, p. 136.

[63] Petty, "A Treatise on Ireland," pp. 561-62.

Petty's process for mathematizing Irish political questions involved forgetting history, forgetting religion, and forgetting language in order to treat the Irish simply as economic beings interested in their own personal profit. In the Down Survey, Petty's men reduced an Irish landscape saturated with local memories and traditions to a series of lengths and angles. With his political arithmetic, he reduced the complexities of Irish society to a set of economic measures, a view that Jonathan Swift would later parody brilliantly in "A Modest Proposal."

The lives of such economic beings could then be reduced to terms of number, weight, and measure, and self-interest, mathematically defined, became a tool that the government could use to engineer a more peaceful and more profitable nation. In this way, political arithmetic was not simply a means to describe the order already inherent in human society; it was a tool for the better imposition of such order. To that end, Petty recommended the creation of a General Registry of Lands, Commodities, and People.[64] This new office, he hoped, would make land a commodity, secure titles and boundaries, and increase trade and manufacture. The Down Survey in itself, he realized, was not enough to serve as a foundation for the new Irish state. An entire administrative system unknown, and perhaps unneeded, in England was essential. It was not enough merely to conquer the natives; a complete system of governance had to be instituted. Political Arithmetic was the ideal tool for this. Petty explained, "finally when wee have a cleere view of all persons and things, with their powers & familyes, wee shall be able to Methodize and regulate them to the best advantage of the publiq and of perticular persons."[65]

After two decades of revolution, factionalism, and sectarian strife, Petty thought that he had found in mathematics a new foundation for political consensus and effective government. Petty's

---

[64]See *Petty Papers*, I: 77-90.
[65] *Petty Papers*, I: 90.

project was closely related to that of his sometime associate Thomas Hobbes. Hobbes offered a mechanical view of a society populated by people operating according to their own rational self-interest. Consensus had to be mandated by the royal authority. Moreover, Hobbes claimed that "The skill of making, and maintaining commonwealths consisteth in certain rules, as doth arithmetic and geometry...."[66] As Petty's biographer put it, "What Hobbes laid down in theory, Petty sought to apply in practice."[67] Political arithmetic, Petty claimed, would provide an ideal tool for the sovereign, allowing "puzling and preplext Matters" to be phrased in "Terms of Number, Weight and Measure, and consequently be made demonstrable."[68] Taken from the *Wisdom of Solomon*, the phrase, "According to Weight, Measure and Number did God create everything," was a commonplace in works on mathematical science. But in Petty's hands it represented something new. Mathematics had become a technique to manage what Hobbes famously called 'the war of all against all.'[69]

While the survey was a phenomenal success, political arithmetic, in Petty's lifetime, was a failure. None of his policy suggestions were taken seriously, and, as Petty himself admitted, his arguments fell seriously short of the standards of mathematical proof. Part of the problem was that the very collection of the numbers that would support Petty's claims required a certain level of government commitment. Petty was well aware of the degree to which he had to stretch the available data to make his arguments. His mathematical arguments, he admitted, "are either true, or not apparently false, and which if they are not already true, certain, and evident, yet may be made so by the Sovereign Power."[70] As with

---

[66] Hobbes, *Leviathan*, quoted in Buck, "Seventeenth-Century Political Arithmetic," pp. 79-80.
[67] Fitzmaurice, *Life of Sir William Petty*, p. 188.
[68] Petty, "A Treatise on Ireland," p. 554.
[69] Buck, "Seventeenth-Century Political Arithmetic," p. 76.
[70] William Petty, "Political Arithmetick," pp. 244-45.

the survey, in the end the truth of Petty's mathematics depended
on the government's power and interest in making them true.

1.6   The Mathematical Principles of Natural Philosophy

While Petty invoked the authority of mathematics to bolster his
political claims, he knew that he would never achieve the certainty
of astronomy or mechanics in his political arithmetic. His sources
were few and his tools inadequate. He begged forgiveness in the
dedication to his *Political Arithmetick*, "for having presumed to
practice a Vulgar Art upon Matters of so high a nature."[71] (Imagine an age when mathematics was thought more vulgar than politics!) Petty's was the mathematics of the craftsman, not of Euclid.
"I know these are not so perfect Demonstrations as are required in
pure Mathematicks;" he explained in his *Treatise on Ireland*, "but
they are such as our Superiors may work with, as well as Wheel-
wrights and Clockmakers do work without the Quadrature of a
Circle."[72] Absolute certainty was not Petty's aim, usefulness was.
As with the Down Survey, Petty's primary goal was cost effectiveness and practicality, not testing the limits of precision. At the
same time, Petty realized that the mathematics he used as a tool
in surveying and political arithmetic could also play an essential
role in the new sciences.

Perhaps the greatest achievement of seventeenth-century science
from Galileo to Newton was the mathematization of the laws of
nature, and Petty offered his own statement on the importance of
mathematics to natural philosophy in his 1674 *Discourse Made Before the Royal Society*.[73] Petty had long been familiar with the new
mechanical philosophy. After studying medicine at the famous European schools of Utrecht, Leiden, and Amsterdam, he had moved

---

[71]William Petty, "Political Arithmetick," p. 239, n. 1.
[72] Petty, "A Treatise on Ireland," p. 611.
[73]See Robert Kargon, "William Petty's Mechanical Philosophy," *Isis* (1965) 56: 63-66.

to Paris in 1645 when it was a hotbed of scientific activity. In Paris, Petty was personally involved with many of the greatest minds of early seventeenth-century science—Mersenne, Gassendi, Hobbes, Descartes, and Roberval. Together these men laid the foundations of the mechanical philosophy and the advanced mathematics that came to be an essential part of the new science.

The new mechanical philosophy was highly amenable to mathematical treatment, despite the fact that some of its most important proponents, like Bacon and Boyle, discouraged the use of mathematics.[74] Scholastic philosophy in the Aristotelian tradition offered four types of causes or explanations in science: the formal, the material, the efficient, and the final. The mechanical philosophy, by contrast, based all explanations on matter in motion. It reduced the sensible qualities of objects (such as color, taste, and texture) to the underlying motions of atoms of different shapes. Reality, for the mechanists, consisted of the collisions of atoms; and atoms could, at least in principle, be described mathematically in terms of size, shape, speed, and direction. Behind the complicated world of appearances, the mechanists claimed, lay a fundamentally simple world of mathematically combining atoms. Petty's *Discourse*, like the work of many others, represented in some sense the union of the practical mathematical tradition with the natural philosophical tradition.

Echoing his work in political arithmetic, Petty chose as the motto for his *Discourse*, 'According to Weight, Measure and Number did God create everything.' In it he described how a square law could explain an enormous variety of phenomena. Petty's account ranged from the velocity of ships, the strength of timbers, the motion of horses, and the velocity of mill wheels, to the effects of gunpowder, the distance at which sounds may be heard, the durations of the lives of men, water pumps, the prices of commodities,

---

[74] See Steven Shapin, "Robert Boyle and Mathematics: Reality, Representation, and Experimental Practice," *Science in Context* (1988) 2: 23-58.

and the compression of wool and other elastic bodies. The application of science, he argued, depends on its expression in terms of number, weight, and measure. Petty wanted to show more than the application of mathematics to natural philosophy, however; he wanted to prove the broader relevance of mathematics to the Royal Society. For Petty, the importance of the new science lay not so much in its ability to discover the laws of nature as in the practical (and profitable) benefits it offered.

Figure 1.4: Sir William Petty (1623–1687), Physician, Surveyor, Founder of Political Economy. Portrait by John Closterman ©National Portrait Gallery, London

Newton's *Mathematical Principles of Natural Philosophy (Principia mathematica)* is often seen as the apotheosis of seventeenth-century science. It represents the university scholar developing complicated new forms of mathematics in order to explain the most basic laws of nature. In many ways, however, Petty's mathematics was much more characteristic of the period. The seventeenth century was an age of conquest, colonization, and capitalism. Mathematics rose to importance because of the role it could play in these activities, through navigation, surveying, and bookkeeping. For Petty, mathematics was a tool with which he could serve the interests of government and pursue his own financial interests. Its beauty or purity was always secondary to its usefulness and profitability. He saw its power more in terms of the transformation of society rather than simply the description of nature. Shortly before his death, Petty was given a copy of Newton's *Principia*. He expressed his tremendous admiration for the book in typically economic terms. "I would give £500 to have been the author of it;" he wrote, "and £200 that [my son] understood it."[75]

---

[75] Petty to Southwell (July 9, 1687) in Fitzmaurice, *Sir William Petty*, p. 306.

Chapter 2

Newtonianism and Natural Theology

The Infidel Mathematician

In January 1719, Sir Samuel Garth lay on his deathbed in his London townhouse. His friend the poet Joseph Addison came to pay his last respects. Garth was not only physician to King George I, he was also a published poet, a wit, and frequenter of the coffee houses popular at that time. In addition to Addison, his literary friends included John Dryden, Alexander Pope, Sir Richard Steele, and three men educated at Trinity College Dublin: William Congreve, Jonathan Swift, and George Berkeley.[1] Concerned for Garth, Addison had come to "discourse with him seriously about preparing for his approach dissolution."[2] Garth, however, refused the holy sacraments. To Addison's great horror, Garth explained, "Surely,

---

Portions of this chapter appeared as the introduction to D.A. Attis, P. Kelly and D. Weaire, "Helsham and the Rise of Newtonian Physics," pp. 1-18 in D. Weaire, P. Kelly and D.A. Attis (eds.), *Richard Helsham's Lectures on Natural Philosophy* [1739] (Dublin: Trinity College Dublin Physics Department, Institute of Physics Publishing, Verlag MIT, 1999)

[1] Daniel McCue, "Samuel Garth, Physician and Man of Letters," *Bulletin of the New York Academy of Medicine* (1977) 53: 368-402.

[2] Joseph Stock, *Memoirs of George Berkeley, Late Bishop of Cloyne*, Second Edition (London 1784), p. 30.

Addison, I have good reason not to believe those trifles, since my friend Dr. Halley who has dealt so much in demonstration has assured me that the doctrines of Christianity are incomprehensible and the religion itself an imposture."[3] The Dr. Halley in question was none other than Edmond Halley, astronomer, mathematician, and associate of Sir Isaac Newton. (It was Halley who had encouraged and funded the publication of Newton's *Principia Mathematica*.) Halley would be appointed Astronomer Royal the following year, and to this day his fame returns every 75 years when the comet that bears his name approaches the earth. Halley, however, was also rumored to be an atheist.[4]

Addison's story apparently made a powerful impression on the young George Berkeley. Berkeley was a fellow of Trinity College Dublin who had come to London in 1713 to seek a broader audience for his works on philosophy. Jonathan Swift had introduced him to the leading lights of the London literary scene, and once there, Berkeley became increasingly concerned about the rise of a covert group of free thinkers (who subjected established religion to rational critique), going so far as to infiltrate their secret clubs in order to prove that they were in reality atheists.[5] Addison's account of Garth's deathbed denial of religion is said to have moved Berkeley to write a book exposing the dangerous tendencies of free thinking and attacking the very mathematics on which they based their arguments.[6] Berkeley was terrified that free thinkers and atheists were rapidly gaining power—ascending to the post of Astronomer Royal, destroying the faith of the king's physician, and publish-

---

[3] Stock, *Memoirs of George Berkeley*, p. 30.

[4] Simon Schaffer, "Halley's Atheism and the End of the World," *Notes and Records of the Royal Society of London* (1977) 32: 17-40.

[5] See David Berman, *George Berkeley: Idealism and the Man* (Oxford: Clarendon, 1994), p. 78.

[6] Given that Addison died in 1720 while Berkeley was on a tour of the Continent and Berkeley did not write *The Analyst* until the 1730s, this exchange is most likely apocryphal. See Alexander Campbell Fraser, *Life and Letters of George Berkeley*, Vol. 4 (Oxford: Clarendon, 1871), pp. 224-25.

ing blasphemous works about the irrationality of revealed religion. The very soul of the kingdom was at stake, and the greatest threat, he believed, lay with 'infidel mathematicians' and purveyors of the mechanical philosophy.

The magnitude of Newton's achievement and the shadow that Newton casts over the next hundred years obscures the fact that Newton's theories were quite controversial at first. In fact, Newton's work was full of inconsistencies, problematic assumptions, and dangerous implications, and the two figures who most clearly demonstrated this were both Irish—George Berkeley and John Toland. Toland was the first and the most dangerous of the free thinkers and the specter that haunts Berkeley's entire body of work. Toland argued that the science of Newton and the philosophy of John Locke invalidate the claims of established religious authorities. The revealed truths of Christianity, the doctrines of the Anglican and Catholic churches, he declared, are all meaningless cant. Matter itself, he argued, is active and requires no external intervention from God. And the universal laws of science, Toland decreed, demand a world where all men are equal before the law. For Toland, Newtonianism supported deism, pantheism, and republicanism, and invalidated Anglicanism and monarchy.

Given the intimate relationship between church and state in eighteenth-century Britain and Ireland, religious questions had definite political implications. In Ireland in particular, the political establishment depended on doctrinal distinctions between Anglicanism, Presbyterianism, and Catholicism. When radicals like Toland questioned those distinctions on the basis of Newtonian natural philosophy, they threatened to destroy the metaphysical foundations of Anglican rule. Newton's followers responded by clarifying the philosophical and religious bases of natural philosophy as a support to the establishment, but others, like George Berkeley, argued that Newtonianism itself was flawed and dangerous to religion.

As the children of the Cromwellian adventurers became landed Irish gentlemen, their tastes shifted from practical science to the pursuit of rigorous knowledge. Nobles, bishops, and government officials actively debated questions of mathematics and natural philosophy along with theology, politics, and literature. In fact, all were intimately related. The Civil War and the Glorious Revolution had not only disrupted political, social, and religious life, they made it clear that the old truths of Aristotelian philosophy, respect for the established Church, and the divine right of kings no longer had the power to contain the disruptive forces of new ideas. The search for social order became a search for methods of generating consensus on what we know and how we know.[7] Mathematics in particular was widely acknowledged as the model for right reasoning, and so arguments about the proper way to think about mathematics had immediate repercussions for philosophy, theology, and even politics.

Berkeley's critique of materialism would become the touchstone for a long tradition of Irish idealism. Generations of Irish scientists and poets, from William Rowan Hamilton to George Francis Fitzgerald to William Butler Yeats, shared Berkeley's distrust of materialism. Berkeley's vision of the power of ideas resonated strongly with the intellectual elite in a country riven by economic, political, and religious tensions. While Petty and his military colleagues valued mathematics as a rough but practical tool, Berkeley and his literary and philosophical friends saw mathematics as the most important front in the battle for the hearts and minds (and souls) of the English and Irish ruling class. Newton's calculus may have been a powerful method for solving problems, but for Berkeley the more important question was whether it would strengthen belief in God and support the Anglican church. Garth's cautionary tale showed just how high the stakes were.

---

[7]Steven Shapin and Simon Schaffer, *Leviathan and the Air-Pump: Hobbes, Boyle, and the Experimental Life* (Princeton: Princeton University Press, 1985).

2.1   A Science Fit for Palaces and Courts

Trinity College Dublin opened its first scientific laboratory in 1711.[8]
The two-story brick building constructed at a cost of one hundred
pounds contained a chemical laboratory, a lecture theater, a dis-
secting room, and a museum. The Board of Trinity intended it to
serve as part of the new medical school and accompanied it with the
establishment of lectureships in anatomy, chemistry, and botany.
That day Richard Helsham gave the first in a series of lectures on
Newton's natural philosophy. William Thompson, a bachelor of the
college, composed a Latin poem for the opening of the laboratory
whose rhetoric would have been familiar to the previous generation
of men like Hartlib, Petty, and Symner. At the beginning of the
age of enlightenment, Thompson's vision of science still reflected
the brutalities of the seventeenth century.

> See, Physics, bearing trophies from the foe
> Now vanquished, moves imperious on her way,
> Leading in chains the crowds of errors past
> And celebrating triumph unperturbed....
>
>
> We follow [Nature] at her heels and track her down
> And capture her, binding her fast with chains;
> Then summon Chemistry's tormenting fires,
> And force her to reveal her mysteries
> Even against her will.[9]

It was almost as if on the occasion of the establishment of a respectable scientific institution, Thompson felt the need to remind

---

[8]See D.E.W. Wormell, "Latin Verses By William Thompson Spoken at the Opening
in 1711 of the First Scientific Laboratory in Trinity College Dublin," *Hermathena* (1962)
96: 21-30, T.P.C. Kirkpatrick, *History of the Medical Teaching in Trinity College Dublin
and of the School of Physic in Ireland* (Dublin: Hanna and Neale, 1912), pp. 76-99, and
Wesley Cocker, "A History of the University Chemical Laboratory, Trinity College, Dublin,
1711-1946," *Hermathena* (1978) 124: 58-76, pp. 58-59.

[9]Wormell, "Latin Verses," pp. 25-26.

his audience of the less than respectable origins of modern science in Ireland.

By the end of the seventeenth century, Petty's view of practical science in the service of empire was gradually being replaced by another approach to science. The children of the Cromwellian adventurers presented themselves as cultured gentlemen, and cultivating the new experimental philosophy was just as important for them as building their country manors. No longer just the concern of practical men, science became a pastime for the elite in new gentlemen's clubs like the Dublin Philosophical Society. By the late seventeenth century, those with an interest in natural philosophy were more often gentry and clergymen than mechanics or navigators. Though they declared science to be independent of all divisive questions in religion and politics, they sought in science a philosophical authority that might justify their claims to religious and political authority rather than a practical tool to exercise military power.

The key figure in the establishment of science in Ireland during this period was William Molyneux.[10] Molyneux's father had served as master gunner in Cromwell's army, and Molyneux would later serve as joint surveyor general and chief engineer of Ireland. But Molyneux also had a strong interest in philosophy. When he entered Trinity College Dublin in 1671, he found the university's Aristotelian curriculum was not to his taste. He later recalled, "I could never approve of that verbose philosophy there professed and taught, but still procured the books of the Royal Society..., Descartes' writings, Dr. Bacon's works..."[11] Independently wealthy from the land acquired by his forebears, Molyneux lived the life of a gentlemen, demonstrating his scholarly interests by translating Descartes' *Meditations* as well as Galileo's *Discourses*

---

[10]See J.G. Simms and D.H. Kelly, *William Molyneux of Dublin, 1656-98* (Blackrock: Irish Academic Press, 1982).

[11]Quoted in Simms and Kelly, *William Molyneux*, p. 19.

on Two New Sciences. He also collected an impressive array of instruments and built his own small observatory.

Figure 2.1: William Molyneux (1656–1698), Founder
of the Dublin Philosophical Society. Portrait by Sir
Godfrey Kneller ©National Portrait Gallery, London

In 1682 Molyneux and his friends began meeting weekly in a private room in a coffeehouse, "merely to discourse of philosophy, mathematicks, and other polite literature."[12] Eventually they decided to formalize the group on the model of the Royal Society of London (itself less than twenty years old).[13] They elected the aging William Petty as their first president, while Molyneux served as secretary and treasurer. The group was primarily Anglo-Irish rather than English. Of the fourteen original members, nine had

---

[12]Quoted in K.T. Hoppen, *Common Scientist in the Seventeenth Century: A Study of the Dublin Philosophical Society, 1683-1708* (London: Routledge & Kegan Paul, 1970), p. 23.

[13]P. Kelly, "From Molyneux to Berkeley: The Dublin Philosophical Society and Its Legacy," pp. 23-31 in D. Scott (ed.), *Treasures of the Mind* (London: Sotheby's, 1992).

studied at Trinity College Dublin and three at Oxford University.[14]
The group soon contained seven present or future Anglican bishops, and only one Roman Catholic. Seven members were or became
fellows of the Royal Society of London, and both the Royal Society
and the Oxford Philosophical Society (founded in 1683) encouraged the infant Dublin Society through official correspondence and
informal ties.

K.T. Hoppen, historian of the society, has argued that the members devoted little time to the mathematical development of natural philosophy.[15] Molyneux, for example, found Newton's *Principia*
nearly incomprehensible when Edmond Halley sent him a copy in
the spring of 1687. After a year's effort, he complained, "I question after all, whether I shall be able to master it...Neither do
I know any mathematical head in this place that has thoroughly
considered the whole."[16] Yet Molyneux and his colleagues made
important advances in raising the status of mathematics in a society where practical mathematics was still often considered a trade.
Molyneux may not have been up to Newton's standards, but he
was very much a proponent of the mathematical sciences. In fact,
Molyneux's favorite subject was optics.[17]

In 1692 Molyneux published his magnum opus, *Dioptrica Nova.*
The work was the first of its kind in the English language, offering
Euclidean style proofs regarding the optics of telescopes, microscopes, magic lanterns, and vision. While the work was full of diagrams and proceeded in a Euclidean style, Molyneux also offered
experimental demonstrations for many of his theorems, combining
the theoretical and the practical. Molyneux divided his work into
two parts. The first part confined itself to geometrical optics, trac-
ing mathematically the paths of rays passing through glass lenses

---

[14]See Hoppen, *Common Scientist,* pp. 25-52.

[15]Hoppen, *Common Scientist,* pp. 130-31.

[16]Molyneux to Flamsteed (May 19, 1688), quoted in Hoppen, *Common Scientist,* p. 124.

[17]Molyneux to Flamsteed (April 11, 1682), quoted in Simms and Kelly, *William Molyneux,* p. 61.

of various types. In the second part, Molyneux allowed himself more room to speculate. He offered evidence that light is a body and described the principle of principle of least time, the argument, which went back to Hero of Alexandria but had been elaborated by Pierre de Fermat, that the laws of reflection and refraction can be derived from the assumption that light always takes the fastest path between two points. Molyneux also offered practical advice regarding optical instruments such as telescopes, spectacles, sextants, and quadrants. He even described the grinding of lenses and recounted debates within the astronomical community about the best kinds of telescopes. While the previous generation had used optics for navigation and gunnery, their children would use it for astronomy and geometrical optics.

The study of optics also led to important questions about perception. As philosophers began to marshal evidence from telescopes and microscopes to support new theories of astronomy or biology, questions arose about the degree to which these phenomena represented reality or were merely a creation of the instrument itself. This led to deeper questions about sense perception, which according to Locke was the source of all knowledge. In a letter to John Locke in 1693, Molyneux proposed what became known as 'the Molyneux Problem'.[18] If a person born blind learns to distinguish shapes by touch (say a cube and a sphere) and then suddenly gains sight, would he immediately be able to recognize the shapes visually? In other words, do our senses need to be trained to recognize a shape? Locke addressed the problem in Book II of the second edition of his *Essay Concerning Human Understanding*. He and Molyneux agreed that the shape of an object is a primary quality that causes our different sense perceptions of that object, but they also agreed that we cannot infer from the ideas of one sense to another. The blind man, they predicted, would not be able to

---

[18]Berman, *George Berkeley*, pp. 10-11, Menno Lievers, "The Molyneux Problem," *Journal of the History of Philosophy* (1992) 30: 399-415.

distinguish the shapes by sight. This idea would later become an important part of Berkeley's philosophy.

St. George Ashe, Molyneux's friend and associate in the creation of the Dublin Philosophical Society, further emphasized a view of mathematics as the study of rigorous truths fit for cultured gentlemen. Ashe was appointed Donegall Lecturer in Mathematics at Trinity College on the death of Miles Symner in 1686. He was also instrumental in getting telescopes for the observatory established at Trinity in 1685. Ashe became a very public promoter of mathematics and the new science. In October of 1685, the Earl of Clarendon (brother-in-law of James II) became Lord Lieutenant of Ireland (the head of the Irish administration), and the Dublin Philosophical Society hoped to recruit him as a patron. To that end, Ashe presented Clarendon with an address on behalf of the society.[19] The rise of the experimental philosophy, he argued, represented more than simply a new way for university scholars to study the natural world. Ashe explained, "philosophy was admitted into our palaces and our courts, began to keep the best company, to refine its fashion and appearance, and to become the employment of the rich and the great."[20] For Ashe (himself a landed gentleman) science was not, as it had been for Petty, primarily a means to get rich. In fact, Ashe praised men like Clarendon, whose wealth "secures [them] from the sordid considerations of gain and profit" and allows them "to reason and philosophize above the common rate of man-kind [or] at least to protect those who do."[21] For Ashe, as for Molyneux, science and mathematics were as much about respectability as they were about utility. As they distanced themselves from the rough manners of their colonizing ancestors, they

---

[19] St. George Ashe, "St. George Ashe's Speech to Lord Clarendon, 25 January 1685/86," pp. 275-78 in I. Ehrenpreis (ed.), *Swift: The Man, His Works, and the Age* (London: Methuen, 1962). See Hoppen, *Common Scientist*, pp. 97-98.

[20] Ashe, "Speech to Lord Clarendon," p. 276.

[21] Ashe, "Speech to Lord Clarendon," p. 275.

also distanced themselves from the rough mathematics of the surveyors and instrument makers.

While he had a great respect for experimental facts, for Ashe deductive mathematics represented the ideal form of knowledge. Mathematics, Ashe explained, "rejects all trifling in words and Rhetorical schemes, all conjectures and authorities, prejudices and passion."[22] He continued, "no proposition is pretended to be proved, which does not plainly follow from what was before demonstrated, as is manifest in Euclid's *Elements*."[23] Of course, this was no new argument. For over a thousand years Euclid's *Elements* had been the paradigm of indisputable truth. But while Petty was content to sacrifice such certainty for the sake of practical results, Ashe was more interested in presenting mathematics as a model of certainty. Ashe claimed that there are few or no controversies between professors of mathematics. He wrote, "most other Knowledge (how diverting soever) is yet still conjectural and litigious, whereas Peripatetick and Cartesian, Catholick and Heretick, do all agree in a Mathematical demonstration...."[24] In a world of warring factions in philosophy and religion, mathematics offered a vision of truth that transcended party, a world where consensus could be reached and prejudice abolished.[25]

The vision presented by Molyneux and Ashe of polite society and rational consensus was ultimately a fragile one. The members of the Dublin Philosophical Society depended on the establishment of Protestantism and the military force of the crown for their land and therefore for their ability to pursue their scientific interests. When

---

[22] St. George Ashe, "A New and Easy Way of Demonstrating Some Propositions in Euclid," *Philosophical Transactions* (1684) 14: 672-76, p. 672.

[23] Ashe, "A New and Easy Way," p. 672.

[24] St. George Ashe, *A Sermon Preached in Trinity College Chappell Before the University of Dublin, January the 9th, 1693/4. Being the First Secular Day Since Its Foundation by Queen Elizabeth* (Dublin, 1694), p. 13.

[25] This was a common claim for mathematics, see Peter Dear, *Discipline and Experience: The Mathematical Way in the Scientific Revolution* (Chicago: University of Chicago Press, 1995).

James II, a Roman Catholic, became king of England in 1685, they consoled themselves with the belief that he would not upset the Protestant establishment; but when his new Lord Lieutenant Tyrconnell began appointing Catholic judges and military officers in Ireland in 1687, they began to fear the worst. Molyneux was suspended from his position as surveyor general, and the Dublin Philosophical Society ceased its meetings. In November of 1688 William of Orange landed in England, and the following month James fled to Ireland. Tyrconnell raised an army and negotiated assistance from the French, preparing for war. Molyneux evacuated to Chester, and Ashe left to London and then to Vienna as chaplain to the English ambassador. In 1689, the provost of Trinity fled, and Tyrconnell replaced him with Michael Moore, a Catholic and an Aristotelian trained in Paris.[26]

On the first of July 1690 the Williamites defeated the Jacobites at the Battle of Boyne, and James fled to France. The Jacobites eventually surrendered, but with a Stuart pretender still alive in France, Irish Catholics were seen as rebels waiting for an opportune moment to strike. Over the next few decades the Irish parliament instituted a series of penal laws restricting their rights. Roman Catholics were forbidden from carrying arms, attending foreign universities, or owning horses greater than a set value. In 1697 all Catholic bishops and regular clergy were banished. In 1704 Catholic inheritance rights were restricted, as was their ability to purchase and lease land. All holders of public office were required to take the sacraments of the Anglican Church, and soon Catholics were also deprived of the right to enter the professions or to vote. As before, Catholic scholarship, suppressed in Ireland, continued to flourish on the continent. Michael Moore returned to France and became Rector of the University of Paris.[27] Patrick D'Arcy, a

---

[26]Thomas Duddy, *A History of Irish Thought* (London: Routledge, 2002), pp. 78-81.

[27]Liam Chambers, *Michael Moore, c. 1639-1726: Provost of Trinity, Rector of Paris* (Dublin: Four Courts Press, 2005).

former Jacobite, gained fame at the French Academy of Sciences for his mathematics.

After the war, the Williamite government appointed Molyneux commissioner for army accounts, and Ashe became Provost of Trinity College. Meetings of the Dublin Society resumed in April of 1693 at the Provost's House, and by the end of that year the Society had grown to 49 members, larger than ever. As Molyneux grew more interested in politics than in science, however, the society began to wane. Molyneux's son Samuel revived the society in 1707. Though it was only briefly lived, the second coming of the Dublin Philosophical Society was a symbolic passing of the torch. Richard Helsham, George Berkeley, and Bryan Robinson were all involved with the society during this final period. Together they explored the educational, scientific, philosophical, and religious consequences of Newtonianism, fast on its way to becoming scientific orthodoxy.

2.2 The Newtonian Establishment

Newtonianism in many ways represented the culmination of the work of men like Molyneux and Ashe. It connected the practical interests of navigators and mechanics with the scholarly interests of mathematicians and astronomers. Newton's *Philosophiae naturalis principia mathematica* [*Mathematical Principles of Natural Philosophy*] (1687) and his *Opticks* (1704) marked an epoch in the history of science.[28] They not only introduced the calculus and solved the problems of the motion of the planets, falling bodies, the tides, and the relation of colors to white light, they also gave simple mathematical laws and methodological precepts that promised to lead

---

[28]See Simon Schaffer, "Newtonianism," pp. 610-26 in R.C. Olby, G. N. Cantor, J. R. R. Christie and M. J. S. Hodge (eds.), *Companion to the History of Modern Science* (London: Routledge, 1990) and Betty Jo Teeter Dobbs and Margaret C. Jacob, *Newton and the Culture of Newtonianism* (Atlantic Highlands, NJ: Humanties Press International, 1995).

to even more discoveries. Newtonian science was so successful, in fact, that it soon came to impact literature, philosophy, and religion as well. Enlightened readers everywhere wanted to understand what it was all about, but (as Molyneux had discovered) Newton's works were written for trained mathematicians. So, beginning in the 1720s and 1730s, popular accounts of Newtonian science began to appear.

Richard Helsham's lectures at Trinity College Dublin were among these early attempts to explain Newton's science in a simple fashion.[29] Helsham began lecturing at Trinity College in 1721, and in 1724 he was appointed as the first holder of the new Erasmus Smith Professorship of Natural and Experimental Philosophy.[30] The new chair, the College's first professorship in the sciences, was funded from the Irish estates of Erasmus Smith, a London merchant who had adventured money in Cromwell's conquest. Despite his title, however, Helsham's main interest was in medicine. He served as president of the Royal College of Physicians as well as personal physician to Jonathan Swift, who described Helsham as "the most eminent Physician of this City and Kingdom."[31]

Helsham's approach was typical of early eighteenth-century popular Newtonianism.[32] On the very first page of the printed text he describes natural philosophy as "entertaining and delightful," in other words, suitable for young gentlemen. Unlike many other early Newtonian texts, however, Helsham's was aimed specifically

---

[29]Kirkpatrick, *History of the Medical Teaching*, pp. 79-82, and Norman D. McMillan, "Richard Helsham M.D. (1683-1737), Medical Man, Virtuoso and Educationalist: Author of the First Purpose-Written Student Textbook on Natural Philosophy in the Vernacular," *Science (Journal of the Irish Science Teachers Association)* (October 1988), pp. 13-18.

[30]See Denis Weaire, "Experimental Physics at Trinity," pp. 270-79 in C. H. Holland (ed.), *Trinity College Dublin & the Idea of a University* (Trinity College Dublin Press: Dublin, 1991).

[31]H. Williams (ed.), *The Correspondence of Jonathan Swift* (Oxford: Clarendon, 1965), Vol. IV, p. 360.

[32]Schofield chose it as one of three representative textbooks. See Robert Schofield, *Mechanism and Materialism: British Natural Philosophy in an Age of Reason* (Princeton: Princeton University Press, 1970).

at university students rather than at a general educated audience. In fact, Helsham's *Lectures* was one of the first books to be published by the new Dublin University Press whose printing house had been completed in 1734. Students were far from qualified to read Newton's original work, so Helsham glossed over Newton's complicated mathematical proofs and added simple experiments on nearly every page. Helsham's *Lectures* was written in a style that reassured students that all important questions had been answered and that modern science rested on secure foundations. As the purpose of science education in the eighteenth-century university was to appreciate examples of God's workmanship in the natural world, Helsham's text was a perfect element in a liberal education designed primarily for the training of Anglican clergymen.

In many ways, Newton's natural philosophy is quite similar to the modern view of physics as the science of particles and forces. Helsham (following Newton) claimed that there exist "certain principles, forces, or powers, wherewith all parts of matter of what kind soever, so far as experience reaches, seem to be endued; and whereby they act upon one another for producing a great part of the phaenomena of nature."[33] While the followers of Descartes claimed that matter is entirely passive and only interacts through collisions, the Newtonians believed that matter has certain powers, such as attraction, whose cause may be unknown but whose existence can be demonstrated experimentally. In the *Opticks*, for example, Newton speculated that God created solid massive particles and endowed them with certain powers that allow them to act at a distance on each other. These forces, Newton claimed, produce the familiar phenomena of nature, such as the refraction and reflection of light, gravity, magnetism, and electricity. Moreover,

---

[33]D. Weaire, P. Kelly and D.A. Attis (eds.), *Richard Helsham's Lectures on Natural Philosophy* [1739] (Dublin: Trinity College Dublin Physics Department, Institute of Physics Publishing, Verlag MIT, 1999), p. 2.

Newton was able to demonstrate that at least one force, gravity, acts according to a simple mathematical law.

Not everyone was convinced by the Newtonian position, however. Many objected that Newton's forces were 'occult' or mysterious. How can one body act on another without touching it? Throughout the *Lectures*, however, Helsham presents these phenomena as obvious and in no need of elaborate justification. The experiments, he seems to say, speak for themselves. This non-polemical style, common to later textbooks but not as common during this period, may be one of the reasons for the great success of the book and its appeal to both beginning students and teachers anxious to present Newtonian science as the bulwark of certainty and rationality. This explains why it was listed as a required text for students up through 1849, over a century after its original publication.

2.3   Robinson and the Aether

While Helsham's lectures exhibited a simplified and matter of fact style avoiding any hint of controversy, Helsham's colleague at Trinity (and the editor of his lectures) Bryan Robinson proposed elaborate theories and developed highly mathematical treatments of a wide range of phenomena.[34] Robinson's work illustrates how the attempt to extend Newton's theories could lead to difficult philosophical and theological problems. Specifically, Robinson's most important contribution was the revival of Newton's theory of the aether.[35] While Newton's immediate successors abandoned his speculative aether theory, new developments in chemistry, electric-

---

[34] Kirkpatrick, *History of the Medical Teaching*, pp. 109-12.

[35] See Schofield, *Mechanism and Materialism*, pp. 106-14 and Arnold Thackray, *Atoms and Powers: An Essay on Newtonian Matter-Theory and the Development of Chemistry* (Cambridge: Harvard University Press, 1970), pp. 135-41, and Peter Heimann, "Nature is a Perpetual Worker, Newton's Aether and Eighteenth Century Natural Philosophy," *Ambix* (1973) 20: 1-25.

ity, and physiology led to a renewed interest in the 1740s. Robinson stood out by his success in transforming Newton's cautious hints into a fully developed mathematical 'theory of everything' that he believed could explain all physical, chemical, and physiological phenomena.

Soon after the publication of Newton's *Principia*, a group of physicians applied Newton's methods to physiology and medicine.[36] Newtonian physicians attempted to find experimentally verifiable mathematical laws of animal and plant physiology. They saw human and animal bodies as complicated webs of vessels and explained the phenomena of physiology by the dimensions of the vessels and the velocities of the fluids within them. Robinson was one of the leaders in the mathematization of physiology.[37] In his 1732 *Treatise on the Animal Oeconomy*, Robinson explained muscular motion, blood flow, respiration, and digestion in mechanical and mathematical terms.[38] In order to explain the motion of the blood, for example, he began with an account of the motion of fluids through cylindrical pipes proven by experiments on brass pipes of different lengths and diameters.

Robinson first became interested in the aether, he claimed, because of Newton's speculation that the vibration of the aether along the nerves causes sensation.[39] In 1743 he offered an elaborate account of the aether as the cause of gravity, attraction and repulsion, elasticity, the phenomena of light, heat, muscular motion, sensation, and fermentation. The aether allowed Robinson to avoid the 'occult' forces that formed a troubling philosophical problem at the center of the Newtonian worldview. Following Newton, Robinson

---

[36]See Theodore M. Brown, "Medicine in the Shadow of the Principia," *Journal of the History of Ideas* (1987) 48: 629-48, and Anita Guerrini, "James Keill, George Cheyne, and Newtonian Physiology," *Journal of the History of Biology* (1985) 18: 247-66.

[37]Schofield, *Mechanism and Materialism*, p. 114.

[38] Bryan Robinson, *A Treatise of the Animal Oeconomy* (Dublin, 1732).

[39]Isaac Newton, *Opticks: Or a Treatise of the Reflections, Refractions, Inflections & Colours of Light-Based on the Fourth Edition London, 1730* (Dover, 1952), Queries 12-17, 23-24, pp. 345-48, 353-4.

imagined the aether as a fluid composed of very small particles
of equal size. Each particle, he hypothesized, acts on its nearest
neighbors with a repulsive force that causes the fluid to expand. In-
troducing forces between the particles of the aether itself, however,
required Robinson to explain the causes of these inter-particulate
forces. This led him into a hornet's nest of complicated issues,
particularly the question of whether force is inherent in matter or
imposed from outside.

As Robinson knew well, the answer to this question had important religious implications. Materialism, or the self-sufficiency of
matter, had long been seen as an attack on God's role in the universe. Robinson, therefore, concluded that the cause of the forces
between the particles cannot be material, "for Matter is in its own
Nature inert, and has not any Activity in itself."[40] The cause,
he decided, must be some spirit, which, "ordains and executes the
Laws by which Aether and Bodies act mutually on one another."
His aether theory, therefore, led him to an explicitly religious conclusion. Robinson later wrote, "The Aether is the most general of
all material causes, from which all particular causes derive their
power; and it derives its great power and force immediately from
God, who by it governs the material world."[41] While it began as
a mechanical explanation and a means to mathematize the phenomena of nature, in the end, Robinson argued that the aether
demonstrates the inability of matter itself to be a cause and the
continuing spiritual activity of God in nature. This view would
hold a fascination for Dublin scientists up through the early twentieth century.

While Helsham believed that Newtonian natural philosophy supported religion by providing evidence of God's design of the physical world, Robinson's claims were much more specific and much

---

[40] Robinson, *Dissertation on the Aether of Sir Isaac Newton* (Dublin, 1743), p. 122.

[41] Bryan Robinson, *Sir Isaac Newton's Account of the Aether, With Some Additions by way of Appendix* (Dublin, 1745), Preface.

more open to dangerous interpretations. Despite his pious conclusions, Robinson's work could easily be read as providing mechanical explanations for all phenomena and therefore denying the role of God in the universe. If mathematical and mechanical laws govern all of nature, some wondered, what role is there for divine providence? Such scientific accounts appeared to lead to determinism and materialism, long the enemies of revealed religion.

2.4 Materialism and Atheism

The dangerous implications that Robinson worked so hard to avoid, John Toland embraced. Toland is one of the most perplexing figures in Irish history, and in the eighteenth century he was regarded as one of the most dangerous. Born in 1670 into a Gaelic-speaking Roman Catholic family in the Inishowen peninsula in Donegal, he went on become alternately a Presbyterian, a freethinker, and an Anglican.[42] A learned Biblical scholar, he was often accused of atheism; and though he associated with eminent philosophers such as Locke and Leibniz, he managed to offend all of them. Toland's most influential work *Christianity Not Mysterious* (1696) utilized Locke's philosophy of language and the example of Newtonian natural philosophy to attack the notion that true Christian doctrine contains anything that cannot be understood rationally. The book initiated what became known as the deist controversy. Deists favored a Christianity based on rational principles, transcending the doctrinal differences between Christian sects, and Toland was widely read by French deists including Voltaire, Diderot, d'Holbach and Montesquieu.[43] When Toland visited Ireland in 1697, however, he encountered only animosity. The book was condemned by the Irish

---

[42]See the critical essays in Philip McGuinness, Alan Harrison and Richard Kearney (eds.), *John Toland's Christianity Not Mysterious* (Dublin: Lilliput, 1997), Robert Sullivan, *John Toland and the Deist Controversy* (Cambridge: Harvard University Press, 1982).

[43]Philip McGuiness, "*Christianity Not Mysterious* and the Enlightenment," pp. 231-42 in McGuinness, *Christianity Not Mysterious*, p. 237.

Parliament and burned by the common hangman. Toland had to flee Ireland to avoid prosecution himself.

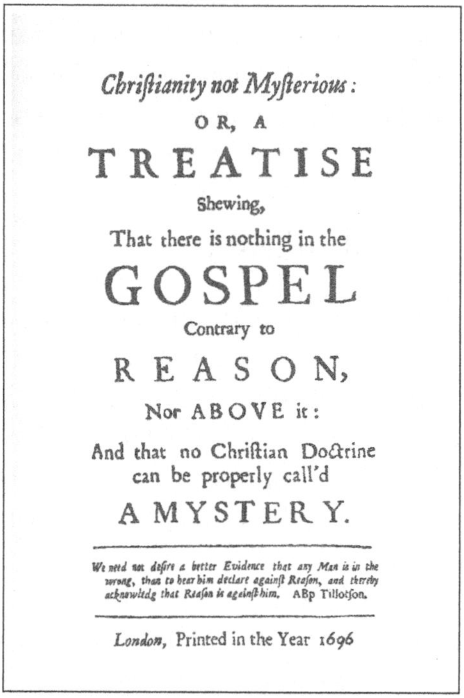

Christianity not Mysterious:
OR, A
TREATISE
Shewing,
That there is nothing in the
GOSPEL
Contrary to
REASON,
Nor ABOVE it:
And that no Christian Doctrine can be properly call'd
A MYSTERY.

We need not desire a better Evidence that any Man is in the wrong, than to hear him declare against Reason, and thereby acknowledge that Reason is against him. ABp Tillotson.

London, Printed in the Year 1696

Figure 2.2: Title Page to John Toland's *Christianity Not Mysterious* (1696), which used the ideas of Newton and Locke to attack established religion

Toland argued that reason has the power to penetrate the mysteries of religion just as it had penetrated the mysteries of the physical world. He ridiculed theological mysteries as "Scholastick Jargon" and "metaphysical Chimeras."[44] "Reason is the only Foundation of all Certitude;" he claimed, and "nothing reveal'd, whether as to its Manner or Existence, is more exempted from its Disquisitions, than the ordinary Phenomena of Nature."[45] Essentially,

[44]McGuinness, *Christianity Not Mysterious*, p. 8.
[45]McGuinness, *Christianity Not Mysterious*, p. 17.

Toland argued for the supreme authority of reason, even in questions of revelation. Locke had argued that if a word does not signify a distinct idea then it must be meaningless, and Toland applied this to the revealed truths of Christianity.[46]

While Toland presented his work as an attempt to save Christianity from the errors of selfish theologians, many saw it as an outright attack on religion. He argued that priests had invented religious mysteries as a means to control the masses. "They gravely tell us we must adore what we cannot comprehend,"[47] he fumed, and invent mysteries, "that we might constantly depend upon them for the Explication."[48] Toland wanted instead to democratize religious belief. Since everyone has reason, he argued, religion should be intelligible to all. While Protestants had long used such claims to criticize the Catholic emphasis on the infallible authority of the Church, such an argument could be used to attack any religious establishment, including Anglicanism. So while Toland attacked Catholic mysteries like transubstantiation, he also claimed that there is "no Difference between Popish Infallibility, and being oblig'd blindly to acquiesce in the Decisions of fallible Protestants."[49] Toland condemned all religious intolerance, preferring, "true Christian Liberty to Diabolical and Antichristian Tyranny."[50]

Toland's attack on Christian mysteries represented a fundamental challenge to the Anglican establishment. As historian David Berman explains, "if there were no Christian mysteries then there could be nothing to separate the rival Christian religions or sects. And then there could be no basis for the Penal Code."[51] A range of Irish philosophers and clergymen felt compelled to defend the

---

[46]Berman, *George Berkeley*, p. 148.

[47]McGuinness, *Christianity Not Mysterious*, p. 15.

[48]McGuinness, *Christianity Not Mysterious*, p. 34.

[49]McGuinness, *Christianity Not Mysterious*, p. 9.

[50]McGuinness, *Christianity Not Mysterious*, p. 11.

[51]David Berman, "The Irish Counter-Enlightenment," pp. 119-40 in R. Kearney (ed.), *The Irish Mind* (Wolfhound Press: Dublin, 1984), p. 137.

faith against Toland's claims. In fact, Berman sees Toland as "the father of Irish philosophy,"[52] giving birth to an Irish tradition in philosophy that lasted from the 1690s to the 1750s. Some writers, like Robert Molesworth (politician and anti-clerical writer), Thomas Emlyn (a Dublin Dissenting minister), and Francis Hutcheson (teacher at a Dissenting Academy in Dublin) reasserted Toland's radical conclusions. While others like Peter Browne (Provost of Trinity College and Bishop of Cork and Ross), Edward Synge (Archbishop of Tuam), William King (Archbishop of Dublin), Edmund Burke, and George Berkeley (Bishop of Cloyne) erected new philosophical supports for orthodoxy.

Toland's arguments against mysteries sabotaged the rational foundations of established religion by comparing them unfavorably to the methods of natural philosophy. But Toland went even further. While the Newtonians claimed that matter is passive and requires an external spirit to move it, Toland argued that matter itself is alive and active.[53] Toland's view, which he called pantheism, threatened to destroy all hierarchies in nature as well as in religion and politics. God is within nature, he claimed, rather than above it. Historians Margaret Jacob and James Jacob explain, "If spirit lay within man and nature, this was a strong argument against organized churches that were supported by tithes and learned ministries that claimed superior spiritual wisdom and separate spiritual authority."[54] Newton's own aether theory (like Robinson's) looked dangerously similar to Toland's pantheism, and Newton seems to

---

[52]See Berman, "The Irish Counter-Enlightenment," David Berman, "Enlightenment and Counter-Enlightenment in Irish Philosophy," *Archiv für Geschichte der Philosophie* (1982) 64: 148-65, David Berman, "The Culmination and Causation of Irish Philosophy," *Archiv für Geschichte der Philosophie* (1982) 64: 257-79.

[53]See John Toland, *Letters to Serena* [1704] (New York: Garland, 1976).

[54]J.R. Jacob and M. Jacob, "The Anglican Origins of Modern Science: The Metaphysical Foundations of the Whig Constitution," *Isis* (1980) 71: 251-67, p. 256.

have censored his own statements about the activity of matter after Toland's book appeared, fearing for their religious consequences.[55]

The Newtonians responded to arguments like Toland's by elaborating what they believed to be the proper metaphysical and religious implications of Newtonianism. A key venue for such claims was the Boyle Lectures, established by the great Irish chemist Robert Boyle on his death in 1691, "to prove the truth of the Christian religion against infidels." Boyle lecturers expounded on the pious implications of true Newtonianism while attacking competing systems of natural philosophy like those of Descartes, Hobbes, Spinoza, and Toland. A spiritually governed moral order, the Newtonians argued, depends on human and divine liberty, which in turn depend on inanimate matter. To deny that God actively rules the physical world, they thought, is to deny that he actively governs the moral world. Newton even believed that the irregularities in planetary motion would destroy the solar system without God's continual intervention.

Many of the earliest supporters and popularizers of Newtonian science were latitudinarian Whigs anxious to find ways to justify the new balance in the English Church and constitution following the Glorious Revolution.[56] They sought to ground order, hierarchy, and human authority in the natural order of the universe rather than the divine right of kings or revelation. These Low Church Whigs found that Newtonianism could be used to attack materialist republicans like John Toland as well as High Church Tories more interested in faith than reason. While Low Church Whigs tended to tolerate dissenters and to see the Church as subordinate to state authority, High Church Tories (including Berkeley and Swift) raged

---

[55]See Philip McGuinness, "'Perpetual Flux': Newton, Toland, Science and the Status Quo," pp. 317-29 in McGuinness, *Christianity Not Mysterious.*

[56]Margaret C. Jacob, *The Newtonians and the English Revolution, 1689-1720* (New York: Gordon and Breach, 1976).

against dissenters and placed the authority of the Church on the same level as the state.

The Newtonians were in a precarious position. On the one hand, they wanted to assert the reasonableness of religion and the importance of tolerance against the claims of the High Churchmen. On the other hand, they wanted to distance themselves from the radical claims made by the freethinkers in the name of reason.[57] The Newtonians felt attacked on both sides, fearing a universe both entirely spiritual and a universe entirely material.[58]

2.5   Immaterialism and Religion

Toland's attack on the mysteries of revealed religion and the power of its ministers showed that Newtonianism could be as much a threat as a support to the establishment. George Berkeley, who entered Trinity College Dublin as a student in 1700, just after the publication of Toland's work, would ultimately attempt to dismantle the very foundations of the Newton-Locke universe in order to rebuild a philosophy that he believed would guide people to God and away from skepticism and atheism. In his defense of revealed religion Berkeley attacked nearly all of the fundamental tenets of Newtonianism—absolute space and time, infinitesimal quantities, abstract ideas such as force, and even the very existence of matter.

While Toland came from a Catholic, Gaelic family, Berkeley was the son of an English commissioned officer and gentleman farmer.[59] He was born near Kilkenny and attended Kilkenny College, the same school that counted Jonathan Swift and William

---

[57] J.R. Jacob and M. Jacob, "The Anglican Origins of Modern Science: The Metaphysical Foundations of the Whig Constitution," *Isis* (1980) 71: 251-67.

[58] See Richard G. Olson, "Tory-High Church Opposition to Science and Scientism in the Eighteenth Century: The Works of John Arbuthnott, Jonathan Swift, and Samuel Johnson," pp. 171-204 in J. G. Burke (ed.), *The Uses of Science in the Age of Newton* (University of California Press: Berkeley, 1983).

[59] Berman, *George Berkeley.*

Congreve among its recent graduates. After Berkeley matriculated at Trinity he quickly came to know Molyneux and Ashe, two men who (along with Jonathan Swift) would help Berkeley navigate the complicated politics of the Dublin and London social worlds.[60] After completing his bachelor's degree in 1704, Berkeley remained at Trinity to prepare for the fellowship examination, renowned as one of the most rigorous and comprehensive of any institution of higher learning. In particular, it required Berkeley to master the latest mathematical techniques in addition to philosophy, history, and divinity. Berkeley became a fellow in 1707 and wrote most of his great works while at Trinity, including *An Essay Towards a New Theory of Vision* (1709), *A Treatise Concerning the Principles of Human Knowledge* (1710), and *Three Dialogues Between Hylas and Philonous* (1713).

When Samuel Molyneux revived the Dublin Philosophical Society in 1707, Berkeley became an active participant. Molyneux and Ashe were still the society's leading lights, but the members now included much of Dublin's Anglican elite such as Peter Browne, the provost of Trinity (and one of Toland's attackers), William King, the archbishop of Dublin, and the Earl of Pembroke, the Lord Lieutenant of Ireland (to whom Locke had dedicated his *Essay Concerning Human Understanding* seventeen years earlier). Science provided an entrée for Berkeley into the Dublin power elite.

Berkeley's notebooks from this period show that he was already immersed in philosophical questions around mathematics. His first published paper, presented to the Dublin Philosophical Society in 1707, investigated whether the infinitesimally small quantities used in the calculus have a sufficiently rigorous conceptual underpinning. Berkeley adopted Locke's position that "we ought to use no sign without an idea answering it," but then he was forced to ask, If

---

[60]On Berkeley and Irish politics, see Richard Kearney, "George Berkeley: We Irish Think Otherwise," pp. 145-56 in R. Kearney (ed.), *Postnationalist Ireland: Politics, Culture, Philosophy* (Routledge: London, 1997).

we are unable to conceive of infinitely small quantities, how can we reliably reason with them? While Berkeley did not dispute that the new techniques of the calculus have enabled "prodigious advances," he was concerned that they have also given rise to disputes "to the great scandal of the so much celebrated evidence of Geometry."[61] His goal in the paper was to show that the results of the calculus can be obtained rigorously using classical geometrical methods.

There was a danger in the approach, however. If we ban all concepts not based on clear ideas, how can we continue to use religious ideas, such as the Trinity, grace, or the afterlife, that refer to mysteries that are incomprehensible. If we hold mathematics to this standard, must we not also hold religion to it? This is what Toland had done. Apparently it was not until he presented his first paper to the Dublin Philosophical Society that Berkeley realized the dangers in arguing that it is meaningless to talk about infinity.[62] After the meeting, the implications seem to have been clear to him and a few months later in his first sermon he went out of his way to use the term 'infinite' in describing the mystery of heaven. A critical question for Berkeley became—if we accept with Locke that all of our ideas come from sense impressions, how can we form ideas of things that we cannot sense directly? And this led him to study the sense of vision.

Berkeley's philosophical development began with a critique of geometrical theories of vision.[63] Optics not only represented one of the most successful areas of applied mathematics, but it also raised questions about perception and hence epistemology. Berkeley started with Molyneux's work in optics, but he came to believe that Molyneux, like other writers on optics, had mistakenly as-

---

[61] Douglas M. Jesseph, *Berkeley's Philosophy of Mathematics* (Chicago: University of Chicago Press, 1993), p. 168.

[62] Berman, *George Berkeley.*

[63] See Geoffrey Cantor, "Berkeley, Reid, and the Mathematization of Mid-Eighteenth-Century Optics," *Journal of the History of Ideas* (1977) 38: 429-48 and C. Turbayne, "Berkeley and Molyneux," *Journal of the History of Ideas* (1955) 16: 339-55.

sumed that the mathematical laws of optics explain how we see. Descartes, for example, argued that we have an innate sense of geometry that enables us to calculate the distance of objects by measuring the angles between the rays of light that come from visible objects, much as a surveyor measures distances through a process of triangulation. In his *Essay Toward a New Theory of Vision* (1709), Berkeley argued that we do not actually perceive distance and magnitude directly but rather through learned associations. We do not see rays of light but rather colors, and so we do not 'see' distance; we judge distance based on visual clues such as faintness of color or smallness of magnitude. And sometimes these clues are misleading. Berkeley noted the phenomenon of the Harvest Moon, for example, which appears larger near the horizon than it does when it is high in the sky even though its actual size does not change. Berkeley explained, "those lines and angles, by means whereof some men pretend to explain the perception of distance, are themselves not at all perceived."[64] Geometrical rays, he continued, "have no real existence in nature, being only an hypothesis framed by the mathematicians, and by them introduced into optics, that they might treat of that science in a geometrical way."[65] While geometry may be a useful calculating device for the solution of certain problems in optics, this does not mean that the human mind actually perceives distance by calculation.

As further evidence, Berkeley brought up the Molyneux problem. Recall that Molyneux had hypothesized that a person born blind and given sight would not be able to recognize the difference between a cube and a sphere at first sight. In fact, Berkeley would later supply experimental support for this hypothesis. A 1728 report in the *Philosophical Transactions* described the case of a boy

---

[64]George Berkeley, "An Essay Towards a New Theory of Vision," pp. 1-70 in Michael Ayers (ed.), *George Berkeley: Philosophical Works Including the Works on Vision* (London: Everyman, 1975), p. 9.

[65] Berkeley, *New Theory of Vision*, sect. 14.

blind from infancy and made to see. Berkeley relates, "When he first saw, he was so far from making any judgment about distances that he thought all objects whatever touched his eyes (as he expressed it) as what he felt did his skin...He knew not the shape of anything, nor any one thing from another, however different in shape or magnitude."[66] This proved, he believed, that we have no innate geometrical ability to measure distance or compare shapes.

Berkeley would follow this train of thought to undermine the distinction between primary and secondary qualities that underlay the Newtonian-Lockean vision of the world. Newton, like most mechanical philosophers, imagined a world of atoms of different sizes, shapes, directions, and velocities. It is this underlying reality that causes our perceptions of secondary qualities such as color, taste, or smell. The secondary qualities are not essential to matter because they are subjective—different people experience them differently—but their underlying cause is objective. This view was critical for the mathematical description of nature, for it made the qualitative world of experience depend on an underlying quantitative reality that could be completely described in mathematical terms (at least in theory). According to Berkeley, however, since we cannot sense these underlying atoms, we can have no real idea of them. Moreover, even the so-called primary qualities, he believed, are themselves subjective. What is small to a human, for example, is large to a fly. Number itself is subjective: "We call a window one," he explains, "a chimney one, and yet a house in which there are many windows and many chimneys hath an equal right to be called one."[67] Numbers cannot exist independently of the perceiving mind: "number...is nothing fixed and settled, really existing in things themselves. It is entirely a creature of the mind."[68] In

---

[66] Berkeley, *New Theory of Vision*, sect. 71.

[67] Berkeley, *New Theory of Vision*, sect. 109.

[68] Berkeley, *New Theory of Vision*, sect. 109.

Berkeley's radical empiricism, the only things that exist are objects of sensation.

So how do we explain the fact that when we see a sphere and reach out and touch it we can also feel the sphere? Common sense says that this is simply due to the fact that there is a single object causing both sensations. Berkeley disagrees. Sense perceptions, he argues, are better thought of as signs. As Berkeley puts it, "visible figures represent tangible figures much after the same manner that written words do sounds."[69] Similarly, Berkeley believes that there is no necessary connection between visual sensations and tangible sensations. We must learn from experience that one goes with the other just as Molyneux's newly sighted man must learn that the sphere he sees is correlated with the sphere he can touch. We associate the two in a conventional manner. In Berkeley's view, we do not see a house; we see colors and shapes and recognize in them the sign for a house. The coincidence of these signs becomes so familiar that it feels as though we perceive the house directly, just as when we hear words in a familiar language we "act as if we heard the very thoughts themselves."[70] It was not until the third edition of the *New Theory of Vision* that Berkeley revealed the nature of this visual language. "The proper objects of vision," he explained, "constitute an universal language of the Author of Nature."[71]

In his *Principles of Human Knowledge* (1710), Berkeley made another great leap—the one that would ensure his inclusion in the history of philosophy. Berkeley's radical conclusion was that *nothing* exists independently of mind. The table, for example, does not exist as an independent structure of atoms; it exists only as the collection of our various perceptions of color, texture, and shape.[72]

---

[69] Berkeley, *New Theory of Vision*, sect. 143.

[70] Berkeley, *New Theory of Vision*, sect. 51

[71] Berkeley, *New Theory of Vision*, sect. 147.

[72]See Daniel Garber, "Locke, Berkeley, and Corpuscular Scepticism," pp. 174-93 in C. Turbayne (ed.), *Berkeley: Critical and Interpretive Essays* (University of Minnesota Press: Minneapolis, 1982), Nancy L. Maull, "Berkeley on the Limits of Mechanistic Explanation,"

These perceptions or ideas can exist only in a perceiving mind. Therefore the universe is composed only of ideas and minds. Ideas, however, being inert, passive, and dependent on mind for existence, cannot be causes. Berkeley elaborates, "the connexion of ideas does not imply the relation of *cause* and *effect*, but only of a mark or *sign* with the thing *signified*. The fire which I see is not the cause of the pain I suffer upon my approaching it, but the mark that forewarns me of it."[73] Nature, far from being an independent entity, is rather a divine language that we perceive directly. Therefore we are constantly in the presence of God, and all of our sensations are evidence of His existence. Berkeley explains, "all the choir of heaven and furniture of the earth, in a word all those bodies which compose the mighty frame of the world, have not any subsistence without a mind ... their *being* is to be perceived or known."[74]

Natural philosophy, on Berkeley's view, becomes the interpretation of God's natural language, just as theology is the interpretation of God's Biblical language. Natural philosophers, according to Berkeley, should look for harmonies and agreements such as Newton found between a falling stone, a planet in its orbit, and the tides, but they should not attribute these effects to the existence of gravity as a quality inherent in all bodies. "There is nothing necessary or essential in the case [of mutual attraction]," claimed Berkeley, "but it depends entirely on the will of the *governing spirit*, who causes certain bodies to cleave together, or tend towards each other, according to various laws."[75] Berkeley hoped to end the dominant focus on the properties of matter, an emphasis that he believed led men away from a proper view of God. "[T]he

---

pp. 95-107 in C. Turbayne (ed.), *Berkeley: Critical and Interpretive Essays* (University of Minnesota Press: Minneapolis, 1982), and Margaret Atherton, "Corpuscles, Mechanism and Essentialism in Berkeley and Locke," *Journal of the History of Philosophy* (1991) 29: 47-67.

[73] George Berkeley, "A Treatise Concerning the Principles of Human Knowledge," pp. 71-154 in Michael Ayers (ed.), *Philosophical Works*, pp. 114-15.

[74] Berman, *George Berkeley*, p. 43.

[75] Berkeley, "Principles of Human Knowledge," p. 131.

doctrine of matter or corporeal substance," he explained, is "the main pillar and support of *scepticism*, [and] upon the same foundations have been raised all the impious schemes of *atheism* and irreligion."[76] Geometrical optics and concepts of force and gravitation, he explained, may "serve the purpose of mechanical science and reckoning; but to be of service to reckoning and mathematical demonstrations is one thing, to set forth the nature of things is another."[77]

Even mathematics itself had the potential to lead men astray. Berkeley believed that it was essential to differentiate between those mathematical practices that were rigorous and well founded and those that were not.[78] Geometry, he felt, offered a paradigm of rational thought, and from it "there is acquired an habit of reasoning, close and exact and methodical: which habit strengthens and sharpens the mind, and being transferred to other subjects is of general use in the inquiry after truth."[79] The calculus, however, was a different story altogether. Newton's fluxional calculus and Leibniz's differential calculus, he argued, are full of questionable claims and vitriolic debates.

While Ashe had testified that there are no disputes in mathematics, Berkeley was not only prepared to acknowledge them, he was willing to abandon the façade of the purity of mathematics to protect theology. In fact, Berkeley became the most incisive and important critic of modern mathematical techniques, unmasking a host of conceptual issues that would not be resolved by mathematicians until the nineteenth century. His goal was not to take issue

[76] Berkeley, "Principles of Human Knowledge," p. 125.

[77] George Berkeley, "De Motu," pp. 253-76 in Ayers (ed.), *Philosophical Works*, p. 259.

[78] See Douglas M. Jesseph, *Berkeley's Philosophy of Mathematics* (Chicago: University of Chicago Press, 1993), Geoffrey Cantor, "Berkeley's *The Analyst* Revisited," *Isis* (1984) 75: 668-83, Helena M. Pycior, "Mathematics and Philosophy: Wallis, Hobbes, Barrow and Berkeley," *Journal of the History of Ideas* (1987) 48: 265-86.

[79] Berkeley, *The Analyst*, pp. 65-66.

with the results of the calculus but to remove the grounds for any argument that mathematics is more rigorous than religion.

Newton and Leibniz had, at about the same time, introduced new methods that they found useful in describing nature mathematically. Their techniques, known as the calculus, were essentially an extension of Descartes' application of algebra to geometry. Descartes described geometrical curves with algebraic equations. The calculus was a method for finding the tangents to such curves or the area underneath them. Tangents and areas had important physical interpretations. The tangent represents the rate of change of a variable. So given an equation for velocity over time, the tangent or differential of the curve represents the acceleration at any given point in time. The integral, or area under the curve, is the inverse. So the area under the velocity curve represents the distance traveled. Crucial to the early method of the calculus was the use of infinitesimal quantities. To find the rate of change in a given moment of time, one considered a finite period of time and gradually reduced the time period to zero.

Berkeley mocked the infinitesimals as "the ghosts of departed quantities."[80] While extremely useful, the method lacked the traditional rigor of Euclidean geometry. Berkeley's critique of the calculus was closely related to his anti-materialism. Only "one who thinks the objects of sense exist without the mind," he explained, could believe "that a line but an inch long may contain innumerable parts really existing, though too small to be discerned."[81] If only objects of sense exist, and infinitesimal quantities cannot be seen, then, he concluded, they cannot exist. In his 1734 text *The Analyst*, Berkeley offered a detailed critique of the calculus, demonstrating the apparently contradictory assumptions made in both Newton's fluxional calculus and Leibniz's differential calculus. His critique hinged on the incomprehensibility of infinitesimal quantities, ex-

---

[80] Berkeley, *The Analyst*, sect. 35.
[81] Berkeley, *Treatise*, p. 140.

hibited in the fact that in the course of a proof they must first be assumed to have some small length and later be assumed to have no length at all.

The difficulties of the calculus not only sabotaged the mathematicians' claims to mastery of reason, they also invalidated their attempt to criticize revealed religion as mysterious. The mysteries of the fluxional calculus, Berkeley argued, are such that "he who can digest a second or third fluxion, a second or third difference, need not, methinks, be squeamish about any point in divinity."[82] While Toland argued that Christianity is no more mysterious than Newtonian natural philosophy, Berkeley retorted that the calculus is at least as mysterious as theology. Even the apparently sedate, stable, and rational world of mathematics was riven by questions about its metaphysical foundations, its applicability to natural philosophy, the role of mystery, and the importance of arguments from authority. Mathematicians, Berkeley realized, laid claim to rationality and used that authority to define rationality in religion as well as science.

Far from denying rationality as some of his critics charged, Berkeley went to the heart of their claims, exposing the new mathematics itself as irrational and mysterious. Interestingly, Berkeley did not believe that the results of the calculus were false. In fact, he showed that it is possible to use the more rigorous method of exhaustion within a traditional Euclidean approach to reach many of the same results. But while the method of exhaustion was rigorous, it was significantly more onerous to apply. The infinitesimal approach is a kind of shortcut, but the price for taking that shortcut was too great for Berkeley. He explained, "I am not concerned about the truth of your Theorems, but only about the way of coming at them; whether it be legitimate or illegitimate, clear or obscure, scientific or tentative."[83] The usefulness, even the truth

---

[82] Berkeley, *The Analyst*, sect. 7.
[83] Berkeley, *The Analyst*, sect. 20.

of mathematics, was less important to Berkeley than the process
for reaching truth and ultimately the impact of that process on a
person's relationship to the divine.

Berkeley had systematically destroyed the foundations of the
materialistic and mechanistic view, removing anything that might
stand in the way of a direct relationship between man and God.
But he still had to rebut Toland's critique of religious mysteries.
The calculus may be as mysterious as the Trinity, but how can we
truly believe in the Trinity if it is not based on sense perceptions?
Berkeley admitted that in fact we can have ideas that do not stand
for sense impressions. Just as a blind man can form no distinct
idea of color but can still acknowledge its existence, so can we talk
about the Christian mysteries even though we cannot base them
on sense perceptions. Moreover, these concepts are useful. Words
can do more than simply refer to distinct ideas; they can also ex-
cite certain passions. As Berman explains, "Berkeley was the first
modern philosopher to formulate and support the theory that words
have legitimate uses which do not involve informing or standing for
ideas."[84] Berkeley argued that the Holy Trinity, original sin, grace,
and heaven are all emotive and not cognitive terms. Non-cognitive
words concerning the afterlife, for example, can evoke emotions, actions, and dispositions. Matter is also one of those emotive words,
but it has a negative tendency towards skepticism, atheism, and
idolatry. It leads men to act as if the causes of their sensations
were material rather than spiritual. The problem, therefore, was
not merely correct belief but also proper conduct, "the apprehen-
sion of a distant Deity, naturally disposes men to a negligence in
their *moral* actions, which they would be more cautious of, in case
they thought Him immediately present..."[85]

[84]Berman, *George Berkeley*, p. 144.

[85]George Berkeley, "Three Dialogues Between Hylas and Philonous," pp. 155-252 in Ayers
(ed.), *Philosophical Works*, p. 247.

Ideas and words, Berkeley concluded, matter because they change the way that people experience the world and lead them to act in different ways. His goal was not a logical deduction— 'I believe in God therefore I must follow His laws'—but rather a feeling—'I experience all of reality as thoughts in God's mind therefore I must follow His laws.' In other words, how one experiences the activity of doing mathematics is more important than any mathematical result. For this reason, Berkeley felt that it was essential to preserve mathematics as unassailably rigorous, for that was the only way to produce the proper emotional response and therefore the best support for religion. As he explained, "it is downright impossible, that a soul pierced and enlightened with a thorough sense of the omnipotence, holiness and justice of that Almighty Spirit, should persist in a remorseless violation of his laws."[86]

Berkeley was willing to sacrifice usefulness for intellectual rigor, and the power of his critique forced the mathematical community in Ireland and England to reconsider their own positions. His arguments elicited more than a dozen responses from mathematicians between 1734 and 1750. Every text on fluxions had to react to Berkeley's charges. In fact, some historians of mathematics blame Berkeley for the fact that mathematics in Britain and Ireland stagnated in the mid to late eighteenth century, held back by an undue focus on rigor and geometrical approaches while mathematicians on the Continent, less concerned with the theological implications of their work, forged ahead in developing a range of sophisticated and powerful new mathematical techniques.[87] It would not be until the 1790s that mathematicians at Cambridge and Trinity College

---

[86]George Berkeley, *A Treatise Concerning the Principles of Human Knowledge*, 1st edn, sect. 155.

[87]Philip Kitcher, *The Nature of Mathematical Knowledge* (Oxford: Oxford University Press, 1984), pp. 239-40. David Sherry, "The Wake of Berkeley's Analyst: Rigor Mathematicae?," *Studies in History and Philosophy of Science* (1987) 18: 455-80.

Dublin would adopt the Continental version of the calculus and once again begin to contribute to the progress of science.

2.6 Swift's Satire on Science

While Berkeley attacked the vanity of the modern mathematicians and the impracticality of their discoveries through detailed philosophical and mathematical analysis, Jonathan Swift took another tack. Through parody and satire, Swift exposed the mathematicians, mechanistic physicians, political economists, and assorted projectors as misguided, incompetent, and even dangerous.

In Book III of *Gulliver's Travels*, Swift created what is perhaps literature's most famous parody of science. At this point in his journey, Lemuel Gulliver visits the flying island of Laputa. The Laputans, Gulliver finds, are a strange race, clothed in gowns covered with astronomical and musical symbols with one eye turned inward and one up towards the sky. Servants follow them everywhere, flapping their ears and mouths. As Gulliver relates, "the minds of these people are so taken up with intense speculations, that they neither can speak, nor attend to the discourses of others, without being roused by some external taction upon the organs of speech and hearing...."[88] In fact, Gulliver finds to his amazement that, "the husband is always so rapt in speculation, that the mistress and lover may proceed to the greatest familiarities before his face, if he be but provided with paper and implements...."[89]

The Laputans are obsessed with pure mathematics and music. At dinner Gulliver is given a shoulder of mutton cut into an equilateral triangle, and a pudding cut into a cycloid. The tailor sent to outfit Gulliver in more suitable clothing takes his measurements with a quadrant and sketches Gulliver's dimensions on paper with

---

[88] Jonathan Swift, "Gulliver's Travels [1726]," pp. 21-278 in Miriam Kosh Starkman (ed.), *Gulliver's Travels and Other Writings* (New York: Bantam Books, 1981), p. 157.

[89] Swift, *Gulliver's Travels*, p. 163.

a ruler and compass. Unfortunately, six days later the clothes arrive "very ill made, and quite out of shape" due to a mistake in the calculation. In fact, for all their expertise in pure mathematics, the Laputans are incapable of doing anything practical. "Their houses are very ill built," Gulliver observes, "the walls bevil, without one right angle in any department; and this defect ariseth from the contempt they bear for practical geometry, which they despise as vulgar and mechanic...."[90]

The flying island uses an enormous magnet (described by Gulliver in great detail) to stay aloft and to move around the kingdom of Balnibarbi over which it rules. As Laputa travels over the kingdom, the people below send up petitions and sometimes wine and food in baskets that are lowered down for the occasion. If the people rebel, the king keeps the island over a town, blocking out the sun and rain until they relent. In extreme circumstances the king can even drop the island directly on their heads.

---

[90] Swift, *Gulliver's Travels*, p. 161.

Figure 2.3: The Flying Island of Laputa from Jonathan Swift's *Gulliver's Travels* (1726). Illustration by Milo Winter (1936)

Annoyed with the abstractness of his Laputan hosts, Gulliver decides to visit the capital city of Lagado in the kingdom of Balnibarbi. There he finds a stark contrast to the wealth of Laputa: "I never knew a soil so unhappily cultivated," he notes, "houses so ill contrived and so ruinous, or a people whose countenances and habit expressed so much misery and want."[91] Lord Munodi, apparently the only sane person in the kingdom, explains the cause of Balnibarbi's ruin:

> [A]bout forty years ago, certain persons went up to Laputa... and after five months continuance came back with a very little smattering in mathematics, but full of volatile spirits acquired in that airy region. ... [T]hese persons upon their return began to dislike the manage-

---

[91] Swift, *Gulliver's Travels*, p. 173.

ment of every thing below, and fell into schemes of putting all arts, sciences, languages and mechanics upon a new foot.[92]

"[N]one of the projects are yet brought to perfection;" Munodi relates, "and in the mean time the whole country lies miserably waste, the houses in ruins, and the people without food or clothes."[93]

Gulliver then decides to visit the Grand Academy of Lagado where he encounters a bewildering array of ill-conceived projects. First he meets a man who has spent the last eight years "upon a project for extracting sunbeams out of cucumbers, which were to be put into vials hermetically sealed, and let out to warm the air in raw inclement summers."[94] Next he encounters a man working on an operation to transform human excrement back into its original food. Gulliver meets a man born blind employed to mix colors (without much success), and when he finds himself with a colic, he is taken to a physician who cures his patients using a bellows to mechanically withdraw or introduce air into the anus. And these are the practical projectors! Among the speculative projectors, one has devised a means to rid language of all words, "since words are only names for things, it would be more convenient for all men to carry about them such things as were necessary to express the particular business they are to discourse on."[95] He has great hopes for his method, despite the fact that the particularly verbose must carry enormous bags full of the objects of their discourses. Another projector is working on a new method of mathematics education where propositions and demonstrations are written on thin wafers and fed to students so that they will be digested and ascend to the brain. A political projector suggests, based on the "strict universal resemblance" between the natural and the political body, that

---

[92] Swift, *Gulliver's Travels*, p. 175.
[93] Swift, *Gulliver's Travels*, p. 175.
[94] Swift, *Gulliver's Travels*, p. 177.
[95] Swift, *Gulliver's Travels*, p. 183.

physicians should be sent to the senate to feel the senators' pulses and prescribe medicines to solve political problems.

The bizarre and senseless projects that Gulliver witnesses in Lagado are all exaggerated versions of actual proposals from the pages of the *Philosophical Transactions* and other scientific journals.[96] Swift parodies the uselessness of abstract mathematics, mechanical theories of the human and animal body, political economy, Locke's theory of language, and agricultural improvement. Balnibarbi, generally read as Ireland, is ruined by its attempt to mimic the innovations of Laputa, generally read as England. The Laputans have been seen as Cartesians or followers of Hobbes,[97] while the Academicians of Lagado are the Baconians. In other texts like *Tale of a Tub, The Battle of the Books,* and *A Modest Proposal,* Swift further makes a mockery of the scientific pretensions of Baconians, Newtonians, natural theologians, and political economists.

Swift's jabs, however, were no more successful than Berkeley's critique in uprooting the steadily growing Newtonian orthodoxy. The phenomenal popularity of Helsham's Newtonian textbook, indeed its familiarity even today, represents the tremendous success of the Newtonian worldview. But these attacks and alternatives demonstrate that even Newtonian natural philosophy, the paradigm of scientific achievement, involved contentious claims about the goals of natural philosophy, the relationship between natural philosophy and theology, and the political implications of various philosophical claims. As Newtonian natural philosophy

---

[96]See Marjorie Nicolson and Nora M. Mohler, "The Scientific Background to Swift's 'Voyage to Laputa'," *Annals of Science* (1937) 2: 299-334, and "Swift's 'Flying Island' in the Voyage to Laputa," *Annals of Science* (1937) 2: 405-30.

[97]See M.K. Starkman, *Swift's Satire on Learning in 'A Tale of a Tub'* (Princeton: Princeton University Press, 1950), Robert H. Hopkins, "The Personation of Hobbism in Swift's *Tale of a Tub* and *Mechanical Operation of the Spirit,*" *Philological Quarterly* (1966) 45: 372-78, Colin Kiernan, "Swift and Science," *The Historical Journal* (1971) 14: 722, Clive T. Probyn, "Swift and the Physicians: Aspects of Satire and Status," *Medical History* (1974) 18: 249-61, David Renaker, "Swift's Liliputians as a Caricature of the Cartesians," *PMLA* (1979) 94: 936-44, Robert P. Fitzgerald, "Science and Politics in Swift's Voyage to Laputa," *Journal of English and Germanic Philology* (1988) 87: 213-29.

made its way into the universities and philosophical societies, its implications for religion and politics could not be ignored. In fact, Newtonianism was valued precisely because people believed that it could teach valuable religious lessons; and once its radical implications were suitably dismissed, it became an essential component of Trinity's theological education.

2.7   We Irishmen ...

While Molyneux, Berkeley, and Swift were all deeply engaged with questions in natural philosophy and mathematics, their participation in the Irish political establishment led them all unavoidably to questions of political philosophy and governance as well. The previous generation had developed Ireland as a colony. Their descendants, however, began to chafe under the political restrictions placed on Ireland by the English parliament. Historian Roy Foster explains, "Those who in the 1690s called themselves 'the Protestants of Ireland' or even 'the English of this kingdom' could see themselves as 'Irish gentlemen' by the 1720s."[98] This 'colonial nationalism' as it was called by later historians represented the increasing self-awareness and self-confidence of the Irish Anglicans.

Molyneux, who represented Trinity College Dublin in parliament, made perhaps the most famous contribution with his pamphlet, *The Case of Ireland's being Bound by Acts of Parliament in England, Stated* (1695). Using Locke's principle of the social contract, he argued that no laws could be imposed on Ireland by the English parliament without the consent of the Irish. He even talked about "the Common Rights of all Mankind." Ireland, he argued, is a separate kingdom, not a colony, and should be governed by the same laws as England.[99] The work was so controver-

---

[98]Roy Foster, *Modern Ireland, 1600-1972* (New York: Penguin, 1988), p. 178.
[99]Duddy, *History of Irish Thought*, pp. 771-74.

sial that it was condemned as seditious and ceremonially burned by the public hangman. When Molyneux talked about common rights, however, he was concerned only with the rights and liberties of his fellow Anglicans. He excluded the vast majority of the inhabitants of Ireland. His pamphlet, however, would become 'the Manual of Irish Liberty' for other advocates of rights, including revolutionaries like Wolfe Tone. (It also caught the attention of the American colonists as they moved toward independence.) Swift similarly questioned the right of the English parliament to make laws binding the kingdom of Ireland, going so far as to argue that "all Government without the consent of the Governed is the very Definition of Slavery."[100] When in 1782 Ireland gained parliamentary independence, Henry Grattan, speaker of the Irish House of Commons, famously declared, "Spirit of Swift, spirit of Molyneux, your genius has prevailed! Ireland is now a nation!"[101]

William Petty's political arithmetic had sought ways to dismantle Ireland and to exploit its resources, but Molyneux and Swift worked to create a vision of a successful Ireland, politically and economically independent. Berkeley's philosophy was also seen by some as a struggle for intellectual independence from England. Berkeley presented his idealist approach as distinctly Irish. Chastising English materialists who say 'the wall is not white' or 'the fire is not hot' because they believe that these secondary characteristics are caused by invisible primary causes, Berkeley countered, "We Irish men cannot attain to these truths."[102] And after critiquing the idea that there can be no such thing as lines without width, Berkeley retorted, "We Irish men can conceive no such lines."[103] For Yeats, this moment represented "the birth of the [Irish] national intellect" and Berkeley's attack on English mate-

---

[100] Berkeley, "The Commonplace Book."

[101] Quoted in J.C. Beckett, *The Anglo-Irish Tradition* (London: Faber and Faber, 1976), p. 9.

[102] Cited in Duddy, *History of Irish Thought.*

[103] George Berkeley, *Philosophical Commentaries*, p. 393.

rialism was "the Irish Salamis," a reference to the battle where a small Greek force defeated the much larger Persian army, protecting (in Yeats' and many others' view) the future of Western Civilization.[104] We see here the beginning of an Anglo-Irish tradition, distinct from the Catholic and Gaelic intellectual tradition, and still politically and intellectually dependent on the English ruling class, yet striving to define a role for themselves in the future of Ireland. Mathematics and science as much as literature and theology were central to defining the Anglo-Irish identity, and they would remain so over the next two centuries.

---

[104]Quoted in Berman, *George Berkeley*, p. 208.

Chapter 3

Enlightenment and Revolution

A Foreign Invasion

On December 11, 1790, Trinity College Dublin appointed John Brinkley as Andrews Professor of Astronomy and director of the new Dunsink Observatory. Having graduated from Cambridge with honors and having assisted the Astronomer Royal at Greenwich Observatory, he was perfectly suited for the job. Yet the members of the college were outraged. The Board of Senior Fellows had vetoed Brinkley's appointment in favor of John Stack, a fellow of the college, but the provost had overruled their veto.[1] The daily papers lamented the appointment of an Englishman over an Irishman.[2] Conscious of the situation, on his arrival Brinkley challenged Stack to a battle of wits and promised to resign immediately if Stack could prove his superior knowledge of astronomy. But, as one observer put it, "Stack was far too wise to accept the challenge."[3]

---

[1] On Brinkley's appointment, see Patrick A. Wayman, *Dunsink Observatory, 1785-1985: A Bicentennial History* (Dublin: The Royal Dublin Society and the Dublin Institute for Advanced Studies, 1987), pp. 21-25, 286-87.

[2] *Freeman's Journal* (December 14 and 22, 1790); *Dublin Evening Post* (December 14 and 16, 1790).

[3] Wayman, *Dunsink Observatory*, p. 289.

There was never really a question as to whether Brinkley was the more scientifically qualified candidate, but in the eyes of the fellows, a competent Trinity man was better than an accomplished 'foreigner.' At the end of the eighteenth century, Trinity remained essentially an ecclesiastical institution, inward looking and tradition bound. The Newtonian natural philosophy that had stirred so much controversy on its introduction was now sanctioned by nearly a century of works designed to show its harmony with Anglican theology. The fellows saw their role as the preservation of this 'Holy Alliance' between Newtonianism and Anglicanism rather than the extension of the bounds of scientific knowledge. Brinkley took a different approach. While he carefully indicated the way in which mathematical science supports religion, he felt the new mathematical techniques of French mathematicians such as Laplace and Lagrange showed this better than Newton's more cumbersome approach. Moreover, as professor of astronomy rather than a fellow, Brinkley had been appointed first and foremost to do research, not to tutor students. Brinkley was one of the first mathematicians in the British Isles to master the French version of the calculus, and he won accolades from scientific societies in Ireland, England, and France for his work.

The fact that Brinkley was English was even more infuriating than the fact that he preferred advanced research to elementary teaching. A long peace, a booming economy, and the softening of sectarian tensions had given Irish Protestants the confidence to press for a degree of independence from England in the 1770s and 1780s. Following in the path set out by Molyneux and Swift, the patriot movement, as it came to be called, resented the fact that the English parliament controlled Irish trade and had the power to veto Irish legislation. Science became an expression of the patriots' aspirations for Ireland. The patriots believed that science would lead to enlightenment, prosperity, and international respect.

Brinkley's appointment sat uneasily with this movement. Clearly patriots would have preferred an Irishman, but they also realized that building an international reputation in science required appointing researchers skilled in the latest methods. And Brinkley did raise the new observatory to international fame. But there was another tension lurking in Brinkley's appointment. While he saw the new French science as a further support for theology, the French themselves associated their science with arguments for revolution and democracy. In the 1790s Irish radicals began to echo the claims of French revolutionaries, proclaiming that science would lead to the end of oppressive political and religious establishments. Their confidence shaken, Irish Anglicans abandoned patriotism for an ideology of Protestant Ascendancy, the belief that as the supporters of education and scholarship only they were qualified to rule. Science justified Ascendancy, they argued, not democracy. The rebellion of 1798 and the Union with Great Britain in 1801 put an end to the enlightened dreams of the patriot movement. Trinity's model of scholarship in the service of the Anglican Church survived, while the patriot's hope that science would lead to industrial and cultural revival foundered in an atmosphere of renewed sectarianism.

3.1   Science, Patriotism, and Polite Culture

Brinkley arrived in Dublin in May of 1791 to find a booming metropolis. Dublin was the second largest city in the British Empire with twice the population of Manchester or Edinburgh. Sir Jonah Barrington recalled, "its progress was excessive—the locality of Parliament—the residence of the nobility and commons—the magnificence of the Viceregal court—the active hospitality of the people—and the increasing commerce of the Port—all together gave a brilliant prosperity to that splendid and luxurious capital."[4] Nowhere was Dublin's self-confidence more apparent than in the magnificence of its new buildings. This was the heyday of the Georgian architecture that still characterizes Dublin today. Trinity College received its now famous West Front and lavish provost's house in 1759 and 1760. The Parliament House gained its eastern portico in 1785, the Custom House was completed in 1791, and the Four Courts opened in 1803. Merrion Square and St. Stephen's Green were now ringed with the Georgian townhouses of the wealthy, while aristocrats erected mansions like Leinster House and Powerscourt House.[5]

Figure 3.1: The Front Gate of Trinity College Dublin (1799).
Illustration by James Malton ©National Library of Ireland

[4]Sir Jonah Barrington, *Historic Memoirs of Ireland*, 2nd Ed. (London, 1835), Vol. 1, p. 7.

[5]Robert Poole and John Cash, *Views of the Most Remarkable Public Buildings, Monuments, and Other Edifices in the City of Dublin* [1780] (Shannon: Irish University Press, 1970).

Much of the money for this explosion of buildings came from the Irish Parliament, swelled with the revenues of Ireland's growing commerce. Lord Chancellor Clare proclaimed, "there is not a nation on the habitable globe which has advanced in cultivation and commerce, in agriculture, and in manufactures, with the same rapidity in the same period."[6] The revival of Irish commerce led to the growing political self-confidence of the Irish gentry. When English troops were withdrawn from Ireland to keep order in America, the Irish gentry formed the Volunteers to defend the country from the threat of a French invasion. The Volunteers, however, soon began demonstrating for free trade and a free parliament. In 1780 they succeeded in passing new legislation allowing Ireland to export wool and to sell to the colonies for the first time, and in 1782 Ireland gained parliamentary independence.

The Irish Parliament drew aristocrats, politicians, and innumerable hangers-on. As a result, Dublin had the reputation of being second only to London in the English-speaking world as a center for upper-class entertainment.[7] Members of Irish high society attended the theatres, balls, assemblies, card-parties, and concerts typical throughout Europe, and like their fellow Londoners and Parisians they also attended scientific lectures, popular experimental demonstrations, intellectual conversazioni, and scientific clubs. Science was one of their many pastimes.[8] Ireland's newfound self-confidence and wealth was not only responsible for the Georgian building boom, it also lay behind the origins of many of Ireland's most important scientific institutions. Throughout Europe, scientifically and industrially-minded gentlemen were coming together

---

[6]Constantia Maxwell, *Dublin Under the Georges, 1714-1830* (London: Faber and Faber, 1956), p. 90.

[7]Tighearnan Mooney and Fiona White, "The Gentry's Winter Season," pp. 1-16 in David Dickson (ed.), *The Georgeous Mask: Dublin, 1700-1850* (Dublin: Trinity History Workshop, 1987), p. 1.

[8]See Máire Kennedy, "The *Encyclopédie* in 18th-century Ireland," *Book Collector* (1996) 45: 201-13; Graham Gorgett and Geraldine Sheridan (eds.), *Ireland and the French Enlightenment* (London: Macmillan, 1999).

to form academies, museums, and societies for the promotion of knowledge. The Royal Academies of London, Paris, Berlin, and St. Petersburg had all been founded by 1724, but in the late eighteenth century provincial societies were founded in Manchester (1781), Edinburgh (1783), and Newcastle (1793).[9] The Irish also played their part in this trend. The 1780s and 1790s saw the expansion of the Royal Dublin Society (founded in 1731), the establishment of the Royal Irish Academy (1785), and the building of Dunsink Observatory (1785). In Ulster, the Belfast Academy opened its doors in 1786 and the Belfast Reading Society was founded in 1788, becoming the Belfast Library and Society for Promoting Knowledge in 1792. Armagh Observatory was erected in 1790.[10]

Figure 3.2: Dunsink Observatory near Dublin. R. Clayton, *Dublin Penny Journal* (August 15, 1835), Vol. 4, p. 49.

---

[9]See J.E. McClellan, *Science Reorganized: Scientific Societies in the Eighteenth Century* (New York: Columbia University Press, 1985).

[10]James Bennett, *Church, State and Astronomy in Ireland: 200 Years of Armagh Observatory* (Armagh: Armagh Observatory, 1990).

These new institutions were directly related to the political and economic revival taking place in Ireland. The Royal Dublin Society, for example, combined science, patriotism, and improvement from the beginning.[11] Central to its mission was the diffusion of practical knowledge and the encouragement of Irish industry. Its members, primarily improving landlords, believed that science was the key to economic progress. They hoped that the same industries beginning to flourish in England could be established in Ireland, so they appointed an itinerant mineralogist to search Ireland for minerals of economic significance and offered prizes for improved farming and manufacturing techniques.

The Royal Irish Academy was founded in 1785 for the promotion of science, antiquities, and polite literature.[12] The original members included an archbishop, four bishops, three lords, the provost, and ten fellows of Trinity. Despite the interest of some participants in Irish industrial development, the members viewed the importance of science as primarily cultural rather than economic. They saw science as benefiting Ireland in much the same way that polite literature and antiquities might benefit Ireland, namely through the creation of an Irish high culture that might stand proudly alongside English models. The members were predominantly Anglo-Irish gentlemen (only two of the eighty-eight foundation members were Catholics), many of whom were involved with the Volunteer movement. The Earl of Charlemont, for example, the first president of the Academy, was also Colonel-in-Chief of the Volunteers. Unlike the popular publications of the Royal Dublin Society, the *Transactions of the Royal Irish Academy* was aimed at an international audience of scholars and savants. In science they emphasized

[11]Henry F. Berry, *A History of the Royal Dublin Society* (London: Longmans, Green and Co., 1915); James Meenan and Desmond Clarke (ed.), *RDS: The Royal Dublin Society, 1731-1981* (Dublin: Gill & Macmillan, 1981).

[12]R.B. McDowell, "The Main Narrative," pp. 1-92 in T. O'Raifeartaigh (ed.), *The Royal Irish Academy: A Bicentennial History, 1785-1985* (Dublin: The Royal Irish Academy, 1985), pp. 1-22.

astronomy and abstract mathematics rather than agriculture and
animal husbandry.

The first volume of the *Transactions* appeared in 1787. In the
introduction, Robert Burrowes, a fellow of Trinity College Dublin,
commemorated the occasion by offering a social history of Irish
science. Burrowes wanted to explain both Ireland's previous lack
of scientific activity and the reasons for its recent flowering. "To the
several advantages which Europe has within these latter centuries
experienced from the cultivation of science and polite literature," he
began, "this kingdom unfortunately has remained in great measure
a stranger."[13] The cause, he explained, was not a lack of capacity on
the part of Irishmen but rather certain 'local peculiarities.' He then
proceeded with a laundry list of political and social impediments to
Irish science.[14] Invasion, sectarianism, the emigration of talented
people, and Ireland's peripheral status in both a geographical and
a political sense, he argued, had all prevented the Irish from taking
their rightful place in the Republic of Letters. Burrowes laid out
these deficiencies, however, not so much to excuse Ireland's past
failures, as to explain the sources of her newfound success. He
continued,

> Peculiar circumstances have now given Ireland an impor-
> tance in the political scale, which habits of well-directed
> industry alone can establish and maintain. Whatever
> therefore tends, by the cultivation of useful arts and sci-
> ences, to improve and facilitate its manufactures; what-
> ever tends, by the elegance of polite literature, to civilize
> the manners and refine the taste of its people; whatever
> tends to awaken a spirit of literary ambition, by keeping

---

[13] [Robert Burrowes], "Preface," *Transactions of the Royal Irish Academy* (1787) 1: ix-
xvii., p. ix.
[14] Burrowes, "Preface," pp. ix-x.

> alive the memory of its antient reputation for learning,
> cannot but prove of the greatest national advantage.[15]

Science, literature, and antiquities, the three subjects promoted by
the new Royal Irish Academy, were to play a role in the promotion
of 'national' advantage. Burrowes believed that a cultural, political, and economic revival was taking place, and the creation of
the Academy represented the central role that science had to play
in such a revival. As historian Jim Bennett observes, Burrowes'
preface "was a challenge to a new order in Ireland to accept the
responsibilities that fell to the cultural elite of a truly European
nation."[16]

Mathematics, moreover, held a privileged role. Striking a familiar note, Burrowes explained that "The researches of the mathematician are the only sure ground on which we can reason from
experiments." But he went on to argue that mathematics and
theoretical science were also the foundation of England's growing
industrial sector, "where the hand of the artificer has taken its direction from the philosopher." "Every manufacture," he claimed,
"is in reality a chemical process, and the machinery requisite for
carrying it on but the right application of certain propositions in
rational mechanics."[17] Like all improving gentlemen, Burrowes was
aware of the enormous changes taking place in the new manufacturing districts of England and Scotland. Despite the decidedly
non-theoretical origins of Britain's growing industrial power, however, Burrowes identified mathematical science as the key to technological progress. Ireland's intellectual elite, represented by the
members of the new Academy, had an essential role to play in
economic and national development, he argued, because abstract
science lay at the root of industrial progress. Science and mathe-

---

[15] Burrowes, "Preface," pp. xi.

[16] Bennett, *Church, State and Astronomy in Ireland,* p. 1.

[17] Burrowes, "Preface," pp. xii.

matics, according to Burrowes, were essential accomplishments for the enlightened and patriotic Irishman.

Ireland's newly important position also brought back scientific talents who might otherwise have made their careers outside of Ireland. Richard Lovell Edgeworth, for example, father of the novelist Maria Edgeworth, was one of the best-connected scientific men in Ireland. He was a member of the Birmingham Lunar Society, which included scientifically and industrially minded men like Erasmus Darwin, Matthew Bolton, and Josiah Wedgewood. In London, he joined the group of men who met weekly in Young Slaughter's Coffee House, including Sir Joseph Banks, president of the Royal Society, and Nevil Maskelyne, the Astronomer Royal. Edgeworth also lived in France for a time where he entertained Rousseau and met other famous scientists.[18] In short, Edgeworth's gentlemanly status and interest in science and enlightenment connected him to the scientific elite in both Britain and France.

Edgeworth returned to Ireland in 1782 to improve his estate and soon became involved with the Volunteers. He carried on research into a semaphore telegraph system, early railways, and improvements to carriages and roads. While such activities were more typical for the practical members of the Royal Dublin Society, Edgeworth was also one of the founding members of the Academy. It is said that when Edgeworth visited the Paris Academy of Science he was entitled to take a seat as a member of the Royal Society of London. He asked instead to take his seat as a member of the Royal Irish Academy, to the puzzlement of his French hosts and the great delight of his Dublin colleagues.

The most famous Irish man of science at this time was Richard Kirwan, a chemist, geologist, and mineralogist of European fame.[19]

---

[18]See Richard L. Edgeworth and Maria Edgeworth, *Memoirs of Richard Lovell Edgeworth Begun By Himself and Concluded By His Daughter, Maria Edgeworth* (London, 1820)

[19]F.E. Dixon, "Richard Kirwan, the Dublin Philosopher," *Dublin Historical Record* (1971) 24: 52-64; Michael Donovan, "Biographical Account of the Late Richard Kirwan," *Proceed-*

Born into a Roman Catholic family in Galway, Kirwan later renounced his Catholicism in order to become a member of the Irish Bar. He moved to London and throughout the 1770s and 1780s was an important member of the Royal Society. In London he socialized with Cavendish, Priestley, Banks, and other scientific notables, and his defense of the phlogiston theory brought him into contact with all of the great European chemists including Scheele and Lavoisier. Kirwan returned to Dublin in about 1787 and was a major contributor to the Royal Irish Academy, becoming its president in 1799. He also held conversazioni every Wednesday night that attracted fellows of Trinity College, legal luminaries, bishops, aristocrats, and even the Lord Lieutenant himself. Like Burrowes, Edgeworth and Kirwan seem to have shared a sense that Ireland was on the verge of greatness. It was crucial for the creation of a scientific culture in Ireland that emigrants like Edgeworth and Kirwan returned to Ireland with scientific knowledge and contacts gained elsewhere. Neither Edgeworth nor Kirwan held university positions; they saw science as the prerogative of all cultured and public-spirited gentlemen rather than simply the domain of scholars.

3.2 The University and Society

As Dublin's polite society grew more interested in science and intellectual culture, Trinity College found itself losing its traditional role as Ireland's source of knowledge. By the late eighteenth century, it was beginning to look terribly out of date. Its main purpose lay in training clergymen for the Church of Ireland (the destination of two thirds of its graduates), and it was run as a religious institution. All fellows had to be in holy orders and were required to remain celibate. In the new enlightened age, however, parents wanted their

ings of the Royal Irish Academy (1847-50) 4: lxxxi-cxviii; J. Reilly and N. O'Flynn, "Richard Kirwan: An Irish Chemist of the 18th Century," *Isis* (1930) 13: 298-319.

children trained to be gentlemen, not monks. Education was now about social position, not theological disputation. One pamphlet for students written in 1798 described the new ideal of the Trinity graduate as "a man of urbanity of manners,—courteous and beneficent to his dependents, select in his friendships, temperate in his life, and useful in society."[20] Provost Hely-Hutchinson struggled "to adapt the course to the education of men of rank as well as men of science," much to the dismay of the clerical fellows.[21]

The changes at Trinity were part of a broader wave of educational reforms in late eighteenth-century Britain. Three new educational models replaced the monastic rigor of the medieval universities: the mathematical (Cambridge), the classical (Oxford) and the practical scientific (Scotland). The Scottish universities had rapidly established themselves as the providers of the most advanced scientific education in the British Isles. Their excellence in specialized scientific topics resulted from their institutional structure, which differed radically from the English universities and Trinity College Dublin. Scotland had five universities controlled by local town councils rather than the established Church. They were non-ecclesiastical, non-residential, non-exclusive, inexpensive, and professional. Professors were paid by the students, and students were free to take any classes they desired. Dissenting academies, created to meet the educational needs of those excluded from the Anglican universities, operated along similar lines. Ulster received its first Dissenting Academy in 1785, followed in 1814 by the Belfast Academical Institution. The emphasis was on the type of practical knowledge that would be of use in a growing industrial center like Belfast.[22]

---

[20]Eumenes, *An Address to a Young Student, on His Entrance into College* (Dublin, 1798), p. 11.

[21]Sir Jonah Barrington, *Personal Sketches of His Own Times* (London, 1827), p. 60.

[22]David Kennedy, "The Ulster Academies and the Teaching of Science (1785-1835)," *Irish Ecclesiastical Record* (1944) 63: 25-38. See also Helena C.G. Chesney, "Enlightenment and

In England, Oxford continued to teach the classics, while Cambridge became increasingly mathematical. By the mid-eighteenth century, the Senate House or Tripos examination at Cambridge had become more important than traditional disputations. It also became increasingly competitive. In 1753 graduates were ranked, and placing high (the surest route to a fellowship) increasingly required a mastery of mathematics. In 1791 the examiners began printing some of the problems, an innovation that would revolutionize education and further stimulate the competitive, mathematical nature of the examination.[23]

By the mid-eighteenth century, however, the course followed by the undergraduates at Trinity College Dublin had changed little in over a century. All students followed the same course and read the same books. The first two years included Aristotelian logic, basic mathematics, and the Greek and Roman classics. Euclid's *Elements* was not a part of the undergraduate course until 1758 and algebra not until 1808.[24] The third year introduced Newtonian natural philosophy (through Helsham's textbook), and the fourth year emphasized natural theology and ethics. Natural philosophy served as a key step in the progression from logic and mathematics to natural theology and ethics, and hence a key element in the training of future clergymen.

As the general cultural importance of science grew, however, so did the need to change its role in the undergraduate course. The most detailed account of the course in the late eighteenth century comes from our friend Robert Burrowes, who in 1792 published *Observations on the Course of Science Taught At Present in Trinity*

Education," pp. 367-86 in J. W. Foster and H. C. G. Chesney (eds.), *Nature in Ireland: A Scientific and Cultural History* (Dublin: Lilliput Press, 1997).

[23]John Gascoigne, "Mathematics and Meritocracy: The Emergence of the Cambridge Mathematical Tripos," *Social Studies of Science* (1984) 14: 547-84.

[24]John William Stubbs, *The History of the University of Dublin, From Its Foundation to the End of the Eighteenth Century* (Dublin, 1889), p. 197. See the Preface to Theaker Wilder, *Newton's Universal Arithmetick* (London, 1769).

College, Dublin, With Some Improvements Suggested Therein. The time had come, Burrowes explained, to reform the course in order to "render our Academic Education more respected and more useful."[25] He contrasted what is currently taught at most universities with "the knowledge most valuable to the Community and most reputable in the World."[26] As the character of the student body was changing, so must the course of study. The location of the college in a metropolis, he explained, where it must educate many 'persons of the middle rank' with no desire to enter the learned professions, must shape the teaching.

Extolling the new interest in practical science in the service of the nation, Burrowes recommended that each student "should readily apprehend the construction of any machine, and account satisfactorily for its movements...and should be able to point out the respective advantages, and disadvantages of projects for improving the Manufactures, or extending the Commerce of his Country."[27] Clearly, Burrowes did not simply have in mind the best education for clergymen. He wanted to prepare students to play a useful role in society. In mathematics, for example, he advocated the teaching of algebra, "the branch of Mathematics most useful for the purposes of ordinary life, and essentially necessary to every science."[28]

A new course was introduced at Trinity the following year in 1793. The general subjects remained the same, but they were now taught using textbooks recently written by Trinity fellows. A system of gold medals was also instituted in which prizes were given to the best students in an effort to encourage greater academic exertion.[29] A printed version of the course from 1799 gives the detail

---

[25] Robert Burrowes, *Observations on the Course of Science Taught At Present in Trinity College, Dublin, With Some Improvements Suggested Therein* (Dublin, 1792), p. 7.

[26] Burrowes, *Observations*, p. 6.

[27] Burrowes, *Observations*, p. 18.

[28] Burrowes, *Observations*, p. 35.

[29] See Stubbs, *History of the University of Dublin*, pp. 257-58.

of the changes.[30] While chemistry, natural history and economics were nowhere to be seen, despite Burrowes' pleas, and the science books were not the most advanced, the fact that so conservative an institution as Trinity adapted at all indicates the extent of the pressures it faced.

In fact, Trinity faced a crisis of credibility in the 1780s and 1790s. Even a friend of the university admitted to students in 1782 that "although outwardly the University may wear a more splendid appearance than ever... yet, it neither commands the respect which it has done in former times, nor can you hope for the same advantages from having gone through a course of education in it, which your predecessors have enjoyed."[31] Burrowes was personally involved in a particularly embarrassing exchange in the 1790s.[32] He failed a student named Deane Swift, a distant relative of Jonathan Swift, who had written an insulting Latin epigram on a fellow. Deane's father, Theophilus Swift, took the opportunity to write a number of scathing pamphlets attacking Trinity's education and its scholarly reputation. "[T]he honor, the interest, the safety of the country," Swift declaimed, "demand the interposition... of the crown itself, to amputate the rotten members, and cut away the proud flesh of a proud body, that will shortly gangrene the whole state of letters."[33] If science was now important to the nation, as Burrowes argued, then Trinity had an obligation to promote science. Trinity's record in this regard, however, was particularly poor.

---

[30] Beresford Papers TCDMS 2770-4/1. See also Richard Eaton, *An Abridgement of Astronomy and Natural Philosophy as Read in the Under-Graduate Course of Trinity College, Dublin* (Dublin, 1797).

[31] *Thoughts on the Present State of the College of Dublin; Addressed to the Gentlemen of the University* (Dublin, 1782), p. 5.

[32] See Studiosus, *Alma's Defence, Or The Critic Refuted: A Poem With Explanatory Notes* (Dublin, 1790) and Theophilus Swift, *Prison Pindarics; Or, a New Year's Gift from Newgate Humbly Presented to the Students of the University* (Dublin, 1795).

[33] Theophilus Swift, *Animadversions on the Fellows of Trinity College* (Dublin, 1794), p. 61.

Swift's most trenchant criticism involved Trinity's reputation for scholarship. He asked, "how comes it to pass, that this enlightened university, ever since its foundation for more than two centuries, hath produced so few men of wit and learning. Let the logicians of Trinity College give some good reason, not a syllogism," he demanded, "why the foreign universities have termed the Dublin Muse the 'SILENT SISTER.'"[34] His claim was more than simply slander; historians McDowell and Webb were unable to find any publications by current fellows of the college during the thirty-year period from 1722 to 1753.[35]

In part the problem lay with Trinity's traditional goal, namely the education of clergymen. Burrowes himself argued that the fellows and professors "are maintained not for solitary study, but for Public Instruction: Not so much that they may write books extending the boundaries of Science, as that they may teach well what is already known."[36] Fellows saw themselves as teachers and clergymen. They were supposed to be men of wide learning and high moral character, not specialized researchers. Neither did the route to fellowship encourage creativity. Fellowship at Trinity (similar to tenure in today's academic system) was obtained through a highly competitive examination, the culmination of what one pamphlet described as "a course of study, which is allowed to be the severest and most extensive, that has been made necessary in any seminary in Europe."[37] The rewards of fellowship, however, were well worth the trouble. It was not only a job for life but a job with increasing salary and decreasing duties. While junior fellows were consumed with teaching undergraduates, senior fellows were not, and they shared in the profits of Trinity's extensive estates. Fellowship was also a stepping-stone to a college living or a bish-

---

[34] Swift, *Animadversions*, p. 11.

[35] R.B. McDowell and D.A. Webb, *Trinity College Dublin, 1592-1952: An Academic History* (Cambridge: Cambridge University Press, 1982), p. 39.

[36] Burrowes, *Observations*, p. 19.

[37] *Thoughts on the Present State of the College of Dublin*, p. 14.

opric. Specialized study rose in importance only to the degree that it improved one's chances in the fellowship examination. A professorship in mathematics, for example, was created in 1762 funded (like the Professorship of Natural and Experimental Philosophy) from Erasmus Smith's estates. But the professor (always a senior fellow) lectured only to the fellowship candidates.[38]

The fellows of Trinity College were clearly scholars, but they were not researchers in the modern sense of the word. They rarely published papers in scientific journals, and though they did publish textbooks and even mathematical works, they spent just as much time writing sermons and studying Biblical languages. In general they published tracts designed to better explicate and support Newton's work and to place it clearly in a natural theological context. This required very little in the way of new experiments or even new mathematics. However, as Irish society changed in the late eighteenth century, and Trinity's role within it changed, natural philosophy assumed a new importance. Perhaps these obscure proofs and simple experiments might be the key to Ireland's industrial success. Certainly, scientific fame was increasingly important for Trinity's reputation and Ireland's reputation more generally. The age of the dilettante scholar was slowly coming to an end, and John Brinkley represented the future.

3.3   The Threat of French Mathematics

Trinity's lack of scientific vigor was symptomatic of a larger British trend. In the wake of Berkeley's critique of the calculus, mathematics in Britain and Ireland had greatly diverged from mathematics on the Continent. Not only had the British and Irish continued to use Newton's fluxional calculus while mathematicians on the Continent used Leibniz's differential notation, but they seemed less

---

[38]TCDMS Mun/P/1/1478.

interested in pioneering new methods than their European coun-
terparts. In the Royal Academies of Paris, Berlin, and St. Petersburg, mathematicians like the Bernoullis, Euler, d'Alembert, Lagrange, and Laplace created a thoroughly mathematical physics.
They used mechanics and astronomy as sources for mathematical
problems and developed new tools to solve them, including partial
differential equations, complex analysis, the variational calculus,
probability theory, and the theory of series.[39] Often supported by
royal patronage and published in international scientific journals,
they focused their energies on the solution of increasingly technical
problems. In the process they remade Newton's geometrical natural philosophy into an analytical mechanics. Key works were Lagrange's *Analytical Mechanics* (1788) and Laplace's *Celestial Mechanics* (1799-1825). The École Polytechnique, founded in 1794 to
provide the French Empire with mathematically and scientifically
trained engineers, became a key site for teaching this advanced
mathematics with an emphasis on its practical applications.[40]

By the end of the eighteenth century, the British were increasingly beginning to feel their scientific inferiority. It was in the military academies that an appreciation of French mathematics first
became evident.[41] If England's most dangerous foe were training
her officers with mathematics, perhaps English officers should be
learning it as well. At the Royal Military Academy at Woolwich,
Charles Hutton brought the new mathematics to public attention,

---

[39]H.J.M. Bos, "Mathematics and Rational Mechanics," pp. 327-55 in G.S. Rousseau and
Roy Porter (eds.), *The Ferment of Knowledge: Studies in the Historiography of Eighteenth-
Century Science* (Cambridge: Cambridge University Press, 1980); Ivor Grattan-Guinness,
*Convolutions in French Mathematics, 1800-1840: From the Calculus and Mechanics to
Mathematical Analysis and Mathematical Physics* (Boston: Birkhäuser, 1990).

[40]Terry Shinn, *Savior scientifique et pouvoir social: L'Ecole Polytechnique et les polytechniciens, 1794-1914* (Paris: Presse Fondation Nationale des Sciences Politiques, 1980); J.H.
Weiss, *The Making of Technological Man: The Social Origins of French Engineering Education* (Cambridge: Harvard University Press, 1982); Bruno Belhoste, Amy Dahan Dalmedico
and Antoine Picon (eds.), *La Formation Polytechnicienne, 1794-1994* (Paris: Dunod, 1994).

[41]See Niccolò Guicciardini, *The Development of the Newtonian Calculus in Britain, 1700-
1800* (Cambridge: Cambridge University Press, 1989), pp. 108-23.

authoring dictionaries, encyclopedias, and textbooks to explain the new methods. Hutton also edited a fascinating periodical called the *Ladies' Diary, or Woman's Almanac.*[42] It began in 1704 as an almanac with articles of general interest to women; but by 1707, the recipes, sketches of notable women, and articles on health and education were accompanied by enigmas, queries, and mathematical questions—often in verse. Problems included algebra, astronomy, dynamics, fluxions, geometry, harmonics, hydrostatics, navigation, optics, and pneumatics. Many of the contributors and editors, like Hutton and Thomas Leybourn, professor at the Royal Military College at Sandhurst, were associated with the military academies. John Brinkley was also a contributor in his younger days. In 1795 Leybourn launched a more highbrow periodical, *The Mathematical and Philosophical Repository*, and in 1807 it became the first British periodical to publish work using the Continental notation for the calculus, $\frac{dx}{dy}$, rather than the Newtonian notation, $\dot{x}$.

In 1808 John Playfair, professor of mathematics at Edinburgh University, lamented the sorry state of British mathematics in a review of Laplace's *Celestial Mechanics*. "[A] man may be perfectly acquainted with everything on mathematical learning that has been written in this country," he complained, "and may yet find himself stopped at the first page of the works of Euler or D'Alembert."[43] Fewer than a dozen mathematicians in the British Isles, he continued, were capable of reading Laplace. Though mathematics by this time dominated the curriculum at Cambridge, the course there differed radically from the École Polytechnique and did not include any of the advances made after the 1720s.[44] It was not until 1803 that Robert Woodhouse's *Principles of Analytical Calculation* became the first major Cambridge work to use the differential nota-

---

[42]Teri Perl, "The Ladies' Diary or Woman's Almanack, 1704-1841," *Historia Mathematica* (1979) 6: 36-53.

[43][John Playfair], "La Place, *Traité de Méchanique Céleste*," *Edinburgh Review* (1808) 11: 249-84, p. 281.

[44]Guicciardini, *Newtonian Calculus*, p. 124.

tion.[45] But analytical mathematics did not begin to make inroads into the curriculum until a group of precocious undergraduates created the Analytical Society in 1812. While the generation of the 1780s and 1790s publicized the continental work, it would not become established in Britain and Ireland until the next generation (see chapter 4).

John Brinkley was at the leading edge of this new movement.[46] Where Brinkley learned these methods is quite a mystery. He began contributing mathematical problems and solutions to the *Ladies' Diary* in 1782 while still at school.[47] Though he graduated from Cambridge as senior wrangler (the top scorer on the Tripos examination) and first Smith's prizeman in 1788, analytical mathematics formed no part of the course at the time. One might also expect that his work assisting at Greenwich Observatory (1787-88) may have introduced him to the mathematical work of the great French astronomers, but Playfair, who visited the observatory in 1782, remarked that the Astronomer Royal Nevil Maskelyne preferred geometry to the advanced methods of the French analysts.[48] Brinkley's interest in astronomy probably served as his primary motivation for learning the new techniques at a time when most university scholars in Britain ignored them. As early as 1799 he was lecturing on the results of the French astronomers, and he published a textbook on astronomy in 1813.[49] As the director of Dunsink Observatory rather than a fellow or tutor in the College, he was professionally obligated to take an interest in the latest as-

---

[45] See Guicciardini, *Newtonian Calculus*, pp. 124-38.

[46] On Brinkley, see I. Grattan-Guinness, "Mathematical Research and Instruction in Ireland, 1782-1840," pp. 11-30 in J. Nudds, N. McMillan, D. Weaire and S. M. Lawlor (eds.), *Science in Ireland 1800-1930: Tradition and Reform* (Dublin: Trinity College Dublin, 1988), pp. 15-16, Guicciardini, *Newtonian Calculus*, pp. 132-33, Wayman, *Dunsink Observatory*, pp. 21-51, P.A. Wayman, "Rev. John Brinkley: A New Start," in Nudds *et al.* (eds.), *Science in Ireland.*

[47] I thank Shelley Costa for her help in tracking down Brinkley's contributions.

[48] Guicciardini, *Newtonian Calculus*, p. 101.

[49] John Brinkley, *Synopsis of Astronomical Lectures, To Commence October 29, 1799, at the Philosophy School, Trinity College, Dublin* (Dublin, n.d.).

tronomical research, and by the late eighteenth century that meant French mathematical physics. While observers strove to improve instruments and methods of data analysis in order to determine the motions of the planets and their satellites more accurately, French theorists pushed the theory of gravitation and the new mathematical methods to their limits to predict subtle new celestial phenomena.

Figure 3.3: *The Ladies Diary* (1723), the first periodical in Britain to publish mathematical problems using the continental notation for the calculus.

Brinkley's work immediately demonstrated that the provost's faith in him had been well placed. He published numerous articles in the *Transactions of the Royal Irish Academy* and the *Philosoph-*

ical Transactions of the Royal Society of London (he had become
a fellow in 1803), and he won the Cunningham Medal of the Royal
Irish Academy in 1818.[50] In fact, his award-winning essay was the
first in the Academy's *Transactions* to use the Continental differential notation. Then in 1823 Brinkley won the Copley Medal of
the Royal Society. Brinkley's fame was such that he was elected
correspondent of the prestigious French Académie des Sciences in
1820, and on his death, Arago published an obituary, referring to
Brinkley as the "célèbre astronome de Dublin."[51] Ironically, Arago
assumed Brinkley was Irish.

Brinkley's work was a marked contrast to the work of the other
Trinity fellows who all emphasized basic physical ideas and qualitative experimental demonstrations rather than advanced calculus
and precision instrumentation. Yet Brinkley's work did not break
completely with the Trinity tradition. His textbook, for example, which according to the *Gentleman's Magazine* was "generally
considered the best introduction to that science in our language,"[52]
resembled the other Trinity works in style. It avoided the advanced
mathematics that would have been incomprehensible even to the
most advanced students and emphasized basic observational practices in astronomy. It also stressed the religious implications of
astronomy. Laplace and Lagrange had calculated that all the secular inequalities of the solar system are periodic, a result widely
taken to mean that the solar system would continue forever without
the orbits of the planets ever collapsing.[53] (Contrary to Newton's
belief that the solar system required God's regular intervention to

---

[50] J. Brinkley, "Investigations in Physical Astronomy, Principally Relative to the Mean
Motion of the Lunar Perigee," *Transactions of the Royal Irish Academy* (1818) 13: 25-52.

[51] [Francois Arago], "Sur la vie et les ouvrages de M. John Brinkley, correspondant de
l'Académie des Sciences," *Comptes rendus hebdomadaires des séances de L'Académie des
Sciences* (1835) 1: 212-25, p. 225.

[52] "Obituary of John Brinkley," *Gentleman's Magazine* (1835) 2: 547.

[53] See M. Norton Wise, "Mediations: Enlightenment Balancing Acts, or the Technologies
of Rationalism," pp. 207-56 in Paul Horwich (ed.), *World Changes, Thomas Kuhn and the
Nature of Science* (Cambridge: MIT Press, 1993), pp. 233-35.

preserve its motions.) Such a result had a powerful resonance for British natural theologians. In his review of the *Celestial Mechanics,* for example, Playfair highlighted Laplace's proof of the stability of the solar system. According to Playfair, the result proved that the universe "is therefore the work of design, or of intention, conducted by wisdom and foresight of the most perfect kind."[54] Brinkley lectured on this theme, and he ended his astronomy textbook by drawing a similar theological lesson from Laplacian astronomy. Modern astronomy, Brinkley claimed, teaches us that "The simple laws of matter and motion, which the Almighty has been pleased to ordain, are sufficient to preserve the motions of the system for a length of time, to which our bounded intelligence cannot put a limit."[55] He continued,

> By the discoveries of Newton we are permitted, as it were, to understand some of the Counsels of the Almighty. From these we can, by demonstration, overturn the absurd doctrine of blind chance. We see that a Supreme Intelligence placed and put in motion the planets about the sun in the centre, and ordained the laws of gravitation, having provided against the smallest imperfection that might arise from time.[56]

In the end, according to Brinkley, we find that the universe is governed by "laws not mechanical, but moral: laws only obscurely seen by the light of reason, but fully illuminated by that of revelation."[57] Like the fellows of Trinity, Brinkley saw the natural theological context as essential.

The irony is that at the same time that Brinkley and other British mathematicians were promoting Laplace's science as the perfect support for Anglican natural theology, Laplace himself and

---

[54][Playfair], "La Place, *Traité de Méchanique Céleste,*" p. 279.

[55]John Brinkley, *Elements of Astronomy,* 2nd Ed. (Dublin, 1819), p. 268.

[56] Brinkley, *Elements of Astronomy,* p. 284.

[57] Brinkley, *Elements of Astronomy,* p. 284.

the other great French mathematicians were drawing very different conclusions. France, the source of the most advanced science, was also the source of the most radical political ideas, ideas that would shake the British and Irish establishments to their very core.

3.4 Science and Revolution

While in England and Ireland, Laplace's astronomical achievement served as evidence for a universe designed and regulated by a benevolent divine power, Laplace himself and many of his French colleagues did not share this view. A popular story recounted that Napoleon, after reading Laplace's *Celestial Mechanics*, asked the mathematician about the role of God in the universe. Laplace is said to have replied tersely, "I had no need for that hypothesis." While British natural philosophers believed that nature could be understood only as divinely ordained, the French experimented with various forms of materialism and determinism. Laplace, for example, claimed that it was possible in principle for a being that knew the exact position of every particle in the universe at a given point in time to calculate its past or future state. What room was there for divine providence in such a clockwork universe?

The French Revolution has often been seen as a revolution of ideas, and behind many revolutionary ideas lay the mathematical sciences as a model of rationality.[58] Analysis, the philosophes believed, was the perfect tool to solve society's greatest problems. Condorcet and others applied the new mathematics of probability to the moral sciences, looking at the evaluation of legal evidence, the selection of juries, demography, and economics.[59] The

---

[58]Robin E. Rider, "Measures of Ideas, Rule of Language: Mathematics and Language in the 18th Century," pp. 113-40 in Frängsmyr, Heilbron and Rider (eds.), *The Quantifying Spirit in the Eighteenth Century*; Lorraine Daston, "Enlightenment Calculations," *Critical Inquiry* (1994) 21: 182-202.

[59]Lorraine Daston, *Classical Probability in the Enlightenment* (Princeton: Princeton University Press, 1988).

physiocrats, moreover, developed Petty's political arithmetic into the nascent science of political economy. Even Lavoiser, Lagrange, and Laplace all applied their mathematics to social and economic problems in addition to chemistry, mechanics, and astronomy. In Britain radical philosophers like Joseph Priestley and Jeremy Bentham similarly deployed materialism and the moral sciences in an explicit attempt to undermine the establishment. Priestley, chemist and Unitarian minister, regularly connected his work in science with his desire for radical social change.[60] In his *Experiments and Observations on Air* (1790), for example, he wrote that the growth of knowledge would "be the means, under God, of extirpating all error and prejudice, and of putting an end to all undue and usurped authority in the business of religion, as well as of science."[61] The English hierarchy, he threatened, has "reason to tremble even at an air pump or an electrical machine."[62] In his *Reflections on the Present State of Free Enquiry in This Country* (1787), he explained, "We are, as it were, laying gunpowder, grain by grain, under the old building of error and superstition, which a single spark may hereafter inflame so as to produce an instantaneous explosion."[63] This combination of science and sedition gave him the nickname 'Gunpowder Joe.'

Jeremy Bentham, founder of utilitarianism, discussed both chemistry and politics with Priestley. He had read in Priestley's *Essay on the First Principles of Government* that "the good and happiness... of the majority of the members of any state, is the

---

[60] See Isaac Kramnick, "Eighteenth-Century Science and Radical Social Theory: The Case of Joseph Priestley's Scientific Liberalism," *Journal of British Studies* (1986) 25: 1-30; Simon Schaffer, "States of Mind: Enlightenment and Natural Philosophy," pp. 233-90 in G.S. Rousseau (ed.), *The Languages of Psyche: Mind and Body in Enlightenment Thought* (Berkeley: University of California Press, 1990).

[61] Joseph Priestley, *Experiments and Observations on Different Kinds of Air* (Birmingham, 1790), p. xxiii. See Maurice Crosland, "The Image of Science as a Threat: Burke versus Priestley and the 'Philosophic Revolution,'" *British Journal for the History of Science* (1987) 20: 277-307.

[62] Quoted in Crosland, "The Image of Science as a Threat," p. 282.

[63] Quoted in Crosland, "The Image of Science as a Threat," p. 285.

great standard by which every thing relating to that state must be finally determined," a principle that became the foundation of the utilitarian philosophy.[64] In the same spirit, Bentham attempted to replace traditional morality with a 'felicific calculus' in which morality was recast as the calculation of the total quantity of happiness that would result from a given action. For Bentham, such a calculational approach represented the triumph of rationality over the mystifications of the Church and the fictions of the traditional legal system.[65] The result of these trends was a model of the individual as a calculator of self-interest, reacting mechanically to sense impressions. Society itself was merely a collection of individuals interacting according to universal and deterministic laws like Newtonian particles. In this way, human behavior and social change could in theory be described mathematically and, even more importantly, controlled through various forms of social engineering. Bentham's goal, like Petty's, was not simply to describe the regularities of society but to set up institutions that would transform people into better citizens. Bureaucrats and government advisors, he argued, should be trained in science rather than theology, and the law should treat all individuals equally since the laws of human nature are universal.[66] Certain members of the scientific elite claimed both the ability and the right to determine the best way to organize society.

In Britain such an approach clearly threatened traditional beliefs and values. Edmund Burke, a proud member of the Irish conservative intellectual tradition and a graduate of Trinity College Dublin, was appalled by the attempt to apply scientific or mathematical reasoning to the structure of the state.[67]   Burke

---

[64] Joseph Priestley, *Political Writings*, Ed. by Peter N. Miller (Cambridge: Cambridge University Press, 1993), p. 13.

[65] See Elie Halévy, *The Growth of Philosophic Radicalism* [1928] Trans. Mary Morris (London: Faber and Faber, 1972).

[66] Schaffer, "States of Mind," pp. 273-74.

[67] Crosland, "The Image of Science as a Threat."

blamed the French Revolution on a conspiracy of philosophers, referring sarcastically to "the learned academicians of Laputa and Balnibarbi."[68]   Burke's *Reflections on the Revolution in France* also attacked Priestly through thinly disguised references. Burke saw Priestley as merely the most obvious English representative of the dangers of abstract reasoning. And he believed the work of philosophes like Condorcet (mathematician, secretary of the French Academy of Sciences, and politician) to be a dangerous experiment in social planning that denied the dignity of the individual and the traditions of society. "[T]he age of chivalry is gone," Burke proclaimed, "That of sophisters, oeconomists, and calculators, has succeeded; and the glory of Europe is extinguished forever."[69] The French legislators, he argued, reduced people to numbers, homogenizing them into a mass rather than distinguishing them as individuals or classes. "[T]he constitution of a country," he claimed, is not simply "a problem of arithmetic."[70] That kind of mathematical reasoning, he believed, had dangerous political consequences. Burke explained, "on the principles of this mechanic philosophy, our institutions can never be embodied, if I may use the expression, in persons; so as to create in us love, veneration, admiration or attachment."[71]

Underlying the materialist perspective on society was a mechanical theory of the mind. Building on Locke's argument that the human mind begins as a blank slate and that all ideas are built up from sense impressions, some radicals hypothesized that the mind itself is simply matter. John Robison, professor of natural philosophy in Edinburgh University, charged that Priestley has "been preparing the minds of his readers for Atheism by his Theory of

---

[68]Quoted in Crosland, "The Image of Science as a Threat," p. 393.

[69] Edmund Burke, *Reflections on the Revolution in France* [1795] (New York: Anchor Books, 1973), p. 89.

[70]Quoted in Crosland, "The Image of Science as a Threat," p. 296.

[71]Quoted in Schaffer, "States of Mind," p. 251.

the Mind."[72] For Robison, human intelligence was the image of the divine intelligence; to reduce one to mere mechanical vibrations in the ether was to make the other nonsense.

The freethinkers that Berkeley had attacked earlier in the century had grown much more aggressive. In Dublin, William Hales, professor of Oriental Languages and staunch Newtonian described the current times as the "last stage of the most tremendous warfare ever conducted by the united powers of INFIDELITY and ANARCHY, not against the outworks, but the very citadel of our inestimable CONSTITUTION in Church and State, and even the impregnable ROCK OF CHRISTIANITY itself...."[73] The French school, headed by Voltaire, he explained, wants to sweep away the Bible, leaving only the book of nature and the use of unassisted reason for the regulation of our faith and morals. But Hales drew support from "the *Newtonian* ASTRONOMY—that sober handmaid of RELIGION."[74] Even Hales' *Analysis Fluxionum* was more than simply a defense of Newton's century-old fluxional calculus against the methods of the French calculus.[75]

The French Revolution and its supporting philosophies led to a battle for the authority of science, with Anglican natural philosophers championing Newton's theologically informed system and Unitarians and deists propounding materialistic and deterministic accounts of physical, social, and even mental phenomena.

[72] Robison, *Proofs of a Conspiracy*, p. 482. See also J.B. Morrell, "Professors Robison and Playfair, and the Theophobia Gallica: Natural Philosophy, Religion and Politics in Edinburgh, 1789-1815," *Notes and Records of the Royal Society* (1971) 26: 43-63.

[73] [William Hales] *The Inspector, Or Select Literary Intelligence for The Vulgar A.D. 1798, But Correct A.D. 1801, The First Year of the XIXth Century* (London, 1799), p. xvii.

[74] Hales, *The Inspector*, pp. 94n-95n.

[75] Florian Cajori, *A History of the Conceptions of Limits and Fluxions in Great Britain from Newton to Woodhouse* (Chicago: Open Court, 1919), p. 268.

3.5   The Ideology of Ascendancy

In addition to its often subtle scientific implications, the French Revolution, of course, had obvious political consequences for the Irish Protestant Establishment.   The ideals of the French Revolution—liberty, fraternity, equality—were directly opposed to a political system that excluded the majority of the population on the basis of their religious beliefs. And while Irish Catholics feared the end of religion as much as Irish Protestants, they had more to gain by siding with the French. The foremost proponents of republicanism in Ireland, however, were the Presbyterians. Excluded from political participation like the Catholics, they had a tradition of rational dissent going back to Toland. They quickly took up the standard of rationality and used it as a weapon against the established Church. While the 1780s began in Ireland with a renewed sense of patriotism, national unity, and enlightenment, by the end of the decade the balanced façade of Georgian life had crumbled. As patriotism gave way to republicanism and the Volunteers became the revolutionary United Irishmen, Anglicans retreated into an ideology of Protestant Ascendancy and a legislative union with Britain.

The political issue first came to a head with the question of the tithe, a tax on all Irish people (Catholic as well as Protestant) used to support the Anglican Church. In the late 1780s, secret agrarian societies began to fight the payment of the tithe. The issue exposed the way in which the opulence of Trinity College and the Church of Ireland was built by taxes on Ireland's poorest farmers, farmers who were themselves excluded from the benefits of either. Supporters of the establishment, including a number of Trinity fellows, found themselves forced to justify their position and for the first time used the term 'Protestant Ascendancy.'[76]

---

[76]The origins of the phrase 'Protestant Ascendancy' have been a source of debate in Irish historiography. See James Kelly, "The Genesis of 'Protestant Ascendancy': The Rightboy

In 1786 the Bishop of Cloyne, Richard Woodward, published a pamphlet on *The Present State of the Church of Ireland*.[77] Woodward unashamedly pointed to what he saw as the real foundation of the supremacy of Protestants over Catholics in Ireland. "I need not tell the Protestant proprietor of land," he asserted, "that the security of his title depends very much (if not entirely) on the Protestant ascendancy; or that the preservation of that ascendancy depends entirely on an indissoluble connexion between the Sister Kingdoms."[78]   The Dublin Corporation would define Protestant Ascendancy in 1792 as, "a protestant king of Ireland, a protestant parliament, a protestant hierarchy, protestant electors and government, the benches of justice, the army, the revenue, through all their branches and details protestant."[79] But Woodward seems to be referring to a more general condition of economic and political superiority. Protestant wealth, he argued, depends on the political supremacy of Protestantism (more specifically Anglicanism) over Catholicism.

For Woodward, however, it was not simply a matter of might makes right.   The Protestant clergy, for example, justified their right to the tithe through their role as 'national instructors of the people.' Woodward explained, "they are the only sources of whatever Learning exists amongst us, having the province of Education

Disturbances of the 1780s and their Impact upon Protestant Opinion," in Gerard O'Brien (ed.) *Parliament, Politics, and People: Essays in Eighteenth-Century Irish History*(Dublin: Irish Academic Press, 1989), W.J. McCormack, "Eighteenth-Century Ascendancy: Yeats and the Historians," *Eighteenth-Century Ireland* (1989) 4: 159-81, p. 174. See also W.J. McCormack, *From Burke to Beckett: Ascendancy, Tradition and Betrayal in Literary History* (Cork: Cork University Press, 1994); Jacqueline Hill, "The Meaning and Significance of 'Protestant Ascendancy', 1787-1840," in *Ireland After the Union: Proceedings of the Second Joint Meeting of the Royal Irish Academy and the British Academy, London, 1986* (New York: Oxford University Press, 1989)

[77]Richard Woodward, *The Present State of the Church of Ireland: Containing a Description of it's [sic] Precarious Situation; and the consequent Danger to the Public* (Dublin, 1787).

[78] Woodward, *The Present State of the Church of Ireland*, p. 17.

[79]Quoted in R.B. McDowell, "The Age of the United Irishmen: Reform and Reaction, 1789-94," pp. 289-338 in T.W. Moody and W.E. Vaughan (eds.) *Eighteenth-Century Ireland, 1691-1800*, Vol. IV, *A New History of Ireland* (Oxford: Clarendon Press, 1986), pp. 315-16.

left entirely in their hands."[80] A reduction in ecclesiastical income, he argued, would mean the loss of all men of high birth, promising abilities, and liberal education to the church. "The last, and perhaps not the least injurious, consequence of a diminution of the Revenues of the Parochial Clergy," he continued, "would be it's [sic] operation on the great Seminary of Learning, from which this Kingdom derives so much credit and advantage."[81] Trinity College and learning in Ireland, Woodward claimed, both depended on the Protestant Ascendancy.

Other writers drew rather different conclusions from Woodward's argument. One pamphlet attacked Trinity, arguing that Protestant privilege had not led to the advancement of science.[82] The only publications to come out of Trinity, according to this author, were bigoted pamphlets like Woodward's. Trinity, he claimed, "must be allowed on all hands to have shut her gates against the great body of the people, and to have tainted the fair garden of science with the sourness of bigotry."[83] The so-called Republic of Letters was in reality a tyranny, reflecting in the scholarly realm the injustices of the political world. Samuel Barber, moderator of the Presbyterian General Synod of Ulster, presented Woodward's claims to authority and the right to decree religious observation for all Ireland as equivalent to the dogmatism of Roman Catholicism. "To make a monopoly of the Sciences," Barber wrote, "is really carrying things with a high hand."[84]

---

[80] Woodward, *The Present State of the Church of Ireland*, p. 39.

[81] Woodward, *The Present State of the Church of Ireland*, p. 85.

[82] *Strictures on the Bishop of Cloyne's Present State of the Church of Ireland* (London, 1787), p. 55.

[83] *Strictures on the Present State of the Church of Ireland*, p. 62.

[84] Samuel Barber, *Remarks on a Pamphlet, Entitled the Present State of the Church of Ireland* (Dublin, 1787), p. 48.

A number of Trinity fellows responded in Woodward's defense.[85] Like Woodward, Robert Burrowes also appealed to the fears of Protestant landowners.

> [A]lmost all the landed property of this Country is held under grants from the Crown of estates forfeited by original Popish proprietors: tradition has been uncommonly industrious to hand down the memory of this supposed injustice. ... Is it not highly probable, that on the first available opportunity, a Popish ascendancy would annul the grants and avenge the usurpation?[86]

Burrowes argued that the advantages of the establishment outweighed its abuses. He also offered a critique of democracy, claiming that the greater numbers of Roman Catholics are "abundantly overcome by the greater degree of knowledge and property in Protestants," and therefore, "the appointment of a religion to be by law established should be properly lodged with [the Protestants]."[87] George Miller, a fellow at Trinity College Dublin, later made a similar argument against granting voting rights to Catholics. "[T]he influences of property and education," he argued, give "to the Protestants of Ireland an effective, though they did not possess a numerical superiority."[88] The mathematics of democracy, it seems, did not suit the mathematicians of the Ascendancy.

Just as Trinity fellows enlisted science in the battle to rationally demonstrate the truths of religion, they also enlisted their superior intellectual accomplishments to justify the divisions in Irish soci-

---

[85] W.J. McCormack, *The Dublin Paper War of 1786-1788: A Bibliographical and Critical Inquiry* (Irish Academic Press, 1993).

[86] Robert Burrowes, *A Letter to the Rev. Samuel Barber, Minister of the Presbyterian Congregation of Rathfryland, Containing a Refutation of Certain Dangerous Doctrines Advanced in His Remarks on the Bishop of Cloyne's Present State of the Church of Ireland* (Dublin, 1787), pp. 28-29. See also Barber's response to Burrowes, Samuel Barber, *A Reply to the Revd. Mr. Burrowes's Remarks* (Dublin, 1787).

[87] Robert Burrowes, *A Letter to the Rev. Samuel Barber*, p. 32.

[88] George Miller, *The Policy of the Roman Catholic Question Discussed, In a Letter to the Right Honourable W.C. Plunket* (London, 1826), p. 25.

ety. It was no longer enough simply to point to the successful conquest of Ireland as justification for Protestant supremacy. Through their monopoly on science, Protestants could justify the rule of an 'enlightened' elite. At the same time they realized that their accomplishments in the ideal world were founded entirely (and precariously) on a political and religious establishment that excluded the great majority of the population of Ireland. Trinity's income came primarily from land expropriated from Catholics in the seventeenth century. Its clerical graduates as well as its retired fellows were all supported by the tithe. Even its professorships of mathematics and natural philosophy were endowed from the lands that Erasmus Smith received in return for his investments in Cromwell's conquest.

But at the same time that the Trinity fellows deployed their arguments for the rationality of the establishment, radicals adopted rationalistic arguments for the end of an establishment that they associated with tyranny and superstition. Many took their cue from Priestley and invoked the authority of science in their quest for political change. Samuel Barber, for example, rejoiced at "the progress of science, which must ever be favourable to truth and fatal to error. Science enlarges the mind, ascertains the rights of man, and before science, sooner or later, all tyranny must fall."[89]

In 1791, a group of Irish radicals founded the United Irishmen in Belfast, drawing heavily on the Ulster and Scottish tradition of rational dissent.[90] William Drennan, for example, one of the movement's leaders, wrote the original prospectus for the United Irishmen, extolling "the Rights of Men and the Greatest Happi-

---

[89]Quoted in W. Bailie, "Rev. Samuel Barber, 1738-1811: National Volunteer and United Irishman," in J. Haire (ed.), *Challenge and Conflict: Essays in Presbyterian Doctrine* (Antrim: The Greystone Press, 1981), p. 82.

[90]See Philip McGuinness, "John Toland and Irish Politics," pp. 261-92 in Philip McGuinness, Alan Harrison, Richard Kearney (eds.), *John Toland's Christianity Not Mysterious* (Dublin: Lilliput Press, 1997), pp. 266-72.

ness of the Greatest Number."[91] Given this background, it is not surprising that the United Irishmen included many members with scientific interests. Thomas Russell served as the librarian for the Belfast Society for Promoting Knowledge. William James MacNeven, one of the regular attendants of Kirwan's Wednesday night conversazioni, swore Kirwan into the United Irishmen. MacNeven, described by one author as "the father of American chemistry,"[92] later moved to the United States and became a Professor of Medicine and Chemistry in New York. One of MacNeven's students, John Patten Emmet was the nephew of Robert Emmet and the father of Thomas Addis Emmet (both leading figures in the United Irishmen). John Patten Emmet later became a professor of Natural History at the University of Virginia. Robert Adrain also escaped to America after the Rebellion, where he became one of America's first research mathematicians.[93] Drennan, MacNeven, William Lawless, and Archibald Hamilton Rowan (whose solicitor was the father of William Rowan Hamilton, the mathematician) were all members of the Royal Irish Academy, and all but the first were expelled in 1798 for their treasonous activities. Even scientific fellows of Trinity College like John Stack (Brinkley's nemesis) and Whitley Stokes joined. They broke with the society, however, when it became more radical and was eventually outlawed.

In 1792 the United Irishmen began publishing a radical newspaper, the *Northern Star*. Politically and philosophically, it expressed the radical version of the Enlightenment, proclaiming, "The present is an age of revolution... This is the consequence of knowledge, the

---

[91] Quoted in Ian McBride, "William Drennan and the Dissenting Tradition," pp. 49-61 in D. Dickson, D. Keogh and K. Whelan (eds.), *The United Irishmen: Republicanism, Radicalism and Rebellion* (Dublin: Lilliput Press, 1993), p. 49.

[92] See Sean O'Donnell, "Early American Science: The Irish Contribution," *Eire-Ireland* (1983) 18: 134-37.

[93] Edward Hogan, "Robert Adrain: American Mathematician," *Historia Mathematica* (1977) 4: 157-72.

effect of intelligence, the result of truth and reason."[94] In the 1780s Burrowes and the other founders of the Academy would have agreed with such an optimistic view of the growth of knowledge and reform, but by the 1790s revolution threatened their most deeply held beliefs.

By late 1792, French military advances made conciliating Catholics a priority for the British government. Catholics were needed to fight, and the government realized that discontented Catholics made tempting allies for the French. In 1793 Irish Catholics were given the right to vote, to hold civil and military offices, and to take degrees at Trinity College. In 1795 the government established the Royal College of Saint Patrick at Maynooth to train priests in Ireland so that they would no longer have to resort to the Continent.

On May 23, 1798, rebellion broke out in Dublin and quickly spread throughout Ireland. In August, French troops arrived, but by then it was already too late. The rebellion was put down, but, according to historian Roy Foster, it was "probably the most concentrated episode of violence in Irish history."[95] Less than two decades after Grattan's famous declaration of Irish nationhood, the political scene had changed dramatically. The rebellion illustrated the tenuous nature of Protestant Ascendancy and exposed Ireland as a weak link in England's defense. In 1800 the Irish parliament voted itself out of existence, and in 1801 Ireland became part of the United Kingdom. The patriotic self-confidence that had fueled the explosion of scientific activity collapsed. With the passage of the Union, all of the forces that Burrowes thought could no longer prevent the pursuit of science in Ireland had returned: sectarianism, loss of political importance, decline in the economy, and emigration of the most talented.

[94]Kevin Whelan, "The United Irishmen, the Enlightenment and Popular Culture," pp. 269-96 in Dickson *et al.* (eds.), *The United Irishmen* (Dublin: Lilliput, 1993), p. 269.

[95]R.F. Foster, *Modern Ireland, 1600-1972* (New York: Penguin Books, 1988), p. 280.

After the Union the nobility and gentry tended to send their sons to Oxford and Cambridge. There was no longer the money or the culture to support science. Science moved out of the public sphere and remained in the cloisters of Trinity College where, thanks in part to the impetus of Brinkley, students would soon be trained in the latest developments in mathematical physics. The era of the gentleman dilettante was on its way to being replaced by that of the university researcher. Science, briefly the pursuit of patriots seeking to build the Irish nation, remained in the service of an increasingly parochial Ascendancy. The groups that supported practical science had collapsed, while those that supported radical science were expelled.

Chapter 4

Examining the Ascendancy

A Mathematical Competition

It was the end of Trinity term, 1832, and students filed into Trinity's Examination Hall to compete for the Science Medal. They faced ninety-two questions in mathematics and mathematical physics over two days. The topics ranged from geometric proofs and calculations in spherical trigonometry to the solution of algebraic equations and problems in the differential and integral calculus.[1] Among the questions were:

> Expand $\tan^{-1} x$ in powers of x by Maclaurin's theorem; also $\tan^{-1}(x + y)$ in powers of y, by Taylor's theorem.

> Determine, by D'Alembert's principle, the motion of two bodies on two inclined planes, connected by a string, the parts of which are parallel to the planes; and thence deduce the theory of Atwood's machine.

> If an undispersed ray be incident on a prism, prove that the dispersion of the emergent ray is inversely as the prod-

---

[1] *Dublin University Calendar* (1832), pp. viii-xx.

uct of the cosines of the angles of refraction at both sur-
faces.

Forty years after Brinkley's arrival, Trinity now trained her best
and brightest students in advanced mathematics and mathematical
physics. While students were still expected to master the Greek and
Latin classics and to dispute propositions in logic and metaphysics,
only the honors won on the competitive medal examinations really
mattered. A revolution had taken place.

The familiarity of such examinations to the modern student
masks their novelty. The medieval tradition of academic disputations had given way to written examinations at Trinity only in 1816,
and the new scientific standards sat uneasily with Trinity's traditional educational role. Most students still entered the clergy, and
even those who intended to pursue their scientific interests generally took holy orders to become fellows at a university. There were
no graduate science degrees, research assistantships, postdoctoral
fellowships, or professional research positions. The very word 'scientist' was not coined until 1833, indicating, particularly in Britain
and Ireland, the lack of a professional role for those who studied
science.

The contrast between Trinity's educational mission and its new
focus on mathematical science appeared absurd to many commentators. *Blackwood's Edinburgh Magazine* lamented that Trinity
College had become almost entirely devoted to "the barren toils of
mathematics; a science in which not one Irishman out of millions
has ever sought or obtained distinction. "[2] Even the *Dublin University Magazine*, run primarily by Trinity students, complained,
"Mathematics occupy the chief attention of the teachers and the
taught, to the exclusion of studies far more practically useful—
more congenial to our national taste—and, above all, more calculated to forward the objects for which Universities were insti-

---

[2]"Edmund Burke," *Blackwood's Edinburgh Magazine* (1833) 33: 277-97, p. 280.

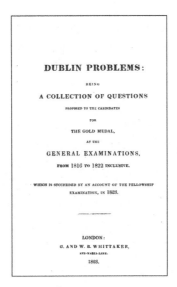

Figure 4.1: Title Page to *Dublin Problems* (1823), which laid out the new rigorous course in science and mathematics for the public.

tuted."[3] To many, mathematics appeared useless in the context of early nineteenth-century Ireland. Another pamphlet explained, "The differential calculus does not in the least promote the understanding of the Bible."[4] Its authors added, "We do not wish the great University of Dublin, founded chiefly to promote the all-important study of Theology, to degenerate into a mere Ecole Polytechnique...."[5]

So why then did Trinity reorganize its curriculum to promote the study of advanced mathematics? At the time there were a number of standard justifications. Some argued that mathematics is the best way to train students to think rationally. Others claimed that mathematics is the subject best suited to competi-

---

[3]"Academical Reform: The Dublin University System of Education Considered in Relation to Its Practicable and Probable Reform," *Dublin University Magazine* (1834) 3: 81-96, p. 84.

[4]Charles Parsons Reichel and William Anderson, *Trinity College Dublin and University Reform* (Dublin 1858), p. 19.

[5]Reichel and Anderson, *Trinity College Dublin*, p. 38.

tive examinations, and many reiterated the traditional argument that mathematical physics best supports natural theology. The fellows of Trinity College promoted mathematics not as training for scientists but rather as the foundation of a liberal education. They valued mathematics primarily as a form of mental exercise. And they could also point to similar developments at Cambridge University, where mathematics had come to dominate the Senate House examination.

Trinity's transformation, however, occurred in a much broader context. In the early nineteenth century, British society faced increasing pressure to remake itself along more secular and democratic lines. It was the Age of Reform, and in both parliament and the popular press, reformers attacked the privileges of the Anglican universities, At the same time, new secular institutions such as the University of London and the Queen's Colleges in Ireland threatened their monopoly on higher education. Trinity's internal reforms in the 1820s and 1830s were intended to avoid parliamentary interference and to compete with the new trends in education. But as the representative of the Protestant Ascendancy, Trinity's status had broader implications. Irish Protestants would defend their ancient privileges by demonstrating their intellectual superiority. When the civil service opened itself to public competition in mid-century, for example, Trinity students outperformed even those from Oxford and Cambridge, proving, they believed, that they had earned the right to govern. In an age of Catholic emancipation, parliamentary reform, and popular education, Trinity felt that rigorous scientific standards for both students and professors offered the best strategy for strengthening the precarious position of the Irish Protestant university and the Protestant Ascendancy it existed to educate.

4.1 The End of the Silent Sister

In the first decade of the nineteenth century, Trinity remained a tradition-bound, ecclesiastical institution. Fellows still saw their duty as the instruction and moral supervision of students rather than research, and students memorized ancient texts for oral examinations. In less than twenty years all of this would change. Provost Bartholomew Lloyd effected an almost complete transformation, restructuring the fellowship examination, the tutorial system, the term examinations, and even the professorships to create an institution designed to foster teaching and research in mathematics.

Lloyd entered Trinity College as a student in 1787.[6] By 1796 he had obtained a fellowship, and he went on to hold professorships in mathematics, natural and experimental philosophy, and Greek as well as a lectureship in divinity. Such variety was common within Trinity College, and Lloyd also served as chaplain to the Lord Lieutenant, president of the Royal Irish Academy, and Provost of Trinity. It is unclear exactly how Lloyd himself mastered analytical mathematics. He may well have studied for the fellowship examination with Brinkley (who arrived in 1792 just as Lloyd completed his BA) or been in contact with some of the young Cambridge mathematicians who were beginning to promote analysis just at this time. While Brinkley's example may have inspired Lloyd to learn advanced mathematics, however, Lloyd himself published only textbooks and made no contributions to mathematical research.

---

[6]J.H. Singer, "Memoir of the Late President," *Proceedings of the Royal Irish Academy* (1836-40) 1: 121-26, p. 17.

Figure 4.2: Bartholomew Lloyd (1782–1837), whose reforms as provost of Trinity College Dublin were critical to the creation of the Dublin School of Mathematics. Drawn by H. O'Neill. Engraved by C. Turner ©National Library of Ireland.

The barriers to research at Trinity in the early nineteenth century were not so much a lack of interest as a lack of time. Trinity had nearly the same number of students as Cambridge University but only a fraction of the staff, meaning that each tutor was responsible for an average of one hundred students. Each tutor, moreover, lectured his students in all aspects of the course from Greek tragedy to geometry. Thomas Romney Robinson, a fellow of the College, complained in 1820 that

> Under the system pursued at present in Trinity College, its fellows can scarcely be expected to devote themselves to any work of research, or even of compilation; con-

stantly employed in the duties of tuition, which harass
the mind more than the most abstract studies, they can
have but little inclination, at the close of the day, to commence a new career of labour.[7]

In 1829 *Blackwood's Edinburgh Magazine* noted that "the government and education of more than fifteen hundred students is con-
fided to a provost, and five and twenty fellows, by whom also various arduous professorships and offices connected with the regimen
of the University are filled."[8] It would be absurd, they added, to
expect "any great exertions in the field of general literature" from
men in such circumstances. Another article found that Oxford University currently held, "a number of endowments for learned men
more than double the number of individuals who have held fellowships and professorships in Dublin since its foundation."[9]

While the professors were overworked, the students faced an
outdated examination system. As at Cambridge, success at Trinity depended upon one's performance on examinations. At Cambridge, however, students took the tripos examination only at the
end of a three-year course, while at Trinity students were required
to pass eleven of sixteen quarterly examinations over the course of
four years. In fact, many Trinity students lived outside the College (many even lived outside of Dublin) and gained their degrees
simply by passing the quarterly examinations. While the Trinity
examinations were oral and designed merely to judge a student's
general competence, at Cambridge students were ranked according to mathematical ability (from the 1760s), examined in English
(from the 1770s) and examined in part on paper (from the 1790s).[10]

---

[7]Thomas Romney Robinson, *A System of Mechanics* (Dublin, 1820), pp. vi-vii.

[8]"The Dublin University," *Blackwood's Edinburgh Magazine* (1829) 26: 153-77, p. 173.

[9]"University of Dublin," *Quarterly Journal of Education* (1833) 6: 201-37, p. 212.

[10]See John Gascoigne, "Mathematics and Meritocracy: The Emergence of the Cambridge
Mathematical Tripos," *Social Studies of Science* (1984) 14: 547-84, Andrew Warwick, "A
Mathematical World on Paper: Written Examinations in Early 19th-century Cambridge,"
Studies in History and Philosophy of Modern Physics (1998) 29: 295-319.

In 1813, when Lloyd was appointed professor of mathematics, the examination system at Trinity had changed little over the past two centuries. Students were examined at the end of each term over two days in randomly chosen divisions of forty. The students sat at a long table while the science examiner proposed questions aloud in Latin to each student and graded his answers on a six degree scale from *optime* down to *vix mediocriter bene* (a failing grade). Of course, in an oral examination each student was asked different questions, and no student received more than ten minutes of questioning while the others sat by waiting and taking turns with the classics examiner at the other end of the table. The severity of the examination depended on the particular tutor assigned to examine a division and perhaps even on his mood that day. The best student in each division was awarded a premium (one for science and one for classics), and a strong record of premiums at the quarterly examinations resulted in a gold medal at the end of the course. Because the divisions were randomly chosen, competition for premiums varied widely. The emphasis was on determining whether or not each student had sufficiently mastered that term's material, not on ranking the students according to merit.

In 1815, soon after his appointment, Lloyd set about reforming the term examination system. Examination sections were now separated according to ability, and those who had won premiums were examined more rigorously on a more advanced and extensive course. Gold medals were awarded for the best answers in classics and science, and special reading courses were required to win them. Also, printed sheets of questions were used for the first time.[11] Now the best students were required to read French mathematics. As the preface to an 1820 Trinity textbook put it, "men who, according to the system pursued two years before the advancement of Dr.

---

[11] *Dublin Problems: Being a Collection of Questions Proposed to the Candidates for the Gold Medal at the General Examinations from 1816 to 1822 Inclusive, Which Is Succeeded by an Account of the Fellowship Examination in 1823* (London, 1823).

Lloyd to the professorship, would be now employed in fathoming the mysteries of Decimal Fractions, are rather more respectably engaged with the *Mecanique Celeste*."[12]

The new medal course represented a definite improvement, but it affected only a small number of students. Since only the best student won the gold medal and all others went unrecognized, only a few students attempted to compete. Richard MacDonnell, however, saw that the same techniques—printed questions and competitive examinations—could reinvigorate all of Trinity's students and rehabilitate Trinity's reputation. In 1828 MacDonnell, former professor of mathematics and future provost, wrote a pamphlet with suggestions for reforming the examinations.[13] The defects of the current system, he explained, tend to "take away [the students'] confidence in the even justice of our Examinations, and weaken their general respect for our Institutions; and I am sorry to say, that such feelings are spreading rapidly from the Students to their parents, and to the Publick at large."[14]

His primary suggestion was the introduction of printed lists of questions, as at Cambridge, in order to supply a uniform standard and to keep all students occupied during the examination. Printed examinations might serve to shield Trinity from charges of bias. "Of the candidates examined under that system," he explained, "though we may meet many who have disappointed themselves, yet I never knew one that did not admit the fairness of the results."[15] Both inside and outside the examination room, Trinity professors faced a crisis of credibility. MacDonnell explained, "The reverence with which a former generation regarded our attainments no longer ex-

---

[12]Dionysius Lardner, *The Elements of the Theory of Central Forces* (Dublin, 1820), p. vii.

[13]Richard MacDonnell, *A Letter to Dr. Phipps, SFTD, Registrar of Trinity College, Concerning the Undergraduate Exams in the University of Dublin* (Dublin, 1828) Beresford Papers, TCDMS Mun/P/1/2770-4/12.

[14] MacDonnell, *A Letter to Dr. Phipps*, pp. 4-5.

[15] MacDonnell, *A Letter to Dr. Phipps*, p. 8.

ists, and the publick are both more competent and more severe in
judging the merits of what we present for their improvement and
instruction."[16] The term examinations, a focal point for claims of
arbitrary judgement, could be transformed into a model of objective standards and meritocratic competition.

When Lloyd became Provost in 1831, he finally had the power
to institute major reforms. As a result of MacDonnell's suggestions and Lloyd's tireless lobbying, Trinity's Board passed a new
resolution completely overhauling the examination system in June
of 1833.[17] Lloyd extended the examinations from two days to four.
On the first day the science examiner questioned the students, and
on the second day the classics examiner evaluated them. All students received written questions as well as oral, and those who
performed well enough were allowed to compete for honors during
the second two days of the examination. Lloyd abolished the system of premiums and gold medals and introduced in its place a
system of moderatorships on the model of the tripos examination
at Cambridge. Students competed to become senior moderators
(first class honors) or junior moderators (second class honors), and
within each group the students were ranked according to merit.
While in Cambridge honors were based almost entirely on mathematical achievement, Trinity instituted three separate moderatorships; mathematics and physics, logic and moral philosophy, and
classics. Moderators followed special courses of reading above and
beyond that required for a mere pass degree.

Once examinations became competitive and specialized, however, students required more specialized training. The existing tutorial system required tutors to teach all subjects, and their salary
depended on the number of students they taught, overburdening
the junior fellows who depended heavily on tutorial fees. Lloyd
persuaded tutors to specialize in a limited range of subjects and to

---

[16] MacDonnell, *A Letter to Dr. Phipps*, pp. 12-13.
[17] See "University of Dublin," pp. 219-37.

admit pupils from other tutors to their lectures. In order to compensate for possible loss of income, he had the tutors pool their tutorial fees and distribute them according to seniority. While fellows were still required to master all the subjects in the undergraduate course to pass the grueling fellowship examination (though this was increasingly dominated by mathematics), they could now specialize their teaching and presumably improve their standards. The most competitive students, however, found even the best tutorial lectures inadequate and often employed private coaches to prepare them for the examinations.[18]

Lloyd's goal was not simply to improve undergraduate teaching, but also to improve the level of professorial research. In order to achieve this, he released the chair of Natural and Experimental Philosophy from tutorial duties, increased its salary to make up for the loss of fees, and required that only a junior fellow hold it. He soon set up the professorship of Mathematics and the Archbishop King's Lecturer in Divinity along similar lines. He explained his goals in a letter to the Vice Chancellor:

> The chief end and aim of [my plan of improvements] is to increase the efficiency of our Establishment in what relates to three of the most labourious professorships by providing for the retirement of the professors after the age of fifty or thereabouts and by quickening the succession in those departments and thereby creating a school of aspirants to those situations to cause those branches of science to be extensively cultivated in our University.[19]

By freeing these key positions from tutorial duties and guaranteeing that energetic young scholars would fill them, Lloyd hoped that Trinity would no longer be known as the Silent Sister. To

---

[18] Andrew Warwick has explored this process at nineteenth-century Cambridge. See his *Masters of Theory: Cambridge and the Rise of Mathematical Physics* (Chicago: University of Chicago Press, 2003).

[19] B. Lloyd to Archbp. Beresford (September 1, 1831) TCDMS Mun/P/1/2770/37.

support this new level of teaching, Trinity fellows produced a number of introductory mathematics and science textbooks designed to introduce students with little background in mathematics to the methods of analytical mechanics.[20] Between 1813 and 1830, Trinity men published eleven works designed specifically for students.

The result of these efforts, explored in more detail in the following chapters, was a series of Trinity graduates who won fame in mathematics and physical science.[21] Bartholomew Lloyd's son Humphrey Lloyd was appointed to the chair of natural and experimental philosophy now shorn of its tutorial duties, and he went on to make important contributions to optics and terrestrial magnetism.[22] James MacCullagh took up the reformed professorship of mathematics. He published important work in geometry and the wave theory of light and through his teaching produced a generation of Trinity mathematicians.[23] William Rowan Hamilton, easily Trinity's most famous mathematician, succeeded Brinkley as professor of astronomy and Royal Astronomer for Ireland in 1826 and gained fame in optics, mechanics, and algebra. Educated under the new medal courses in the 1820s, these three together made Trin-

---

[20]See I. Grattan-Guinness, "Mathematical Research and Instruction in Ireland, 1782-1840," pp. 11-30 in J. Nudds, N. McMillan, D. Weaire and S. M. Lawlor (ed.), *Science in Ireland 1800-1930: Tradition and Reform, Proceedings of an International Symposium held at Trinity College Dublin March 1986* (Trinity College Dublin: Dublin, 1988), pp. 14-15.

[21]See A.J. McConnell, "The Dublin Mathematical School of the First Half of the Nineteenth Century," *Proceedings of the Royal Irish Academy* (1944) 50A: 75-88, Norman D. McMillan, "The Analytical Reform of Irish Mathematics, 1800-1831," *Irish Mathematical Newsletter* (November 1984), pp. 61-75, James G. O'Hara, *Humphrey Lloyd (1800-1881) and the Dublin Mathematical School of the Nineteenth Century* (PhD Thesis: University of Manchester, 1979), Grattan-Guinness, "Mathematical Research," Ken Houston (ed.) *Creators of Mathematics: The Irish Connection* (Dublin: University College Dublin Press, 2000), Raymond Flood, "Taking Root: Mathematics in Victorian Ireland," pp. 103-19 in Raymond Flood, Adrian Rice and Robin Wilson (eds.) *Mathematics in Victorian Britain* (Oxford: Oxford University Press, 2011).

[22]See J. G. O'Hara, *Humphrey Lloyd*; T.D. Spearman, "Humphrey Lloyd, 1800-1881," *Hermathena* (1981) 130: 37-52.

[23]T.D. Spearman, "James MacCullagh," pp. 41-59 in J. Nudds, *Science in Ireland*; B.K.P. Scaife, "James MacCullagh MRIA, FRS 1809-47," *Proceedings of the Royal Irish Academy* (1990) 90C: 67-106.

ity's name synonymous with pioneering research in mathematics and mathematical physics.

Thus, from his appointment in 1813 through the first few years of his Provostship, Lloyd accomplished a major restructuring of the university. The immediate goals of Lloyd's reforms were clear. He wanted to establish a system of incentives to promote the highest standards of scholarship among both students and professors in addition to providing young professors with the time (and money) to devote themselves to research. After Lloyd's reforms, the top portion of the student body were expected to study recent French works in mathematical physics, while tutors were expected to offer specialized lectures to such students as well as pursue their own lines of research.

4.2   Mathematics and Liberal Education

While competitive examination and professorial research may seem like appropriate or even obvious improvements, the choice to focus on mathematics is more puzzling. Trinity could have instead reaffirmed its traditional focus on logic, the classical languages, or theology, and moderatorships in logic and moral philosophy and classics did leave room for these subjects in the new competitive examinations. But advanced mathematics soon dominated all other fields, in part because it had become essential for success in the fellowship examination. Yet mathematics dominated the examinations at a time when the vast majority of students were destined for the clergy, the bar, or the civil service. Such occupations hardly required a facility with the differential calculus. The key to the rise of mathematics lay in its perceived status as a paradigm of rational thought.[24] Mathematics was seen as the most certain and objec-

---

[24]See Florian Cajori, *Mathematics in Liberal Education: A Critical Examination of the Judgments of Prominent Men of the Ages* (Boston: Christopher, 1928).

tive form of knowledge, a model for all of the sciences. Thomas
Romney Robinson, former fellow of Trinity and director of the Armagh Observatory, described mathematics as the noblest branch
of science: "By it alone we reach that which is the great aim of
the good and wise, absolute truth. All the rest of our knowledge is
only probable, varying in degree from the verge of certainty down
to mere shadowy conjecture, and trustworthy exactly in proportion as the intellectual processes which deduce it are trained and
used in analogy to its practice."[25] Mathematics was seen as the
standard by which all other forms of knowledge were to be judged.
It represented absolute certainty and contained truths that all rational humans had to accept regardless of nationality, religion, or
political party.

Mathematics therefore played a role in a liberal education as an
exercise of the mind rather than a useful skill. William Whewell,
master of Trinity College Cambridge and the foremost defender of
the role of mathematics in a liberal education, argued that mathematics provides training in rational thinking applicable to any
future pursuit. "[W]e do not wish our pupils to possess mathematics principally as information, nor even as an instrument...," he
explained, "but as an intellectual discipline."[26] For Whewell the importance of mathematical study lay in its ability to make students
into rational beings.

The intellectual discipline of mathematics could have benefits
in all areas of life, even theology. William Phelan, writing in the
Transactions of the Royal Irish Academy, offered a statement of
the importance of mathematical training for clergymen:

It is the peculiar glory of the Church of England, that
besides giving the most able and irrefragable defense of

---

[25]T.R. Robinson, "Opening Address to Section A," *Twenty-Seventh Report of the BAAS
Held in Dublin in 1857* (London 1858), p. 1.

[26]William Whewell, *On the Principles of English University Education, Including Additional Thoughts on the Study of Mathematics* (London, 1838), p. 42.

those tenets in which she differs from other Christian societies, she has, in every age since the Reformation, produced hosts of zealous and enlightened men, who have
stood forth the champions and protectors of Christianity
in general, and successfully exerted themselves in over-
turning whatever had even the slightest or remotest tendency to weaken the stability of the true faith. . . . Now
almost all those, by whom such inestimable service has
been performed, were. . . carefully disciplined in scientific
reasoning, almost all considerable mathematicians, acute
metaphysicians, and carrying their estimation of logic so
far, as to use it technically and with the most complete
success in their arguments and refutations.[27]

Mathematics had replaced logic as the best discipline for forming
rational minds. And clergymen enlisted it as a valuable support
for their religious arguments. While such arguments had long been
commonplace, their appeal increased with the rise of strident voices
attacking the traditional foundations of church and state.

If the predominance of mathematics at Trinity as part of a liberal education for clergymen and barristers is somewhat puzzling,
the rise of analytical mathematics is even more odd. Berkeley had
argued over a century before that while Euclidean geometry certainly represents the highest standards of reasoning, analysis, particularly the differential and integral calculus, has a much more
tenuous claim to such status. Traditionalists in the early nineteenth century similarly argued against training students in the
calculus rather than geometry or Newton's more intuitive fluxional
calculus. While from the 1780s to the 1800s men like Hutton, Playfair, and Brinkley promoted the use of the differential and integral

---

[27]William Phelan, "An Essay on the Subject Proposed by the Royal Irish Academy,
'Whether, and how far, the pursuits of Scientific, and Polite Literature, assist, or obstruct,
each other'," *Transactions of the Royal Irish Academy* (1815) 12: 3-60, p. 41.

calculus, it was not until the 1810s that analytical mathematics really began to find a home in the universities.

At Cambridge, a key moment was the foundation of the briefly-lived Analytical Society in 1812 by a group of precocious undergraduates including Charles Babbage, John Herschel, and George Peacock.[28] In Dublin there was a similar youthful verve for the exciting new mathematical methods. The *Dublin Philosophical Journal*, established in 1825 and lasting only a few issues, illustrates the trend. Contributors included Trinity professors such as John Brinkley, Bartholomew Lloyd, Dionysius Lardner, and Humphrey Lloyd. The first volume offered an anonymous review of a pamphlet attacking the rise of analysis at Cambridge by the Rev. Arthur Browne. The reviewer quipped,

> Pause, we beseech thee, reader, and reflect on what the English and Irish nations will suffer in reputation, if Cambridge and Dublin will pertinaciously send annually forth into the professions, youths whose minds have been disciplined to think 'in the obscure manner' of D'Alembert! whose ideas are as confused as those of Lagrange! and who can reason no better than Laplace or Poisson![29]

D'Alembert, Lagrange, and Laplace, pilloried in the late eighteenth century as enemies of all true religion and government, were now praised as model reasoners. But while the Dublin mathematicians shared the Cambridge analysts' contempt for the staid traditions of British mathematics, they hardly shared their verve for political reform.

---

[28] See Philip C. Enros, "The Analytical Society: Mathematics at Cambridge University in the Early Nineteenth Century," (PhD Thesis: University of Toronto, 1979), Harvey W. Becher, "William Whewell and Cambridge Mathematics," *Historical Studies in the Physical Sciences* (1980) 11: 1-48, J.M. Dubbey, "The Introduction of the Differential Calculus to Great Britain," *Annals of Science* (1963) 19: 37-48.

[29] "Review: *A Short View of the Principles of the Differential Calculus* by the Rev. Arthur Browne, Fellow of St. John's College, Cambridge (Cambridge, 1824)," *Dublin Philosophical Journal* (1825) 1: 203-11, p. 208.

4.3 The Age of Reform

The 1820s and the 1830s were the Age of Reform in Britain. After the conservative reaction to the French Revolution, Dissenters, industrialists, Catholics, and other groups began to clamor for a voice in parliament. Public opinion was no longer tolerant of aristocratic privilege, and once venerable institutions like rotten boroughs, the established church, and the universities became the targets of radical journals and parliamentary inquiries. Efficiency was the new watchword, and reformers gradually gained enough power to force traditional institutions to adapt to the new spirit of the times. Trinity's improvements came about not simply because of Lloyd's personal mission to introduce high mathematical standards but also because an enormous amount of public pressure was brought to bear to force Trinity to meet new requirements of utility.

In the 1780s and 1790s Irish Protestants had struggled for autonomy from England and the creation of an Irish Protestant nation. In the early decades of the nineteenth century, however, most came to believe that the Union was the only way to guarantee their privileged position in Irish society. But while the Union protected the Anglo-Irish from the overthrow of Protestantism in Ireland, it also exposed them to a British parliament increasingly sympathetic to the claims of Catholics and reformers.[30] Institutions that had been acceptable or tolerable in the eighteenth century were now found in need of reform, and Irish Protestants no longer had the power to prevent such reforms.

Among the foremost proponents of reform in Ireland was Daniel O'Connell. Elected to parliament in 1828 despite the fact that as a Catholic he was barred from taking his seat, O'Connell gained enough power through alliances with the Whigs and radicals to achieve Catholic emancipation (the ability to hold government of-

---

[30] Edward Brynn, *The Church of Ireland in the Age of Catholic Emancipation* (New York: Garland, 1982), p. 3.

fices) in 1829. He then immediately began to campaign for the repeal of the Union with Great Britain with the hope of creating an Irish Catholic nation. When a Whig administration took office in 1831 and carried the first reform bill in 1832, the future did not look bright for the Protestant Ascendancy. One of the reformers' first targets was the Church of Ireland. The Whig *Edinburgh Review* reported that the established Church of Ireland, supported by tithes on Irish people of all religions, represented less than 11% of the Irish population.[31] A Royal Commission began an inquiry into the Church of Ireland that would lead to the end of the tithe and the elimination of two archbishoprics and eight bishoprics. As historian Edward Brynn explains, "the Irish Establishment became the first great historical institution to be measured by the evidence of its usefulness to the nation. When this yardstick was applied, the Church of Ireland was found wanting."[32] While the established Church in England was subject to similar criticisms, its central position in English life made it better able to withstand the attacks. The Irish Church, as a 'national' church that represented only a fraction of the population, had fewer friends and more enemies.

As branches of the established church, the Anglican universities were also subject to the attacks of reformers. They were accused of wasting their incredible wealth for the support of a few privileged and useless dons who failed to promote the pursuit of knowledge. In an 1808 review, Scottish scientist John Playfair had blamed them for the backwardness of British science, accusing particularly "the two great centres from which knowledge is supposed to radiate over all the rest of the island,"[33] namely Oxford and Cambridge. Playfair's review was only the first in a number of attacks published in the *Edinburgh Review*, culminating in a series of essays

---

[31]"State of the Irish Church," *Edinburgh Review* (1835) 61: 490-525.

[32]Brynn, *The Church of Ireland*, p. 5.

[33]John Playfair, "La Place, *Traité de Méchanique Céleste*," *Edinburgh Review* (1808) 11: 249-84, p. 283.

by the philosopher Sir William Hamilton (no relation to the Irish mathematician Sir William Rowan Hamilton). In 1831 Hamilton wrote,

> This is the age of reform—Next in importance to our religious and political establishments, are the foundations for public education. . . . Public intelligence is not, as hitherto, tolerant of prescriptive abuses, and the country now demands—that endowments for the common weal should no longer be administered for private advantage.[34]

Hamilton and others argued that the universities, which had been endowed by the crown for the support of education, were in fact 'public' institutions with certain responsibilities to the nation, responsibilities they had long ceased to fulfill.

The Benthamite *Westminster Review* was even more scathing in its criticism of the universities, accusing them of "a hideous laziness, an enormous and insatiable greediness, and a crapulous self-indulgence."[35] It went so far as to recommend a parliamentary inquiry that would "transfer the entire sovereignty over the realms of learning to the hands of laymen; to rescue these fair domains from the withering sway of ecclesiastics."[36] Not only did these institutions fail to encourage study, the *Review* explained, they actively discouraged it. "[T]he pursuit of learning is the flimsy pretext," the author claimed, "the real aim is to obtain preferment in the church."[37] "[O]ur Universities," he added, "fall far short of that utility which we have an inalienable right to insist upon reaping from our public domains."[38] As an increasing number of groups came

---

[34] Sir William Hamilton, "On the State of the English Universities, with more especial reference to Oxford," pp. 386-434 in Sir William Hamilton, *Discussions on Philosophy and Literature, Education and University Reform* (London 1852), p. 386-7. [Originally appeared in *Edinburgh Review* (June 1831).]

[35] "The Universities of Oxford and Cambridge," *Westminster Review* (1831) 15: 56-69, p. 59.

[36] "The Universities of Oxford and Cambridge," p. 59.

[37] "The Universities of Oxford and Cambridge," p. 61.

[38] "The Universities of Oxford and Cambridge," p. 63.

to recognize the importance of education, they demanded that the universities become more than just Anglican seminaries. Trinity College Dublin, much smaller in terms of faculty and less well endowed, often escaped the explicit notice of such attacks. While Oxford University had 450 ecclesiastical livings, 24 headships of colleges, and 570 fellowships, and Cambridge University had 330 livings, 17 headships of colleges, and 420 fellowships,[39] Trinity College Dublin had only 21 livings, 1 provostship and 25 fellowships. Moreover, while Catholics and Dissenters were not allowed to take degrees at Cambridge or even to matriculate at Oxford, since 1793 they had been admitted to degrees at Trinity. However, Catholics and Dissenters were still excluded from fellowships and scholarships at Trinity; and as one of the largest landholders in Ireland and a representative of elitism, exclusivity, and ancient privilege, Trinity had no friends among the reformers.

In 1827, for example, the *Freeman's Journal* proposed "to lift the veil from the 'Silent Sister,' and submit her to the scrutinizing test of public notice, from which she has too long been screened, both for her own real interest and the general good." The Provost, they reported, earns about £4,000 per year, while the fellow's income (long a secret), they estimated to be £2,000 per year. "What benefit does the public receive at their hands?" asked the *Freeman's Journal*, "Literally none!"

> We have Professors of Mathematics, who deliver about
> twelve lectures in twelve months, on the first twelve pages
> of some elementary treatise of fluxions; while two assis-
> tants, in little more than the same number of lectures,
> labour through a few of the most elementary departments
> of algebra and geometry, *when they have an audience,*

---

[39]"The Irish and English Universities," *Dublin Review* (1836) 1: 68-100.

which, so thoroughly are the students aware of the humbug thus practised on them, does not frequently occur.[40]

(Trinity, it should be said, sued the paper for libel over these claims and received an apology from the proprietor.) Soon afterwards, the Catholic *Dublin Review* mounted its own attack on Trinity's religious exclusivity. While the fellows of the university claimed they were a private organization with their own endowment, the *Review* argued that they were "answerable to the Nation, at large, for every measure which is adopted by them; and are, in fact, but trustees of the interest of the people."[41] Another article found Trinity's religious restrictions intolerable. All fellows at Trinity were required to take an oath to "constantly resist all opinions, which either Papists or others maintain against the truth of sacred Scripture," and to acknowledge the king of England, Scotland, and Ireland "to be subject to the power of no foreign prince or pontiff."[42] At the very least, the article argued, Catholic fellows should be appointed to look after the souls of Catholic students.

Why were calls for educational reform so strident? At the beginning of the nineteenth century, education was becoming an increasingly essential means of social mobility.[43] Catholics and Dissenters, excluded from the universities, were also excluded from those professions that required a university education. At the same time, members of the working class realized that education could mean mobility for them as well. While Whigs and radicals criticized the universities in print, they were also busy promoting alternative schemes of education.

In the late eighteenth century, Dissenting Academies and Literary and Philosophical Societies (like those in Belfast) had served

---

[40]*Freeman's Journal* (January 2, 1827).

[41]"The Irish and English Universities," p. 68.

[42]"Trinity College Dublin," *Dublin Review* (1838) 4: 281-307, p. 296.

[43]Perry Williams, "Passing on the Torch: Whewell's Philosophy and the Principles of English University Education," pp. 117-47 in Menachem Fisch and Simon Schaffer, *William Whewell: A Composite Portrait* (Oxford: Clarendon, 1991), p. 117.

as centers for radicalism. In the early nineteenth century, reformers promoted new scientific organizations like the Royal Institution and the Mechanics' Institutes, particularly in urban and industrial centers like Manchester, Newcastle, Glasgow, and London.[44] The Rev. Thomas Dix Hincks, for example, a Presbyterian minister and former student of Joseph Priestley, founded the Royal Cork Institution in 1803.[45] Hincks published pamphlets on Catholic emancipation and promoted practical scientific education.[46] Some of his major supporters were the brewers Beamish and Crawford. The conjunction of useful knowledge, working class education, and social reform found its greatest proponent in Henry Brougham, a powerful Whig politician with an interest in science. Brougham was a central figure in the creation of new institutions such as the London Mechanics Institution (1823), the Society for the Diffusion of Useful Knowledge (1825), and (together with utilitarians like James Mill) University College London (1827), England's first secular university.[47] Brougham, a major supporter of parliamentary reform and Catholic emancipation, also argued for the right of Dissenters to take degrees at Oxford and Cambridge.

Thus in the early decades of the nineteenth century, the institutions of the Establishment—the Anglican Church and the universities—were submitted to public scrutiny and criticism as

---

[44] See D.S.L. Cardwell, *The Organization of Science in England* (London: Heinemann, 1972), Colin A. Russell, *Science and Social Change in Britain and Europe, 1700-1900* (New York: St. Martin's Press, 1983), Morris Berman, *Social Change and Scientific Organization: The Royal Institution, 1799-1844* (Ithaca: Cornell University Press, 1978), Robert H. Kargon, *Science in Victorian Manchester: Enterprise and Expertise* (Manchester: Manchester University Press, 1977), Steven Shapin and Barry Barnes, "Science, Nature and Control: Interpreting Mechanics Institutes," *Social Studies of Science* (1977) 7: 31-74, Ian Inkster, "Science and the Mechanics Institutes, 1820-50," *Annals of Science* (1975) 32: 451-74.

[45] S.F. Pettit, "The Royal Cork Institution: A Reflection of the Cultural Life of a City," *Journal of the Cork Historical and Archaeological Society* (1976) 81: 70-90.

[46] James Thomson, father of William Thomson (Lord Kelvin) and professor at the Belfast Academical Institution shared similar political and education views, see Crosbie Smith and M. Norton Wise, *Energy and Empire: A Biographical Study of Lord Kelvin* (Cambridge: Cambridge University Press, 1989), pp. 3-19.

[47] See J.N. Hays, "Science and Brougham's Society," *Annals of Science* (1964) 20: 227-41.

never before. Measured by their utility to the public, the universities were seen as a shameful waste of resources and talent, to be dealt with by parliamentary commissions and the removal of clerical control. While local critics like Theophilus Swift and Samuel Barber had criticized Trinity in the late eighteenth century for its religious exclusivity and lack of scholarship, they could only tarnish Trinity's reputation. Powerful parliamentarians like Brougham and O'Connell, on the other hand, had the power to remake Trinity's very foundation.

4.4   The Political Context of Lloyd's Reforms

The attacks of the political reformers formed an important context for the mathematical reforms at Trinity College. Lloyd, for example, was appointed provost in 1831 by the new Whig government. McDowell and Webb explain that the government saw Lloyd as "a man who, though not a committed Whig, could be relied on to carry out a policy of reform consonant with that which the Government was proposing in other fields."[48] Lloyd, however, was a reformer out of necessity, not by choice. His letters to the vice-chancellor of the university, Lord Beresford, Archbishop of Armagh, in the 1830s illustrate his sense that the university was under siege. Lloyd referred, for example, to "the attacks made on our institution, the gross misrepresentations respecting our management of its concerns—and the various libelous paragraphs which from time to time appear in the daily prints."[49] He also spoke of "the Terror with which I should regard any attempt to open this place to the government of R. Catholics."[50] In 1834, Lloyd mentioned a rumor of a commission to overhaul the property and manage-

---

[48]R.B. McDowell and D.A. Webb, *Trinity College Dublin, 1592: An Academic History* (Cambridge: Cambridge University Press), p. 152.

[49]B. Lloyd to Beresford (November 17, 1831) TCDMS Mun/P/1/2770/44.

[50]B. Lloyd to Beresford (April 22, 1836) TCDMS Mun/P/1/2770/151.

ment of Trinity, "in accordance with Brougham's threat against
the senior fellows."[51] "I do think it hard," he wrote, "I might say
insulting, that our Protestant Institution should be placed at the
mercy of its most rancorous enemies."[52] Even Lloyd's obituary in
the *Dublin University Magazine* emphasized "his vigilant guardian-
ship of the interests of the institution from *its external and political*
assailants."[53]

Lloyd explained to the Archbishop of Dublin that he most feared,
"those who seek to convert Trin. Coll. into a popish or, what is still
worse, a Lay College for Learning merely secular like the polytechnic school of our neighbors." He continued, "For I am persuaded
that to transfer the business of Education from the Clergy to the
Laity would be to make this, or any other such Establishment a
nursery for Deists, Atheists and Anarchists."[54] Despite his verve
for science, Lloyd did not believe that Trinity College existed primarily to promote scientific teaching and research. That Trinity
might become another École Polytechnique, the most advanced scientific institution of the early nineteenth century, was his greatest
fear, not his wish. Lloyd's goal was to demonstrate the importance
of religious education in an age of increasing secularism.

Lloyd's reforms, therefore, were designed not to train scientists
but to protect an ecclesiastical institution under siege from those
who wanted to sever the ties between religion and education. His
eulogist saw Lloyd's scientific reforms very much in this context,
remarking,

> The English universities are beginning at length to wel-
> come loudly their Irish sister to the generous strife of sci-
> entific advancement; and, even in the cabinet itself (we

---

[51] On the attempts to reform Trinity's management of its estates see R.B. McCarthy, *The
Trinity College Estates 1800-1923: Corporate Management in an Age of Reform* (Dublin:
Dundalgan Press, 1992).

[52] B. Lloyd to Beresford (January 20, 1834) TCDMS Mun/P/1/2770/106.

[53] "The Late Provost," *Dublin University Magazine* (1838) 11: 111-21, p. 119.

[54] B. Lloyd to Whately (April 24, 1834) TCDMS Mun/P/1/1723/4.

speak from authentic sources) projects of rude inquiry have, within the last few years, been checked by the remark from the *highest* authority—that 'Dublin College is reforming itself!'[55]

If the Whigs and Benthamites could condemn a privileged minority religious institution, they could not so easily attack one that had proven itself capable of promoting the advancement of science. By reforming themselves, the fellows of Trinity hoped to prevent others from reforming them even further.

Lloyd's reforms, however, were seen in an even broader context than just preventing enemies from meddling in the affairs of the College. With the rise of radicalism and utilitarianism, the upper classes, the clergy, and more specifically the Protestant Ascendancy were forced to find new ways to justify their privileged position. The cultivation of science was an ideal way to achieve this. An 1834 essay on academic reform in the *Dublin University Magazine* explained,

> These too, be it remembered, are peculiar times—times when we are persuaded the supremacy of rank can only be maintained by the superiority of intellect. A mighty impetus has been given by the diffusion of information to the energies of the lower classes; and if the middling and upper classes desire to maintain their elevation, they must keep pace with the spirit of improvement.[56]

No longer would the privileges of birth or religion be sufficient to protect the Protestant Ascendancy. In a world of Catholic emancipation and Mechanics Institutes, the Protestant gentry could only maintain their ascendant position through the demonstration of their intellectual superiority. Trinity, the article continued, must

---

[55]"The Late Provost," p. 119.
[56]"Academical Reform," p. 86.

train the young men of the upper classes "to compete with the march of intellect, and to sway the wills of the multitude by the mere moral power of superiority of attainments."[57] While Irish Protestants had merely asserted their intellectual superiority in the face of attacks by Catholics and Dissenters in the 1780s, in the 1830s they would have to prove it (mathematically!).

Among Lloyd's reforms was the decision in 1833 to publish for the first time the *Dublin University Calendar*, which gave a history of the college, the details of the new undergraduate course, and the examination questions themselves. A reviewer in the *Dublin University Magazine* explained the tremendous importance of such a document, "by appealing to which the cavils of the ignorant declaimer may at once be decidedly refuted, and the friends of the establishment supplied with conclusive arguments to defend the interests of the Dublin University."[58] Students competed on mathematical examinations not just to prove their own abilities but also to defend Trinity's reputation against 'the cavils of the ignorant declaimers.'

4.5   Meritocracy

When the dreaded parliamentary commission finally did arrive to investigate Trinity in 1853, the fellows pointed again and again to Lloyd's reforms as evidence that Trinity could govern itself.[59] One witness argued that the change to the professorships of mathematics and natural philosophy (requiring them to be held by junior fellows and removing tutorial obligations) "seems to have been successful in promoting a very high cultivation of the branches of

---

[57]"Academical Reform," p. 86.

[58]"The Dublin University Calendar for 1833," *Dublin University Magazine* (1833) 1: 105-6, p. 106.

[59]Dublin University Commission, *Report of Her Majesty's Commissioners Appointed to Inquire Into the State, Discipline, Studies, and Revenues of the University of Dublin, and of Trinity College* (Dublin, 1853).

science to which the Professorships relate."[60] Another praised the arrangement for giving separate lectures to honors students.[61]

Trinity students, however, won success not only in the honors examinations but also in the new examinations for the civil service and the military. In the 1850s the British government instituted competitive examinations for many of the same reasons Oxford, Cambridge, and Trinity had—to encourage hard work, avoid charges of bias, and offer avenues of upward mobility.[62] Trinity quickly organized special courses specifically to train students for these examinations, and their results were quite impressive. Trinity students won commissions and cadetships in the Ordnance Corps, the Royal Artillery, the Royal Engineers, the Indian Civil Service, and the Royal Military Academy.[63] In fact, Trinity students often beat out graduates of Oxford, Cambridge, and other universities in such competitions.

In 1855–56, for example, Trinity students obtained 48 military commissions, compared to 12 for Cambridge, 5 for Oxford, 4 for the Scottish Universities, 3 for London University, and none for the Queen's University.[64] "The superiority of Dublin and Cambridge over Oxford," one article explained, "was caused by mathematics, and that of Dublin over Cambridge by natural and experimental science."[65] On the Indian Civil Service examination, which emphasized classics and polite literature, Trinity men obtained 33 positions, while Oxford won 52, Cambridge 45, the Scottish Uni-

---

[60] Dublin University Commission, p. 49.

[61] Dublin University Commission, p. 68.

[62] Roy MacLeod (ed.), *Days of Judgment: Science, Examinations and the Organization of Knowledge in Late Victorian England* (Driffield: Studies in Education, 1982), C.J. Dewey, "The Education of a Ruling Caste: The Indian Civil Service in the Era of Competitive Examination," *English Historical Review* (1973) 88: 262-85.

[63] *University of Dublin Military Class: The Conditions of Admission, and the Examination Papers Set at the Entrance Examination in October 1857* (Dublin, 1857).

[64] *University Education in Ireland in the Year 1860* (Dublin, 1861). [Reprinted from the *Dublin Evening Mail.*]

[65] *University Education in Ireland*, p. 39.

versities 16, the Queen's University 15, and London University 13. The article concluded,

> Whether, therefore, the test be applied in literature or science, the ancient seats of learning have fully justified their character, and have taught the advocates of new systems of education a lesson that they cannot soon forget. As for the University of Dublin, it may fairly be asserted that these public examinations have done more to wipe out her reproach as the 'Silent Sister,' than the publication of miles of trite mathematics—after the manner of the Cambridge wranglers—or of tons of theological heresies—in the fashion of the Oxonians—could possibly have done.[66]

The examinations, it seemed, provided an objective measure of Trinity's efficiency. Taking into account the smaller population from which it drew students, Trinity was three times more successful than Oxford or Cambridge. When Trinity once again came under attack in the 1860s, its defenders pointed with pride to its record in competitive examinations.[67] One fellow argued,

> At the present moment, Trinity College may be regarded as a manufactory for turning out the highest class of competitors for success in the Church, at the English Bar, in the Civil Service of India, and in the Scientific and Medical Services of the Army and Navy; and any legislation which would produce the effect of lowering the present high standards of her degrees, would tend to destroy the prospects of the educated classes in Ireland, and become to those classes little short of a national calamity.[68]

---

[66] *University Education in Ireland*, p. 40.

[67] On the Irish university question see, T.W. Moody and J.C. Beckett, *Queen's Belfast 1845-1949: The History of a University* (London: Faber and Faber, 1959).

[68] Samuel Haughton, *University Education in Ireland* (London, 1868), p. 11.

Intellectual superiority, this author and others argued, justified the exclusivity of Trinity College and the privileges of Irish Protestants.

Trinity's success in mathematics and on the civil service and military examinations provided what many saw as concrete evidence of the intellectual superiority of Protestantism over Catholicism. While Protestant scientists no longer derided scholasticism, they rarely failed to mention the banning of Copernicus, the persecution of Galileo, and the poor record of Irish Catholic scholarship in the sciences.

But Catholic scientists were quick to respond. William K. Sullivan, for example, one of Ireland's most prominent Catholic scientists in the nineteenth century, saw the new meritocracy as little more than a cover for old aristocracy. The "ascendancy party," as he called them, "know that a properly educated Catholic middle class would soon deprive them of a monopoly which they formerly defended in the name of conservatism, but which they now propose to maintain in the name of liberalism and enlightenment."[69] In terms of Catholic achievement in mathematics, he asked,

> Is it not a mockery for a member of that ascendancy party, which used in former times such unholy means to crush out every trace of mental culture from amongst us, and which now uses mean calumny and vulgar gibes, to ask us where are our senior wranglers [the top scorers on a mathematical examination]?[70]

In 1847, nearly fifteen years after the creation of the moderatorships, Catholics had won only two of the forty-seven gold medals

[69]William K. Sullivan, *University Education in Ireland: A Letter to Sir John Dalberg Acton* (Dublin, 1866), p. 5. See James Bennett, "Science and Social Policy in Ireland in the mid-Nineteenth Century," pp. 37-47 in P. J. Bowler and N. Whyte (eds.), *Science and Society in Ireland: The Social Context of Science and Technology in Ireland, 1800-1950* (The Institute of Irish Studies-Queen's University of Belfast: Belfast, 1997).

[70] Sullivan, *University Education in Ireland*, p. 15.

awarded in mathematics and physics.[71] As stepping-stones to fellowships that could be held only by Anglicans (most in fact required the holder to actually be a clergyman), there was little incentive for Catholics to compete. And yet their failure to pursue such goals was sometimes offered as evidence of their unsuitability not only for mathematics or science but also for the administration of a university or even for government.

A notable exception was John Casey, a Catholic schoolteacher who came into contact with the mathematical fellows of Trinity through his advanced work in mathematics. Casey took a degree at Trinity and in 1873 was offered the professorship of mathematics.[72] The fellows hoped to prove that Trinity rewarded talent regardless of religious differences. Casey, however, opted instead to take a professorship at the Catholic University. Henry Hennessy, a Catholic physicist, also taught at the Catholic University.[73] Both Casey and Hennessy were members of the Royal Irish Academy and Fellows of the Royal Society. Irish Catholics established their own institutions and produced their own scientific experts in the mid-nineteenth century, putting further pressure on Protestants to compete or risk losing their historic claim to be the source of Irish learning.

On the surface, the introduction of teaching and research in advanced mathematics at Trinity College appears to be simply the inevitable progress of science education. Discarding the medieval traditions that retarded scholarship throughout the eighteenth century, reformers like Brinkley and Lloyd modernized the curriculum, bringing Trinity into the age of the modern research university. The fact that similar reforms were taking place at Oxford and Cam-

---

[71] See "Reform of the Dublin University: The Scholarship Question," *Dublin Review* (1847) 23: 228-51.

[72] "The Late Professor Casey," *The Irish Monthly* (1891) 19: 106-108; "John Casey," *Proceedings of the Royal Society of London* (1890-1891) 49: xxiv-xxv.

[73] "Henry Hennessy, 1826-1901," *Proceedings of the Royal Society of London* (1904-1905) 75: 140-42.

bridge, Paris, and Berlin seems to indicate a global rather than a local phenomenon. Certainly Trinity responded to a broad shift in European attitudes about the aims and responsibilities of a university and modeled its reforms on those adopted in other educational institutions. But the fellows at Trinity were also trying to navigate the dangerous waters of British public opinion and Irish politics. They never lost their sense that Trinity's purpose lay not in the training of mathematicians but in the preparation of the sons of the Protestant Ascendancy to take up their traditional roles in the church, the bar, and the civil service. Their pre-eminence, they claimed, derived not from inherited privilege but from intellectual superiority. Just as an individual student's future prospects came to depend increasingly on his ability to demonstrate his talent on written examinations, so the future of the Ascendancy hinged on the degree to which it could prove that its elevated status was rooted in ability. Trinity transformed itself into a manufactory for mathematicians because its most powerful fellows (themselves chosen by their success on a mathematical examination) believed that this was the best way to demonstrate their intellectual ascendancy.

Chapter 5

Truth and Beauty

The Poet and the Mathematician

It was August 1829, and the eminent poet William Wordsworth was reading aloud from his poem *The Excursion* in the drawing room of Dunsink Observatory just outside of Dublin.[1] His host, the young Royal Astronomer for Ireland, William Rowan Hamilton, had prompted this performance after expressing concern that some passages in the poem seemed to show "a slight reverence for Science." While this was Wordsworth's first visit to Ireland, he and the young Hamilton were already well acquainted.[2] In fact it was Hamilton who had convinced Wordsworth to visit Ireland despite his concerns about the safety of travel during a time of unrest. When the two had first met in September 1827 on a tour of

An earlier version of this chapter appeared as David Attis, "The Social Context of W.R. Hamilton's Prediction of Conical Refraction," pp. 19-36 in Peter J. Bowler and Nicholas Whyte (eds.), *Science and Society in Ireland: The Social Context of Science and Technology in Ireland, 1800-1950* (Belfast: Institute of Irish Studies-Queen's University Belfast, 1997).

[1] Robert P. Graves, *Life of Sir William Rowan Hamilton.* 3 Vols. (Dublin: Hodges, Figgis and Co., 1882, 1885, 1889), 1: 311-314; Thomas Hankins, *Sir William Rowan Hamilton* (Baltimore: Johns Hopkins University Press, 1980), pp. 102-104.

[2] On Hamilton's friendship with Wordsworth see George Dodd, "Wordsworth and Hamilton," *Nature* (1970) 228: 1261-1263; Thomas Owens, "Wordsworth, William Rowan Hamilton and Science *in The Prelude*," *Wordsworth Circle* (2011) 42: 166-169, as well as Hankins, *Sir William Rowan Hamilton* and Graves.

the Lake District, they had become instant friends.[3] Wordsworth wrote to Hamilton just after their initial encounter, "Seldom have I parted—never, I was going to say—with one whom, after so short an acquaintance, I lost sight of with more regret."[4] Hamilton, for his part, was star struck. He immediately composed a poem and sent it Wordsworth.

Thus Hamilton had no compunction in pressing Wordsworth on his negative portrayal of science. The passage in question was apparently in the fourth book of *The Excursion* where the wanderer criticizes men of science who,

> Viewing all objects unremittingly
> In disconnection dead and spiritless;
> And still dividing, and dividing still,
> Break down all grandeur, still unsatisfied
> With the perverse attempt, while littleness
> May yet become more little; waging thus
> An impious warfare with the very life
> Of our own souls![5]

According to the account left by Hamilton's sister Eliza, the great poet defended himself "from the accusation of any want of reverence for Science, in the proper sense of the word—Science, that raised the mind to the contemplation of God in works, and which was pursued with that end as its primary and great object." But Wordsworth stood firm in his opposition to "all science which was a bare collection of facts for their own sake, or to be applied merely to the material uses of life. . . . All science which waged war with and wished to extinguish the Imagination in the mind of man.[6] He railed against "what is disseminated in the present day under the

---

[3] Hankins, *Sir William Rowan Hamilton*, pp. 50-52.

[4] Wordsworth to William Rowan Hamilton (September 24, 1827) Graves 1: 268.

[5] Wordsworth, *The Excursion*, Book 4.

[6] Eliza Hamilton's account of Wordsworth's visit, Graves 1: 311-314, p. 313.

title of 'useful knowledge"' as being of a dangerous and debasing tendency.

As it turned out, Hamilton and Wordsworth were in violent agreement. Both believed that the ultimate goal of science and mathematics should be to raise the mind up from the material to the ideal and divine, and they feared that the British obsession with empiricism, inductive reasoning, and useful knowledge was not only misguided, but dangerous. But Hamilton still believed that it was possible to create a science that accomplished what the best poetry did. For Hamilton, mathematics requires imagination and creativity and, even more importantly, it connects men to the highest of truths.

In his youth, Hamilton aspired to fame in both mathematics and poetry. Hamilton's rapid ascent in the scientific world provided an entrée into Irish polite society. In fact, Hamilton seems to have spent more time socializing with novelists and poets than with other scientists. He was a frequent visitor at Edgeworthstown, home of Maria Edgeworth (daughter of Richard Lovell Edgeworth), one of the most famous novelists of her day.[7] Maria's brother Francis Beaufort Edgeworth became one of Hamilton's closest friends, along with two other literary young men. Aubrey de Vere was the son of the poet and playwright Sir Aubrey de Vere and became a poet in his own right.[8] He as well as Hamilton had close ties to Wordsworth, and it was de Vere who introduced Tennyson to Wordsworth. Viscount Adare, later Lord Adare, was the son of the Count of Dunraven and brought Hamilton into the homes of the best families of Ireland and England. All of these young men shared Hamilton's passion for romantic poetry and metaphys-

---

[7] Michael Hurst, *Maria Edgeworth and the Public Scene: Intellect, Fine Feeling and Landlordism in the Age of Reform* (London: Macmillan, 1969).

[8] See S.M. Paraclita Reilly, *Aubrey De Vere: Victorian Observer*, 3rd Ed. (Dublin: Clonmore and Reynolds, 1956) and Wilfrid Ward, *Aubrey De Vere: A Memoir Based On His Unpublished Diairies and Correspondence* (London: Longmans, Green, and Co., 1904).

ical idealism.[9] Their correspondence is full of the philosophies of
Plato, Berkeley, and Coleridge. In fact, it was Coleridge who had
perhaps the greatest intellectual impact on Hamilton even though
they met only in 1832, two years before the poet's death.

While Hamilton moved easily between the literary and scientific worlds, Wordsworth convinced him that it was impossible to
achieve success in both poetry and mathematics simultaneously—
not because of any inherent contradiction between the two but because each requires a separate set of technical skills that can only
be mastered through complete devotion. On Wordsworth's advice,
Hamilton gave up his aspirations for poetry and committed himself to finding the same beauty in mathematics. In his valedictory
poem, "To Poetry," Hamilton wrote,

> Spirit of Beauty, though my life be now
> Bound to thy sister Truth by solemn vow;
> Though I must seem to leave thy sacred hill,
> Yet be thine inward influence with me still;
> And with a constant hope enquire,
> And with a never sequenced desire
> To see the glory of your joint abode,
> The home and birthplace, by the throne of God.[10]

Hamilton never gave up his desire to connect the search for beauty
with the search for truth. He wrote to Francis Beaufort Edgeworth, "I must say that I believe myself to find in mathematics
what you declare you do not—a formable matter out of which to
create Beauty."[11] And while such proclamations of the beauty of
mathematics are commonplace among mathematicians, for Hamilton they had a very specific meaning. Hamilton's pioneering work

---

[9]Terry Eagleton places Hamilton's idealism in the context of the cultural currents surrounding the Young Ireland movement. Terry Eagleton, *Scholars & Rebels in Nineteenth-
Century Ireland* (Oxford: Blackwell, 1999), esp. pp. 86-89.

[10]Graves 1: 317.

[11]W.R. Hamilton to F.B. Edgeworth (November 20, 1829) Graves 1: 348.

in geometrical optics and in mechanics represented his attempt to remake science in order to achieve the same goals that Wordsworth had for poetry—to raise the mind up to God and to create a feeling within the mathematician of devotion and reverence.

Like Berkeley, what was important to Hamilton—more than the application of mathematics—was the fact that mathematics has the power to transform the practitioner. Mathematical symbols, he explained, "are at once signs and instruments of that transformation by which thoughts become things, and spirit puts on body, and the act and passion of mind are clothed with an outward existence, and we behold ourselves from afar."[12] Writing out equations makes our thoughts visible and encourages us to appreciate our ability to turn thoughts into things. The mathematician for Hamilton has the power to shape the experience of the beauty and transcendence of mathematics in much the same way that poetry, in Wordsworth's view, transforms its practitioners. As literary critic M.H. Abrams explains, "Wordsworth maintains that, instead of telling and demonstrating what to do to become better, poetry, by sensitizing, purifying, and strengthening the feelings, directly *makes* us better."[13] Through the representation of the connection between mind and nature, the symbols of mathematics and poetry have the power to transform their readers. Hamilton, Wordsworth, and Coleridge all believed that the ability to shape the experience of truth in this way could be a powerful force in the shaping of society.

---

[12]W.R. Hamilton, "Introductory Lecture on Astronomy, 1832," Graves 1: 643.

[13]M.H. Abrams, *The Mirror and the Lamp: Romantic Theory and the Critical Tradition* (Oxford: Oxford University Press, 1953).

5.1   The Geometry of Light

William Rowan Hamilton was born in Dublin in 1805.[14]   His father was the estate agent for the exiled United Irishman Archibald
Hamilton Rowan. Hamilton, however, was raised primarily by his
uncle, an Anglican curate in Trim. Hamilton described his father
as "a liberal, almost a rebel," but his uncle had a very different
temper; he "was a Tory to the back-bone, and doubtless taught
me Toryism along with Church of Englandism."[15]   In many ways,
Hamilton met his destiny in August of 1821 when his uncle gave
him a copy of Bartholomew Lloyd's *Analytical Geometry*. By age
sixteen, Hamilton was reading Laplace's *Celestial Mechanics* and
had even discovered an error in the text. He entered Trinity College Dublin the following year, outperforming all competitors in
the entrance examination and on every subsequent examination.
He soon devoted himself to preparing for the grueling fellowship
examination, but his plans changed when John Brinkley left the
Professorship of Astronomy in 1826. Though still an undergraduate (!), Hamilton won the appointment over more experienced
Trinity fellows as well as George Airy, future Astronomer Royal for
England.

---

[14]See Graves, Hankins, *Sir William Rowan Hamilton*, Seán O'Donnell, *William Rowan Hamilton: Portrait of a Prodigy* (Dublin: Boole Press, 1983).

[15]William Rowan Hamilton to Augustus De Morgan (July 26, 1852), Graves 3: 392.

Figure 5.1: Sir William Rowan Hamilton (1805–1865), Royal Astronomer of Ireland, Mathematician. Wikimedia Commons.

The work that gave Hamilton a reputation strong enough to warrant a professorship before his graduation began in the early 1820s.[16] His optical research covered almost ten years in which he went from being the precocious son of a Dublin solicitor to Royal Astronomer of Ireland. Hamilton's breakthrough was to apply the latest techniques in analytic geometry to create a single general so-

---

[16]WRH to Cousin Arthur Hamilton (May 31, 1823) Graves 1: 141.

lution to all optical problems—to "remould the geometry of light."[17]
From the beginning, however, Hamilton's work sat uneasily with
that of his contemporaries. The Royal Irish Academy refused to
publish his first paper in 1824, claiming that it was too abstract,
but Hamilton was undaunted. He immediately began working on
what would become his "Theory of Systems of Rays," a series of
four long essays read to the Academy between 1827 and 1832.[18]
Some of the earliest drafts bear the more explicit title "Application
of Analysis to Optics."[19] Essentially, Hamilton applied new French
techniques in analytic geometry to the old science of geometrical
optics.[20]

For Hamilton, optics was essentially a branch of geometry, and
Bartholomew Lloyd's *Analytic Geometry* and Gaspard Monge's *Application of Analysis to Geometry* were his models. Lloyd had written that one of the most important benefits of analysis was that "a
single formula shall frequently offer to the mind a greater variety
of connected truths than can be collected from many pages of Geometry."[21] Hamilton proposed to do this for geometrical optics, to
demonstrate that its many separate truths could be deduced from

---

[17]W.R. Hamilton to S.T. Coleridge [Not Sent] (October 3, 1832), Graves 1: 592.

[18]W.R. Hamilton, "On Caustics. Part First [1824]," pp. 345-63 in A.W. Conway, and J.L. Synge (eds.) *The Mathematical Papers of Sir William Rowan Hamilton.* Vol. I: Geometrical Optics. (Cambridge: Cambridge University Press, 1931); "Theory of Systems of Rays. Part First [1827]," *Transactions of the Royal Irish Academy* (1828) 15: 69-174, *Mathematical Papers* 1: 1-88; "Theory of Systems of Rays. Part Second [1827]," *Mathematical Papers* 1: 88-106; "Supplement to an Essay on the Theory of Systems of Rays [1830]," *Transactions of the Royal Irish Academy* (1830) 16: 1-61, *Mathematical Papers* 1: 107-144; "Second Supplement to an Essay on the Theory of Systems of Rays [1830]," *Transactions of the Royal Irish Academy* (1831) 16: 93-125, *Mathematical Papers* I: 145-163; "Third Supplement to an Essay on the Theory of Systems of Rays [1832]," *Transactions of the Royal Irish Academy* (1837) 17: 1-144, *Mathematical Papers* 1: 164-329.

[19]See Tsuyoshi Ogawa, "His Final Step to Hamilton's Discovery of the Characteristic Function," *Journal of the Society of Arts and Sciences, Chiba University* (1990) B-23: 45-62. Ogawa reprints Hamilton's manuscript "Application of Analysis to Optics" (TCDMS 1492/11) on pp. 49-62.

[20]See Hankins, *Sir William Rowan Hamilton*, pp. 61-98.

[21][Rev. Bartholomew Lloyd], *Analytic Geometry; or A Short Treatise on the Application of Algebra to Geometry, Intended Chiefly for the Use of Undergraduates in the University of Dublin* (Dublin, 1819), p. vi.

a single analytic expression. Hamilton wanted to find one equation that could express all of the fundamental laws of geometrical optics simultaneously.

Hamilton found such an equation using the principle of least action. As far back as the ancient Greeks, mathematicians had recognized that when light travels in a straight line or is reflected at equal angles, it is following the shortest path between two points. The French mathematician Pierre de Fermat had described this as the principle of least time, and William Molyneux had praised it in his *Dioptrica Nova*. Many mathematicians had interpreted the principle as an expression of God's design. Because the direction of the ray of light at any point depends on the length of the entire path, it seems to imply that the light somehow 'knows' where it is going. Throughout the eighteenth century, many of the great European mathematicians including Leonard Euler, Joseph Maupertuis, Jean d'Alembert, Pierre-Simon Laplace, and Joseph Lagrange took up and extended what became known as the principle of least action (action being the product of momentum and distance).

While Enlightenment thinkers distanced themselves from the explicit metaphysical implications of least action (which reeked of the teleology they so detested in scholasticism), they embraced the mathematics as a way to recast geometrical optics and Newton's mechanics as deductive mathematical sciences where all of the known laws could be derived from a single principle. This was a more aesthetically pleasing form than Newton's collection of three separate laws and enabled them to recast mechanics as a deductive science much like Euclid's geometry. The culmination of rational mechanics was Lagrange's *Analytical Mechanics*, which Hamilton referred to as "a scientific poem."[22] In contrast to English accounts of mechanics (like Helsham's which was still in use at Trinity) that focused on descriptions of mechanical apparatus, the French math-

---

[22]W.R. Hamilton, "Inaugural Address by the President," *Proceedings of the Royal Irish Academy* (1841) 1: 107-120, p. 115.

ematicians made mechanics a branch of mathematics. Their goal was a mathematical unification of the known laws of dynamics expressed as a single equation derived from a more general principle. For Hamilton, this was precisely the type of beauty to which science should aspire.

While Molyneux and other writers on geometrical optics constructed diagrams to determine the directions of rays passing through systems of mirrors and lenses, Hamilton proved that each problem could be solved by a single analytical equation. Just as in the application of algebra to geometry, where a single equation represents a curve, in Hamilton's optics a single function V represents an entire optical system. Hamilton reduced all of geometrical optics to the search for a single function for each optical system.[23] In this way, he accomplished his primary goal, namely to introduce a unity and a beauty to geometrical optics. Hamilton's mathematics did not depend on the physical nature of light. In fact, his equation could be interpreted as describing the paths of particles or the motion of waves. Hamilton soon realized that it could also be used to describe the motions of the planets.[24] In 1833 he began to apply his method to astronomy.[25] Just as the characteristic function summarized all information about a given optical system, it could also be used to summarize all information about a given dynamical system. The mathematical methods Hamilton developed allowed him to solve many problems more easily than Newton's laws or even Lagrange's analytical methods.

[23]See H.A. Buchdahl, *An Introduction to Hamiltonian Optics* (New York: Dover, 1993).

[24]Michiyo Nakane, "The Role of the Three-Body Problem in W.R. Hamilton's Construction of the Characteristic Function for Mechanics," *Historia Scientiarum* (1991) 1: 27-37, Hankins, *Sir William Rowan Hamilton*, pp. 172-98.

[25]Hamilton, "On a General Method of expressing the Paths of Light, and of the Planets, by the Coefficients of a Characteristic Function." *Dublin University Review* (1833), pp. 795-826. Reprinted in A.W. Conway and A.J. McConnell (eds.) *The Mathematical Papers of Sir William Rowan Hamilton.* Volume II: Dynamics (Cambridge: Cambridge University Press, 1940).

The characteristic function eventually came to be known as the Hamiltonian function, and by the late nineteenth century it became the standard form for classical mechanics. In their 1867 *Treatise on Natural Philosophy* William Thomson and P.G. Tait transformed Hamilton's dynamics from an abstract mathematical principle into a description of the flow of energy, establishing the Hamiltonian function as the foundation of mathematical physics. In the early twentieth century it was applied to quantum mechanics as well (see chapter 8), and it is now an essential mathematical method in a wide range of sciences. The desire to reduce complex phenomena to a simple set of equations remains the reigning aesthetic in high energy physics where the Hamiltonian is still an important tool (see chapter 9).

Hamilton's vision of mathematical science, though familiar to many modern physicists, was actually quite peculiar at the time, particularly in Britain and Ireland. Everyone who worked on optics in the British Isles was almost exclusively concerned with the physical nature of light. Even textbooks on geometrical optics moved quickly from the basic geometry of light rays to an examination of telescopes, microscopes, and the structure of the eye (as Molyneux's had). All of the published articles on geometrical optics other than Hamilton's were on the construction of optical instruments.[26] In fact John Herschel complained that the advanced French mathematical work in optics, "though confessedly exact in theory, have never yet been made the basis of construction for a single good instrument."[27] Hamilton took what was generally considered a simple and practical method for improving optical

---

[26]I. Grattan-Guinness, "Mathematics and Mathematical Physics from Cambridge, 1815-40: A Survey of the Achievements and of the French Influences," pp. 84-111 in P.M. Harman (ed.), *Wranglers and Physicists: Studies on Cambridge Physics in the Nineteenth Century* (Manchester University Press: Manchester, 1985) lists six research papers on geometrical optics in Britain during the period 1815 to 1840.

[27]J.F.W. Herschel, "On the Aberrations of Compound Lenses and Object-Glasses," *Philosophical Transactions* (1821) 11: 222-267, p. 222.

instruments and transformed it into an abstract mathematical science. In fact, other British scientists struggled even to understand Hamilton's work. George Airy wrote, "To understand the whole is barely possible."[28] And John Herschel despaired, "Alas! I grieve to say that it is only the general scope of the method which stands dimly shadowed out to my mind amid the gleaming and dazzling lustre of the symbolic expressions in which it is conveyed."[29]

5.2   Meta-Mathematics

Not only did Hamilton's mathematical research differ from his contemporaries', but his philosophy of science differed as well. He once explained, "I differ from my great contemporaries... not in transient or accidental, but in essential and permanent things: in the whole spirit and view with which I study Science."[30] While most British scientists were empiricists looking for new facts about the physical universe, Hamilton was engaged in the search for abstract mathematical laws that stood above any physical theories. A visit with the Englishman George Airy impressed upon him the differences between them:

> On the whole, his mind appeared to me an instance, painful to contemplate, of the usurpation of the understanding over the reason, too general in modern English Science. The Liverpool and Manchester Railway, he said, playfully perhaps, but, I think, sincerely, he considered as the highest achievement of man.[31]

While his fellow scientists followed the philosophies of Bacon and Newton, Hamilton found inspiration in the idealism of the German

---

[28] George B. Airy to Hamilton (July 23, 1827) Graves 1: 274.
[29] John F.W. Herschel to Hamilton (June 13, 1835) Graves 2: 127.
[30] W.R. Hamilton to Aubrey de Vere (February 9, 1831), Graves 1: 519.
[31] W.R. Hamilton to Viscount Adare (August 23, 1831), Graves 1:444.

philosopher Immanuel Kant and the poet Coleridge. While British scientists were beginning to forge connections between science and industry, Hamilton was more interested in the relationship between science and poetry.

In the late eighteenth and early nineteenth century, the loose affiliation of philosophical and aesthetic beliefs known as romanticism not only transformed poetry, music, and art but also intrigued scientists looking for new ways to understand nature and man's relationship to it.[32] Hamilton, for example, saw his search for abstract and ideal laws of mathematical physics as imaginative work of the same level as that of his literary friends.

In his introductory lectures to his astronomy students, for example, Hamilton waxed poetic about the role of the imagination in science.[33] Newton's science, he explained, "seems to me in a greater degree than perhaps is generally admitted to belong to imagination also, and to bear analogy to the products of the arts." In contrast to the standard accounts of Newton's work as a model of inductive reasoning, Hamilton described Newton's intellectual process as a form of creation,

> [A]s the mind of an artist calls up many forms, he meditated on many laws and caused many ideal worlds to pass before him: and when he chose the law that bears his name, he seems to have been half determined by its mathematical simplicity, and consequent intellectual beauty,

---

[32]See Trevor H. Levere, *Poetry Realized in Nature: Samuel Taylor Coleridge and Early Nineteenth-Century Science* (Cambridge: Cambridge University Press, 1981), Andrew Cunningham and Nicholas Jardine (eds.), *Romanticism and the Sciences* (Cambridge: Cambridge University Press, 1990) and Richard Yeo, *Defining Science: William Whewell, Natural Knowledge, and Public Debate in Early Victorian Britain* (Cambridge: Cambridge University Press, 1993), pp. 65-71.

[33]On science and the imagination during this period see, Meyer H.Abrams, *The Mirror and the Lamp: Romantic Theory and the Critical Tradition* (Oxford: Oxford University Press, 1953) and Jonathan Smith, *Fact and Feeling: Baconian Science and the Nineteenth-Century Literary Imagination* (Madison: University of Wisconsin Press, 1994).

and only half with its agreement with the phenomena already observed.[34]

Hamilton saw an essential role for the imagination in science. As he explained to Wordsworth, "I have always aimed to infuse into my scientific progress something of the spirit of poetry, and felt that such infusion is essential to intellectual perfection."[35]

What Hamilton meant by 'the spirit of poetry' becomes clear from his philosophy of science.[36] More important than any particular result, he argued, is the bare fact that mathematics applies to the physical world. As he explained, "the visible world supposes an invisible world as its interpreter, and...in the application of the mathematics themselves there must, if I may venture upon the word, be some thing meta-mathematical."[37] The most important conclusion one reaches when reflecting on the possibility of mathematical science, Hamilton argued, is that the laws of thought must somehow be identical to the laws of nature. This is how it is possible for the mind of man to discover the laws of nature simply through meditating on known facts. Here Hamilton invoked the philosophy of George Berkeley. Facts, he explained, are "but passive states of our own being," states of mind over which we have no control, and the cause of these states is God. Hamilton believed that "the *immediate cause* of all our sensations is the Supreme Spirit, in Whom we live and move and have our being, acting on subordinate minds according to rules which He has allowed them to discover."[38] This explains how the laws of thought can be identical to the laws of nature: "that one Supreme Spirit excites perceptions in dependent minds, according to a covenant or plan, of which the terms or conditions are what we call the Laws

---

[34]"Introductory Lecture on Astronomy 1831," Graves 1: 502.

[35]Hamilton to Wordsworth (February 1, 1830) Graves 1: 354.

[36]Hankins, *Sir William Rowan Hamilton*, pp. 172-80.

[37]Hamilton, "Draft of 1833 Lecture on Astronomy," Graves 2: 68.

[38]W.R. Hamilton to H.F.C. Logan (June 27, 1834), Graves 2: 87.

of Nature." Mathematical physics, then, for Hamilton, involves discovering the internal laws of the mind and finding that they correspond to the external laws of nature.

This harmony of man and nature was a central feature of romanticism. In fact, Hamilton found a nearly identical view of science in Coleridge's first *Lay Sermon*,

> The human mind is the compass, in which the laws and actuations of all outward essences are revealed as the dips and declinations. (The application of Geometry to the forces and movements of the material world is both proof and instance.) The fact therefore, that the mind of man in its own primary and constituent forms represents the laws of nature, is a mystery which of itself should suffice to make us religious: for it is a problem of which God is the only solution.[39]

For Hamilton, then, mathematical science demonstrates the harmony of the mental world and the physical world, a harmony that Hamilton believed to be proof that both were created by the same omnipotent being. Science is imaginative because the human mind is structured so as to be able to imagine the true laws of nature.

Hamilton believed that such divine proof could be realized, however, only through the logical process of deduction. Induction, the gradual ascent from facts to theory, is only the beginning of science, he explained, not its entirety, as many of his contemporaries believed. Hamilton aimed at

> the exhibition of *a deductive rather than an inductive unity* in our contemplation and knowledge of nature, a Kantian rather than a Baconian connexion between the

---

[39] S.T. Coleridge, *The Statesman's Manual or The Bible the Best Guide to Political Skill and Foresight: A Lay Sermon Addressed to the Higher Classes of Society* [1816], pp. 3-114 in S.T. Coleridge, *Lay Sermons*, Ed. by R.J. White (London: Routledge and Kegan Paul, 1972), p. 78.

several parts of physical science—one springing from the
mind itself, than from the things which it beholds, and
in which things are rather viewed as illustrations of one
principle than as materials of one edifice.[40]

Like Berkeley, Hamilton was concerned that contemporary scientists focused too much on material nature and brute facts. It was
the truth and beauty of abstract ideas and simple laws, he believed,
that remind the scientist of his own divine nature and connection
to God. Science, according to Hamilton, must show that truth
comes from the mind and the realm of ideas rather than from the
material world. In this way deduction has a metaphysical goal. It
demonstrates that the unity of science derives from the mind itself
and ultimately from God.

Clearly, Hamilton's work in optics and dynamics fit quite well
with his philosophy. He had proven that all the various phenomena
of these sciences could be deduced from one general mathematical
principle, and he was able to construct an abstract science that
avoided any reference to the physical nature of light. Hamilton's
theory was, in a sense, meta-physical because its truth did not
depend on any physical foundations other than the laws of reflection and refraction. In fact Hamilton's optics bears a remarkable
resemblance to one of Coleridge's expressions of the best kind of
science,

The highest perfection of natural philosophy would con-
sist in the perfect spiritualization of all the laws of nature
into laws of intuition and intellect. The phenomena (the
material) must wholly disappear, and the laws alone (the
formal) must remain. ... The optical phenomena are but
a geometry, the lines of which are drawn by light.[41]

---

[40]W.R. Hamilton, "Memorandum of September 10, 1839," Graves 2: 303-4.

[41]S.T. Coleridge, *Biographia Literaria: or Biographical Sketches of My Literary Life and Opinions* [1817], 2 Vols., Ed. by James Engell and W. Jackson Bate (London: Routledge and Kegan Paul, 1983), p. 292.

In his optical work, Hamilton spiritualized the laws of nature into the laws of the mind. In fact, Hamilton wrote a draft of a letter to Coleridge emphasizing the similarity of their views of science. "My aim has been, not to discover new phenomena, nor to improve the construction of optical instruments," he explained, but "to remould the Geometry of Light, by establishing one uniform method for the solution of all problems in that science, deduced from the contemplation of one central or characteristic function."[42]

5.3   A Scientific Prophecy

Hamilton had built the entire edifice of his research around a mathematical approach that avoided the controversial question of whether light is a wave or a particle. And yet in 1832, Hamilton decided to apply his techniques to the wave theory. Almost immediately, he made an important discovery. He found that a single ray of light could, under the proper circumstances, be refracted into a cone of light within a certain kind of crystal.[43] The phenomenon, known as conical refraction, was not particularly important in its own right, but it was a stunning prediction based entirely on mathematical theory.[44] When experimentally confirmed by Hamilton's Dublin colleague, Humphrey Lloyd, it created a sensation in the

---

[42]W.R. Hamilton to S.T. Coleridge [not sent] (October 3, 1832), Graves 1: 592.

[43]Sarton, "Discovery of Conical Refraction," Hankins, *Sir William Rowan Hamilton*, pp. 88-98; James G. O'Hara, *Humphrey Lloyd (1800-1881) and the Dublin Mathematical School of the Nineteenth Century* (PhD. Thesis: University of Manchester, 1979), ch. 3; James G. O'Hara, "The Prediction and Discovery of Conical Refraction by William Rowan Hamilton and Humphrey Lloyd (1832-1833)," *Proceedings of the Royal Irish Academy* (1982) 82A: 231-257, Lars Gårding, "History of the Mathematics of Double Refraction," *Archive for History of Exact Sciences* (1989) 40: 355-385, pp. 360-62, Jed Z. Buchwald, *Rise of the Wave Theory of Light* (Chicago: University of Chicago Press, 1989), pp. 344-45.

[44]Conical refraction has long been seen as an interesting curiosity with no practical applications. In 2010, however, researchers at Trinity College Dublin found that conical refraction can be used to make an 'optical trap' for microscopic objects suspended in liquid with applications in the manipulation of living cells and driving micromotors. D. P. O'Dwyer, et al., "Conical Diffraction of Linearly Polarised Light Controls the Angular Position of a Microscopic Object," *Optics Express* (2010) 18: 27319-27326.

world of nineteenth-century British physical science. Reaction was
intense. George Airy exclaimed, "Perhaps the most remarkable
prediction that has ever been made is that lately made by Professor Hamilton."[45] The German mathematician Julius Plücker later
wrote, "No physical experiment has made such an impression on my
mind as that of conical refraction."[46] Charles Babbage even used it
as evidence that what appears to be miraculous is actually governed
by scientific laws.[47] In 1835 Hamilton was knighted and awarded
the Royal Medal of the Royal Society for "discoveries in Optics, and
particularly that of Conical Refraction."[48] Even twentieth-century
commentators presented it as a paradigm of scientific prediction.[49]

Hamilton's prediction was taken as important support for the
wave theory of light. While the debate over the wave theory of
light came to a head in Britain only in the 1830s, the theory itself
was at least as old as Newton's particle theory of light. The wave
theory faced a number of difficulties, however, and throughout the
eighteenth century, nearly all British natural philosophers accepted
Newton's theory without question.[50] The revival of the wave theory
of light began with Thomas Young's lectures in London in 1799.[51]

---

[45] G.B. Airy, *Philosophical Magazine* (June 1833), p. 420, quoted in Graves 1: 637.

[46] Julius Plücker, "Discussion de la forme générale des ondes lumineuses," *Crelle's Journal* (1839) 19: 1-44, quoted in Graves 1: 637.

[47] Charles Babbage, *The Ninth Bridgewater Treatise: A Fragment, Second Edition* [1838], Vol. 9 in Martin Campbell-Kelly (ed.), *The Works of Charles Babbage* (New York, 1989), pp. 33-34.

[48] J.W. Lubbock to William Rowan Hamilton (November 30, 1835), Graves 2: 170.

[49] In particular George Sarton, "The Discovery of Conical Refraction by William Rowan Hamilton and Humphrey Lloyd (1833)," *Isis* (1932) 17: 154-70.

[50] Geoffrey Cantor, *Optics After Newton: Theories of Light in Britain and Ireland, 1704-1840* (Manchester: Manchester University Press, 1983), pp. 25-90.

[51] Maurice Crosland and Crosbie Smith, "The Transmission of Physics from France to Britain: 1800-1840," *Historical Studies in the Physical Sciences* (1978) 9: 1-61, pp. 30-48, Geoffrey Cantor, "The Reception of the Wave Theory of Light in Britain: A Case Study Illustrating the Role of Methodology in Scientific Debate," *Historical Studies in the Physical Sciences* (1975) 6: 109-132, Xiang Chen, "Young and Lloyd on the Particle Theory of Light: A Response to Achinstein," *Studies in History and Philosophy of Science* (1990) 21: 665-676, Xiang Chen and Peter Barker, "Cognitive Appraisal and Power: David Brewster, Henry Brougham, and the Tactics of the Emission-Undulatory Controversy During the Early 1850s," *Studies in History and Philosophy of Science* (1992) 23: 75-101, Sir Edmund Whittaker, *A*

Using only minimal mathematics, Young emphasized the analogies between sound and light and introduced the principle of interference. Just as sound waves require air to propagate, Young argued that light waves require a luminiferous ether.[52] The vast majority of natural philosophers in both Britain and France, however, ignored Young's ideas.

The tide began to turn with the work of Augustin Fresnel, a graduate of the École Polytechnique who introduced the first mathematically sophisticated form of the wave theory in 1816. Fresnel's wave theory was just one of the French mathematical theories that began to interest British mathematicians in the second decade of the nineteenth century.[53] As they began to study the French calculus, they also became interested in the new mathematical theories of nature. At Cambridge and at Trinity College Dublin, science now meant analytical science, and the new mathematical theories of physics soon made it on to the undergraduate examinations that formed the core of a Cambridge or Dublin education.

A key player in the rise of advanced mathematics at Trinity College Dublin was James MacCullagh.[54] MacCullagh had gradu-

History of the Theories of Aether and Electricity [1910] (New York: Harper & Brothers, 1960), pp. 100-127, Cantor, _Optics After Newton_, pp. 129-46.

[52]Henry John Steffens, _The Development of Newtonian Optics in England_ (New York: Science History Publications, 1977), pp. 107-36.

[53]Grattan-Guinness, "Survey," Crosland and Smith, "Transmission," P.M. Harman (ed.), _Wranglers and Physicists: Studies on Cambridge Physics in the Nineteenth Century_ (Manchester: Manchester University Press, 1985), Cantor, _Optics After Newton_, pp. 147-72. A similar trend occurred in Germany, see Kenneth L. Caneva, "From Galvanism to Electrodynamics: The Transformation of German Physics and Its Social Context," _Historical Studies in the Physical Sciences_ (1978) 9: 63-160.

[54]"Obituary of James MacCullagh," _Proceedings of the Royal Society_ (1847) 5: 712-18, T.D. Spearman, "James MacCullagh," pp. 51-59 in John Nudds, Norman McMillan, Denis Weaire, and Susan McKenna Lawlor (eds.) _Science in Ireland 1800-1930: Tradition and Reform_ (Dublin: Trinity College Dublin, 1988); B.K.P. Scaife, "James MacCullagh MRIA, FRS 1809-47," _Proceedings of the Royal Irish Academy_ (1990) 90C: 67-106; J. Bennett, "MacCullagh's Ireland: The Institutional and Cultural Space for Geometry and Physics," _The European Physics Journal H_ (2010) 35: 123-132; T.D. Spearman, "James MacCullagh, 1809-1847," _The European Physics Journal H_ (2010) 35: 113-122; Olivier Darrigol, "James MacCullagh's Ether: An Optical Route to Maxwell's Equations?" _European Physics Journal H_ (2010) 35: 133-172.

ated from Trinity in 1828 with honors in mathematics and in 1830
presented his first research papers to the Royal Irish Academy. In
1832 MacCullagh became a fellow of Trinity College, and in 1835
he was appointed to the chair of mathematics on the new terms set
by Bartholomew Lloyd, allowing him to focus on research rather
than just teaching. Though MacCullagh tended to use geometrical
rather than analytical techniques, his work on double refraction
was quite similar to Hamilton's brief foray into the field. So similar, in fact, that MacCullagh claimed that conical refraction was
"an obvious and immediate consequence" of his own work.[55] In one
of these papers MacCullagh offered a number of theorems in the
geometry of ellipsoids that he then used to construct Fresnel's wave
surface in biaxial crystals in a more simple and elegant manner.[56]

Hamilton had found in his own investigation of double refraction that Fresnel's wave surface has four conoidal cusps or dimples. Even more importantly he realized that these dimples would
have an interesting physical consequence. A ray that strikes the
crystal at the right angle should be refracted not into two rays as
usual but into a cone of rays. Hamilton made his prediction at
the Royal Irish Academy on October 22, 1832, and the following
day he asked Humphrey Lloyd to perform the experiment to verify the prediction. By December 14, Lloyd had observed conical
refraction. In January he presented his results to the Royal Irish
Academy.[57] Lloyd also discovered in the course of his experiments
that all of the rays of the cone are polarized in different planes,

---

[55] J.G. O'Hara, "Humphrey Lloyd," pp. 3.114-3.126. For Hamilton's correspondence regarding conical refraction, see Graves 1: 623-638, 685-92.

[56] James McCullagh, "On the Double Refraction of Light in a Crystallized Medium Ac-
cording to the Principles of Fresnel," in John H. Jellett, and Samuel Haughton (eds.) *The
Collected Works of James MacCullagh* (Dublin: Hodges, Figgis & Co., 1880). See also J.G.
O'Hara, "Humphrey Lloyd," pp. 3.92-3.113.

[57] H. Lloyd, "On the Phenomena presented by Light in its Passage along the Axes of Biaxal
Crystals," *Philosophical Magazine* (1833) 37: 112-120; H. Lloyd, "Further Experiments on
the Phenomena presented by Light in its Passage along the Axes of Biaxal Crystals," *Philosophical Magazine* (1833) 37: 207-210; and H. Lloyd, "On Conical Refraction," pp. 370-73
in *Report of the Third Meeting of the BAAS; Held at Cambridge in 1833* (London, 1834).

and he demonstrated that this law of conical polarization was a consequence of Fresnel's principles.

Lloyd's experiments were not simply the confirmation of Hamilton's prediction.[58] Lloyd actually discovered the law of conical polarization himself. Both Lloyd and Hamilton, however, concealed this fact. Before his paper was published, Hamilton added a section to his "Third Supplement" describing conical polarization.[59] Not only did this section make no reference to Lloyd's contribution, but Hamilton failed to note that it had been added after the original presentation of the prediction on October 22. Lloyd, for his part, wrote to Charles Babbage, who was editing his description of the experiment for publication,

> I have proposed to omit the words 'unpredicted by theory.' Because, though I was the first to perceive the law of polarization of the *external* cone, it was an easy consequence of Hamilton's views and could not fail to be observed by him on developing them. I should be sorry that anything were expressed from which it might be inferred that I aimed to be anything but the *interpreter* between him and nature.[60]

Both Hamilton and Lloyd had similar commitments to a philosophy that saw the experimenter merely as the interpreter between the theorist and nature. And they took steps to ensure that the published record reflected this philosophy.

So why did Hamilton decide to descend from the lofty heights of abstract mathematics to the mundane world of physical optics? First of all, conical refraction offered a striking example of the ability of mathematical theory to predict physical reality. Secondly, Hamilton's decision to take up the wave theory of light lay in his

---

[58] O'Hara, "The Prediction and Discovery of Conical Refraction."

[59] W.R. Hamilton, "Third Supplement," pp. 138-41.

[60] H. Lloyd to Charles Babbage (March 5, 1837) Babbage Papers, British Library 37192/49. Emphasis in original.

relationship to the rest of the British scientific community. While
no one would deny the skill with which Hamilton had written about
geometrical optics, no one was really interested in it (and few could
understand it). In the 1830s a battle was taking place between two
scientific factions, the Scottish proponents of the particle or corpuscular theory of light and the Cambridge proponents of the wave
theory. But the dispute was more than just a question of the physical nature of light. The Scotsmen favored an empirical approach
to science based on Bacon's philosophy. The Cambridge men, on
the other hand, promoted a philosophy more similar to Hamilton's
that emphasized the importance of mathematical theory and the
priority of ideas over facts. With conical refraction, Hamilton offered important support both for the wave theory of light and for
the Cambridge program for the mathematization of the physical
sciences.

Not everyone in the British Isles agreed that the new mathematical theories were the best way to study nature. In Scotland
in particular, natural philosophers opposed the use of hypotheses
in general and the wave theory of light in particular. They still
feared the dangerous implications of the materialist ether theories
of the late eighteenth century and claimed to follow Bacon and
Newton in practicing the slow and careful induction from experimental facts to higher-order generalizations and laws of nature. In
the late eighteenth century, the Scotsman Thomas Reid developed
a philosophy of mind that he hoped would defend the religious
beliefs and moral tenets of moderate Presbyterianism from both
Hume's skepticism and Priestley's materialism.[61] Central to his
program was an emphasis on the inductive nature of Newtonian

---

[61]L.L. Laudan, "Thomas Reid and the Newtonian Turn of British Methodological
Thought," pp. 103-31 in R. E. Butts and J. W. Davis (eds.), *The Methodological Heritage of
Newton* (Basil Blackwell: Oxford, 1970). For the broader context see George Elder Davie,
The Democratic Intellect: Scotland and Her Universities in the Ninenteenth Century (Ed-
inburgh: Edinburgh University Press, 1961), Anand Chitnis, *The Scottish Enlightenment:
A Social History* (London: Croom Helm, 1976).

science. He emphasized the gradual generalization of experimental facts and the avoidance of hypotheses, conjectures, and complicated theories. Reid stated that "philosophy has been, in all ages, adulterated by hypotheses; that is, by systems built partly on facts, and much more upon conjecture."[62] Newton himself, as Reid never tired of pointing out, had warned against the use of hypotheses.

Hypotheses not only lead the mind astray, he noted; they also assume that man's reason is capable of understanding the works of God. He explained, "the works of nature are contrived and executed by a wisdom and power infinitely superior to that of man; and when men attempt, by the force of genius, to discover the causes of the phaenomena of Nature, they have only the chance of going wrong more ingeniously."[63] Reid feared that too much confidence in man's abilities could lead to atheism (as it appeared to have done in the case of the French mathematicians) while too little might lead to skepticism (as it had for Hume, who doubted the very notion of causality). Hypotheses were not only bad science, the Scots believed; they were also a threat to religion. The Scot John Robison, outspoken foe of the French Revolution and the ether, explained, "a fancied or hypothetical phenomenon can produce nothing but a fanciful cause, and can make no addition to our knowledge of real nature." Moreover, "Although all the legitimate consequences of a hypothetical principle should be perfectly similar to the phenomenon, it is extremely dangerous to assume this principle is the real cause."[64] Robison doubted that a hypothesis could be proven true by any amount of evidence.

---

[62]Thomas Reid, *Essay on the Intellectual Powers of Man* (1785), p. 236 quoted in Laudan, "Thomas Reid," p. 108.

[63] Reid, *Essay on the Intellectual Powers of Man* (1785), p. 472, quoted in Laudan, "Thomas Reid," p. 111.

[64]John Robison, "Philosophy," *Encyclopaedia Britannica*, 3rd Ed. Vol. II, p. 593b, no. 66, quoted in E.W. Morse, *'Natural Philosophy, Hypotheses and Impiety,' Sir David Brewster Confronts the Undulatory Theory of Light* (PhD Thesis: University of California, Berkeley, 1972), p. 28.

The most blatant and dangerous example of a hypothetical cause, according to the Scottish philosophers, was the ether. Even though it could never be observed, it was used to explain an increasing range of natural phenomena. Those natural philosophers that followed Bryan Robinson, for example, threatened to break down the distinction between mental and physical phenomena by explaining thought in terms of vibrations in an ether. Thomas Reid and John Robison both attacked such physiological ether theories, while other Scots attacked ether theories of gravity and electricity as well as optics.[65] So when Thomas Young attempted to revive the wave theory of light in 1799, Henry Brougham responded with the full force of the Scottish tradition.[66] Brougham had studied at Edinburgh University, and he was a close friend of Robison. Brougham launched his attack in the first issue of the *Edinburgh Review*. Invoking the reputations of Bacon and Newton, Brougham described a hypothesis as "a work of fancy, useless to science."[67] After Brougham's vitriolic attack, the wave theory found few supporters.

David Brewster, another Scot, soon joined Brougham in his crusade against the wave theory, particularly with the arrival of Fresnel's theory in the 1820s.[68] They were also skeptical of the wave

---

[65] Richard Olson, *Scottish Philosophy and British Physics, 1750-1880: A Study in the Foundations of the Victorian Scientific Style* (Princeton: Princeton University Press, 1975), pp. 170-77.

[66] On Brougham see Ronald K. Huch, *Henry, Lord Brougham: The Later Years, 1830-1868* (Lewiston, NY: Edwin Mellen, 1993), Trowbridge H. Ford, *Henry Brougham and His World: A Biography* (Chichester: Barry Rose, 1995). On Brougham's optical research see Olson, *Scottish Philosophy*, pp. 219-24, Geoffrey Cantor, "Henry Brougham and the Scottish Methodological Tradition," *Studies in History and Philosophy of Science* (1971) 2: 69-89; Steffens, *Newtonian Optics*, pp. 107-36.

[67] [Henry Brougham], "Review of Young's Lectures," *Edinburgh Review* (1802-1803) 1: 450-56, quoted in Cantor, "Methodological Debate," p. 116.

[68] On Brewster's view of the wave theory see Morse, '*Natural Philosophy, Hypotheses and Impiety,*' Geoffrey Cantor, "Brewster on the Nature of Light," pp. 67-76 in A. D. Morrison-Low and J. R. R. Christie (ed.), '*Martyr of Science': Sir David Brewster, 1781-1868, Proceedings of a Bicentennary Symposium held at the Royal Scottish Museum on 21 November 1981* (Edinburgh: Royal Scottish Museum, 1981), Olson, *Scottish Philosophy*, pp. 177-88, Steffens, *Newtonian Optics*, pp. 137-51.

theory's heavy reliance on analytical mathematics.[69] Brewster argued that "a very great degree of information in natural philosophy may be obtained, without any knowledge of the higher mathematics."[70] Even as the wave theory of light expanded its ability to account for more phenomena, Brougham and Brewster refused to allow it the status of scientific truth. Brewster accepted the interference of light as a fact and even appreciated the power of the wave theory to account for a wide range of phenomena. But no amount of evidence could prove to him the existence of an unobservable ether pervading all of space. Brewster felt that the proponents of the wave theory made dogmatic claims about its veracity and that such claims were dangerous to the progress of science.

The mathematicians of Cambridge and Dublin, on the other hand, supported the wave theory of light with their own opposing methodological and epistemological arguments.[71] For them it represented the ideal of a mathematical, predictive science in which abstract theories are confirmed by experiment. William Whewell attacked the opposing view in his speech to the British Association for the Advancement of Science in 1833 (a speech in which he praised Hamilton's prediction of conical refraction): "It has of late been common to assert that *facts* alone are valuable in science; that theory, so far as it is valuable, is contained in the facts; and, so far as it is not contained in the facts, can merely mislead and preoccupy men." However, he continued, "it is only through some view or other of the *connexion* and *relation* of facts that we know what circumstances we ought to notice and record."[72] Facts,

[69]Richard Olson, "Scottish Philosophy and Mathematics 1750-1830," *Journal of the History of Ideas* (1971) 32: 29-44.

[70]David Brewster, "Evidence, Oral and Documentary, Taken and Received by the Commissioners for Visiting the Universities of Scotland: The University of Edinburgh," *Parliamentary Papers* (1837) 35: 556, quoted in John Hedley Brooke, "Natural Theology and the Plurality of Worlds: Observations on the Brewster-Whewell Debate," *Annals of Science* (1977) 34: 221-86, p. 245.

[71]Cantor, *Optics After Newton*, pp. 173-87.

[72]William Whewell, "Opening Address," p. xx.

Whewell argued, are meaningless without theories, a view he later expanded into a complete philosophy of science.[73] Of all the British scientists, Whewell's philosophy most resembled Hamilton's. Both emphasized the importance of ideas and theory over bare collections of facts, and both believed that the human mind actively shapes our perceptions.

Supporters of the wave theory admitted that they could not observe the ether directly, but they argued that the ability of the wave theory to account for so many experimental facts in an elegant and precise manner made it highly probable. Moreover, they pointed to the ability of the wave theory to predict new phenomena as important evidence for its truth. Even John Herschel, who shared certain sympathies with the Scottish school, explained, "The surest and best characteristic of a well-founded and extensive induction... is when verifications of it spring up, as it were, spontaneously, into notice. ... Evidence of this kind is irresistible."[74] As evidence accumulated, supporters of the wave theory came to believe that the ether really does exist, that it is not simply a mathematical creation of the imagination. The primary arena for British physical science in the 1830s was the British Association for the Advancement of Science. The BAAS originated in the debates over the decline of science in Britain in the early 1830s.[75] Brewster laid the blame at the feet of the universities claiming,

---

[73]William Whewell, *The Philosophy of the Inductive Sciences* [1840], Vol. V & VI in G. Buchdahl and L.L. Laudan (eds.), *The Historical and Philosophical Works of William Whewell* (London: Frank Cass, 1967). On Whewell see especially Menachem Fisch, *William Whewell, Philosopher of Science* (Oxford: Clarendon, 1991); Menachem Fisch and Simon Schaffer (eds.), *William Whewell: A Composite Portrait* (Oxford: Clarendon, 1991) and Yeo, *Defining Science*.

[74]John F.W. Herschel, *A Preliminary Discourse on the Study of Natural Philosophy* [1830] (Chicago: University of Chicago Press, 1987), p. 170.

[75]Jack Morrell and Arnold Thackray, *Gentlemen of Science: Early Years of the British Association for the Advancement of Science* (Oxford: Clarendon, 1981), J.B. Morrell, "Brewster and the Early British Association for the Advancement of Science," pp. 25-29 in Morrison-Low and Christie (eds.), *'Martyr of Science'*.

Within the last fifteen years not a single discovery or invention of prominent interest has been made in our colleges, and...there is not one man in all eight universities of Great Britain who is at present known to be engaged in any train of original research.[76]

Brewster went on to spearhead the effort to establish a new scientific society, one that would spurn the aristocratic pretensions of the Royal Society and the Anglican universities. Each year it would be held in a different city or town, and it would be open to a very wide membership.

The first meeting took place in York in 1831. Bartholomew Lloyd, Provost of Trinity, was the only one to make the journey from Dublin. In fact, none of the major mathematicians from Cambridge—George Airy, William Whewell, or John Herschel—attended because of the role Brewster had played in organizing the meeting. Whewell explained, "even if other circumstances allowed me I should feel no great wish to rally round Dr. Brewster's standard after he had thought it necessary to promulgate so bad an opinion of us who happen to be professors in universities."[77] Hamilton seems to have avoided the meeting for similar reasons. At the second meeting in 1832, held in Oxford, the Cambridge men finally decided to join in, as did the Irishmen Hamilton, Humphrey Lloyd, and T.R. Robinson.

---

[76][David Brewster], "Review of *Reflexions on the Decline of Science in England, and on some of its Causes* by Charles Babbage," *Quarterly Review* (1830) 43: 305-42.

[77]William Whewell to J.D. Forbes (July 14, 1830), quoted in Morrell and Thackray, *Gentlemen of Science*, p. 66.

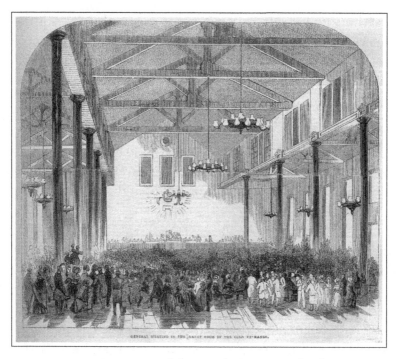

Figure 5.2: Meeting of the British Association for the
Advancement of Science in Cork (1843).
Illustrated London News (August 26, 1843), p. 132.

At the Oxford meeting Brewster received an honorary degree
along with Dalton and Faraday (all Dissenters who were barred
from regular degrees at Oxford, as Brewster pointed out). Brewster
gave a "Report on the Progress of Optics," which summarized the
known experimental facts about light and claimed that the corpuscular hypothesis was better corroborated than the wave theory.[78]
Hamilton was not yet involved in the debate. However, just five
days before leaving for the meeting, he had begun to look at Fresnel's wave theory of light, and by October he had discovered conical

[78] David Brewster, "Report on the recent Progress of Optics," pp. 308-22 in *Report of the
First and Second Meetings of the BAAS; At York in 1831, and at Oxford in 1832* (London,
1833).

refraction.[79] The meeting, therefore, may have been a precipitating event in his interest in physical theories of optics. Conical refraction provided the ideal ammunition to use against the opponents of the wave theory. The bizarre cones of light could never have been discovered through simple generalization from experimental data. Without Hamilton's prediction, no one would have known to look for them. They appeared to offer a striking confirmation, not only of Fresnel's wave theory but also of a scientific method that proceeded through mathematical theory and hypothesis. Lloyd described the discovery as "not only unsupported by any facts hither to observed, but even opposed to all analogies derived from experience."[80] Augustus De Morgan explained that he believed the discovery of conical refraction to be "a most important one, as a *predicted* result, in the very teeth of all former experience... important to the *philosophy* of induction."[81] Conical refraction was the talk of the 1833 meeting of the BAAS, and Whewell used the opportunity of the opening address to praise the wave theory of light. While he conceded that the undulatory theory still faced certain problems, he went on,

> the doctrine will probably gain general acceptance... as prophecies of untried results are delivered and fulfilled. In the way of such prophecies, few things have been more remarkable than the prediction [of conical refraction].[82]

The fulfillment of Hamilton's mathematical prophecy renewed the faith of the Cambridge and Dublin physicists that the wave theory

---

[79]Hamilton's first study of physical optics seems to have been for a review of one of MacCullagh's papers dated July 5, 1830 (TCDMS 1492/307/33). There appears to be no other work on physical optics until January 31, 1832 (TCDMS 1492/307/64), a study of Fresnel's wave and TCDMS 1492/321/1-18 which begins June 11, 1832 and includes the discovery of circles of contact on October 19, 1832.

[80] Humphrey Lloyd, "On the Phenomena presented by Light in its Passage along the Axes of Biaxal Crystals," p. 145.

[81]Quoted in W.R. Hamilton to Helen Bayly (March 14, 1833), Graves 2: 26.

[82] Whewell, "Opening Address," p. xvi.

of light, though still imperfect, was quickly approaching the status of scientific truth. Humphrey Lloyd was asked to deliver a report on optics to the 1834 meeting. The fact that another report on optics was requested so soon after Brewster's indicated the changing status of the wave theory of light and the role the Cambridge mathematicians were playing in the BAAS. While Brewster emphasized the accumulation of experimental data in his report, Lloyd stressed from the very beginning that his report would evaluate the truth of the two rival theories.[83] He continued with a clear statement of the scientific ideal of the wave theory,

> Whatever may be the simplicity of an hypothesis, — whatever its analogy to known laws,—it is only when it admits of *mathematical* expression, and when its *mathematical* consequences can be *numerically* compared with established facts that its truth can be fully and finally ascertained.[84]

Only three years before in his *Treatise on Light and Vision*, Lloyd had concluded that "the question of the nature of light is still a doubtful one."[85] Now, after the example of conical refraction, he went so far as to compare the status of the wave theory of light with Newton's theory of universal gravitation.

Conical refraction, however, was hardly enough to convince Brewster or Brougham to support the wave theory. In response to Hamilton's successful prediction Brewster wrote,

> The recent beautiful discoveries of Professor Airy, Mr. Hamilton, and Mr. Lloyd afford the finest examples of its influence in predicting new phaenomena. The power

---

[83]Humphrey Lloyd, "Report on the Progress and Present State of Physical Optics," pp. 295-413 in *Report of the Fourth Meeting of the BAAS, Held at Edinburgh in 1834* (London, 1835).

[84]Lloyd, "Report on Optics," pp. 295-96.

[85]Humphrey Lloyd, *Treatise on Light and Vision* (London, 1831), p. 7, quoted in Hankins, *Sir William Rowan Hamilton*, p. 133.

of a theory, however, to explain and predict facts, is by
no means a test of its truth.[86]

Brewster disagreed with the very foundation of the new philosophy
of science. No number of successful predictions, no matter how
stunning, would ever convince him of the existence of the ether or
the truth of the wave theory.

Lloyd's report, however, was the final nail in the coffin for the
corpuscular theory of light. Though Brewster and Brougham continued to defend it up through the 1850s, they were ostracized
from the scientific establishment. Brewster did not attend the 1835
meeting of the BAAS. He felt increasingly excluded from the organization that he had helped create. In an 1834 review he claimed
that "the contingent from Cambridge" was taking control of the
BAAS.[87]

There was more at stake in these debates than simply the philosophy of science. In addition to their methodological and philosophical differences, the proponents of the wave and particle theories of
light also differed on social and political issues. Not only were
Brewster and Brougham educated in the Scottish university system, but both made their careers outside of the university. Brewster scraped together a living as an editor of encyclopedias and an
inventor (he invented the kaleidoscope), while Brougham served as
one of the leading lights of the Whig party. Brewster was a Scottish evangelical and was expelled from his post as Principal of the
United College at St. Andrews for his failure to conform to the
established Church. Historian E.W. Morse explains that Brewster
"devoted much of his life to efforts to apply the results of science to
solve human problems, to extend scientific education throughout

---

[86]David Brewster, "Observations on the Absorption of Specific Rays, in Reference to the
Undulatory Theory of Light," *Philosophical Magazine* (1833) 2: 360-61, quoted in Hankins,
Sir William Rowan Hamilton, p. 151.

[87]David Brewster, *Edinburgh Review* (1834-35) 60: 363-94, p. 392, quoted in Hankins,
Sir William Rowan Hamilton, p. 145.

society, to reform the institutions of science, and to elect reform-minded candidates to Parliament."[88] Brougham, for his part, was particularly active in the promotion of non-denominational schools, popular education, legal reform, and Catholic Emancipation. Both Brougham and Brewster opposed all that the Anglican universities stood for and exerted great effort to criticize, reform, and replace them. They attacked the dogmatism of the Anglican establishment as well as the dogmatism of the wave theory of light.

Meanwhile the proponents of the wave theory were almost all professors at Cambridge, Oxford, and Trinity College Dublin. In their history of the BAAS, Morrell and Thackray explain,

> In the 1830s Section A of the British Association became the familiar haunt of a group of Anglican, mainly clerical, gentlemen committed to the placement of physical science within the dominion of mathematical analysis. What may conveniently be called the Cambridge programme had implications of a moral, institutional and career nature, in addition to mathematical and practical aspects. Subscribing to or opposing that programme was not simply a matter of how one interpreted certain experiments in optics, though that was one central question. It was also a matter of commitment to, or rejection of, the growing role of the English universities in British life. And, not surprisingly, commitments of this kind were closely interwoven with wider religious and philosophical views.[89]

In general, the proponents of the wave theory were Anglican clergymen who supported the exclusion of Dissenters from the universities and opposed political reform. For them the wave theory expressed a range of values. It validated the analytical education that domi-

---

[88] Morse, 'Natural Philosophy, Hypotheses and Impiety,' p. 75.
[89] Morrell and Thackray, Gentlemen of Science, p. 479.

nated Cambridge and Trinity, it supported a philosophy that gave priority to ideas, and it even dovetailed well with Anglican natural theology.

5.4   The Benthamites and the Coleridgians

The claims about the nature of science that arose in the debate over the wave theory had much broader philosophical implications than merely the proper form for the physical sciences. Hamilton's friend Aubrey de Vere, for example, saw the philosophical implications of Hamilton's prediction clearly, writing "this sort of *à priori* science seems to me its utmost and ultimate triumph. I confess I like to see experiment occasionally put to the wheel, and reason harnessed as leader in these utilitarian times."[90] In the age-old battle between materialists and idealists, Hamilton and his friends were on the side of idealism, and they saw in utilitarianism everything they despised about the modern age.   Conical refraction offered support for a philosophy of science and of mind that they believed invalidated the claims of the utilitarians.   De Vere summarized his fears in relating to Hamilton an incident in which he confronted a number of utilitarians,

> The three gentlefolks differed in some respects but agreed in these enlightened principles of modern philosophy:—'there is no natural, necessary, or eternal right or wrong; our impressions of those subjects are only associations instilled in us during childhood, for the good of society; the human mind has no natural *principles* of beauty, much less *Idea* of beauty; there is no such thing as conscience; morality is a mere name. ... [T]he only true method of pursuing metaphysical subjects is experience; and Bacon's Inductive philosophy is the key to all philosophy;

---

[90]Aubrey de Vere to W.R. Hamilton (December 29, 1832), Graves 2: 617.

the first desire of every man is and ought to be his own happiness.' These doctrines are, I am afraid, terribly prevalent these days: and if so, what hope is to be entertained for a nation consisting of men who believe them?[91]

De Vere linked utilitarian ethics, associationist psychology, and Baconian empiricism. He felt that their rise represented the moral and philosophical decline of the nation. Nor was de Vere alone in this belief. William Whewell, William Wordsworth, and S.T. Coleridge expressed similar beliefs.[92] They all feared for the state of the nation and felt that the only solution lay in philosophy, particularly the philosophy of science. Because of its importance to the philosophy of science, therefore, the prediction of conical refraction had a significance that reached far beyond the narrow limits of mathematical physics. It could be taken as proof of the priority of ideas over material facts and as support for the entire idealist program.

In an 1840 essay, John Stuart Mill claimed that "every Englishman of the present day is by implication either a Benthamite or a Coleridgian,"[93] and he proceeded to analyze this dichotomy in terms of epistemology, ethics, and political theory. Each side believed that theories of knowledge entail a certain social order.[94]

---

[91] Aubrey de Vere to W.R. Hamilton (October 6, 1832), Graves 2: 616-17.

[92] On the politics of Romanticism (esp. Wordsworth and Coleridge) see Crane Brinton, *The Political Ideas of the English Romanticists* (Oxford: Oxford University Press, 1926); Alfred Cobban, *Edmund Burke and the Revolt Against the Eighteenth Century: A Study of the Political and Social Thinking of Burke, Wordsworth, Coleridge and Southey* (London: George Allen & Unwin, 1929); Charles R. Sanders, *Coleridge and the Broad Church Movement* (Durham: Duke University Press, 1942), John Colmer, *Coleridge: Critic of Society* (Oxford: Clarendon, 1959); R.W. Harris, *Romanticism and the Social Order* (London: Blandford Press, 1969); Ben Knights, *The Idea of the Clerisy in the Nineteenth Century* (Cambridge: Cambridge University Press, 1978); Marilyn Butler, *Romantics, Rebels and Reactionaries: English Literature and its Background 1760-1830* (Oxford: Oxford University Press, 1981); John T. Miller, *Ideology and Enlightenment: The Political and Social Thought of Samuel Taylor Coleridge* (London: Garland, 1987).

[93] John Stuart Mill, "Coleridge [1840]," pp. 177-226 in Alan Ryan (ed.), *Utilitarianism and Other Essays: J.S. Mill and Jeremy Bentham* (London: Penguin, 1987), p. 180.

[94] On idealism vs. utilitarianism see Sheldon Rothblatt, *The Revolution of the Dons: Cambridge and Society in Victorian England* (London: Faber and Faber, 1968), pp. 97-116, and Yeo, *Defining Science*, pp. 176-230.

The Benthamites adopted an empiricist epistemology based on the notion that all knowledge is generalization from experience.[95] Mill explained that Bentham believed that "abstractions are not realities *per se*, but an abridged mode of expressing facts."[96] Ideas are only copies of sensations.[97] Sensation, therefore, is the sole source of knowledge, and metaphysics is simply mysticism. All of our beliefs, he claimed, are built from the mechanical association of sense impressions. Bentham also argued that since there is no such thing as an inherent moral sense, all ethical decisions must be based on empirical facts: in simplistic terms, the greatest good for the greatest number. Mill explicitly linked this philosophy to political reform,

> The practical reformer has continually to demand that changes be made in things which are supported by powerful and widely-spread feeling, or to question the apparent necessity and indefeasibleness of established facts; and it is often an indispensable part of his argument to show, how those powerful feelings had their origin, and how those facts came to seem necessary and indefeasible. There is therefore a natural hostility between him and a philosophy which discourages the explanation of feelings and moral facts by circumstances and association, and prefers to treat them as ultimate elements of human nature.[98]

Utilitarianism also held that like abstractions, political institutions are simply artificial contrivances of convenience. They are nothing more than the sum of their parts and so should reflect the wishes

---

[95]See Elie Halévy, *The Growth of Philosophic Radicalism.* Trans. A.D. Lindsay (London, 1928).

[96]J.S. Mill, "Bentham [1838]," pp. 132-75 in Ryan, *Utilitarianism*, p. 140.

[97]See also James Mill, *Analysis of the Phenomena of the Human Mind,* 2nd Ed. (London, 1878).

[98]J.S. Mill, *Autobiography* [1873] (New York: Bobbs-Merrill, 1957), p. 175.

of the majority of their components. Mill argued that Benthamism was essential in bringing about the parliamentary reforms of the 1830s.

For idealists such as Coleridge, Wordsworth, Hamilton, and Whewell, knowledge is impossible without ideas, which are a precondition for experience. They believed, as Whewell explained, that "an activity of the mind, and an activity according to certain Ideas, is requisite in all our knowledge of external objects."[99] But idealist epistemology also opposed Benthamism on ethical and political issues. Fundamental ideas, they argued, are the basis of ethics just as much as they are the basis of mathematics. Similarly, idealists saw political institutions as based on fundamental ideas and therefore not subject to the desires of the majority. Mill and Whewell engaged regularly and publicly in these debates.[100] Mill observed that the tendency of Whewell's efforts "is to shape the whole of philosophy, physical as well as moral, into a form adapted to serve as a support and a justification to any opinions which happen to be established."[101] And Mill wrote his *System of Logic* explicitly in opposition to Whewell's philosophy as an empirical antidote to idealism, whose political implications he feared.

In this battle between utilitarianism and idealism, mathematics and poetry (Hamilton's two great loves) played a crucial role, for they were extremely difficult for empiricists to account for. An epistemology based solely on sense perception found it almost impossible to explain artistic creativity or mathematical truth (though Mill tried). Conical refraction was a particularly good example of a mathematical prediction that could not be explained on the view that mathematics is simply generalization from experience. It could be seen as proof of the priority of ideas and therefore as

[99] Whewell, *Philosophy*, p. 27.

[100] In addition to Yeo, *Defining Science*, see Perry Williams, "Passing on the Torch: Whewell's Philosophy and the Principles of English University Education," pp. 117-47 in Fisch and Schaffer, *William Whewell*.

[101] J.S. Mill, "Whewell on Moral Philosophy," pp. 228-70 in Ryan, *Utilitarianism*, p. 230.

proof of the entire idealist agenda. Mill recognized the danger when idealism used science as its support in this way. He explained,

> The notion that truths external to the mind may be known by intuition or consciousness, independently of observation or experience, is, I am persuaded, in these times, the great intellectual support of false doctrines and bad institutions. ... And the chief strength of this false philosophy in morals, politics and religion, lies in the appeal which it is accustomed to make to the evidence of mathematics and of the cognate branches of physical science.[102]

Physical science, as the paradigm of certain knowledge in mid-nineteenth century Britain, had become a battleground. In a time before sharp distinctions were drawn between scientist, philosopher, theologian, and political theorist, the evidence of the physical sciences could be utilized in debates as far afield as ethics and parliamentary reform.

5.5 The Clerisy

While Hamilton's prediction had a relevance to British scientific and philosophical debates, it also had implications much closer to home. In the Irish context, conical refraction was important for two primary reasons. First of all, it represented an Irish scientific achievement in an area of tremendous interest to English scientists. In this way, it served to strengthen the political and cultural union with England while at the same time asserting the position of Ireland within that union. Secondly, it supported an idealist philosophy that buttressed the rule of an intellectual elite and the exclusion of Catholics from British politics.

---

[102] J.S. Mill, *Autobiography*, p. 145.

Hamilton always thought of himself as Irish and described his youthful appointment to the professorship of astronomy as "the solemn call of God and my country."[103] He knew that he had been appointed to improve Trinity's reputation through his work in mathematics, and he wrote, "I should like to contribute my mite, or shall I say, my stone to throw upon the pile which hides the buried slander against the 'Silent Sister'."[104] Hamilton's early work in geometrical optics was hardly likely to bring him or his university the renown to which they aspired. This helps explain why Hamilton would have been interested in pursuing research of interest to the mathematicians at Cambridge, for they were the primary audience for Hamilton's science. De Vere recognized this and wrote to Hamilton just after his prediction of conical refraction, "I can most entirely sympathise with the exultation you must feel at the success of your mathematical discovery. I should think from its connexion with Physics, the popular part of Science, it is more likely to enlarge the 'crescent sphere' of your fame than anything else you have done."[105]

Hamilton believed that increasing the reputation of Ireland in this way could have very concrete benefits for daily life in Ireland. When de Vere later wrote to him during the height of the Famine to ask what he was doing to help, Hamilton responded,

> It is the opinion of some judicious friends... that my peculiar path, and best hope of being useful to Ireland, are to be found in the pursuit of those abstract and seemingly unpractical contemplations to which my nature has a strong bent. If the fame of our country shall be in any degree raised thereby, and if the industry of a particular kind thus shown shall tend to remove the prejudice which supposes Irishmen to be incapable of perseverance, some

---

[103]W.R. Hamilton to Eliza Hamilton (September 26, 1827), Graves 1: 271.
[104]W.R. Hamilton to H. Lloyd (January 16, 1836), Graves 2: 177.
[105]Aubrey de Vere to W.R. Hamilton (December 29, 1832), Graves 2: 17.

step, however slight, may be thereby made towards the
establishment of an intellectual confidence which cannot
be, in the long run, unproductive of temporal and material benefits.[106]

And de Vere agreed. In his book *English Misrule and Irish Mis-
deeds*, he traced the problems of Ireland during the Famine not to
the Union (which he strongly supported) but to the English lack
of respect for the Irish. By proving to the English that the Irish
are capable of great accomplishments, Hamilton (and his friends)
believed his work to be crucial to Irish prosperity. Thus science
in Ireland was an important part of the cultural and political relationship with England. There was a belief among the educated
Irish that scientific achievement could strengthen both the Union
and the position of Ireland within the Union.

But while science might strengthen the position of the Irish
Protestants with respect to the English establishment, it could also
support arguments for the exclusion of Catholics. The reforms that
Hamilton and his Irish friends feared in Ireland were simply a more
extreme case of the political reforms Whewell, Wordsworth, and
Coleridge feared in England, and just as their idealist social theory
supported the establishment in England so could it also support
the Ascendancy in Ireland. Philosophy, politics, and religion came
together most clearly in Coleridge's prose writings. Like Berkeley, he blamed the mechanical philosophy for a decline in faith
and morals, and like Burke, he traced the horrors of the French
revolution to bad philosophy. For Coleridge the worst of French
materialism, empiricism, and democracy was represented in England by the utilitarians. Because of the political importance of
fundamental ideas, Coleridge believed that the battle for the fu-

---

[106]W.R. Hamilton to Aubrey de Vere (February 6, 1847), Graves 2: 558.

ture of society was a battle for the philosophy of the upper classes. This is the basis of his theory of the clerisy.[107]

In April of 1816, Coleridge moved to Highgate where he would spend the rest of his life producing conservative defenses of the established government and religion. The first product of his writings in Highgate were the *Lay Sermons*, commentaries on the post-Napoleonic War depression. In the first of these, *The Statesman's Manual or The Bible the Best Guide to Political Skill and Foresight: A Lay Sermon Addressed to the Higher Classes of Society* (1816), Coleridge explained that the problems facing England are moral, not institutional, and therefore their solution lies in the moral rather than the political reform of the country. He traced all social and political problems to bad metaphysics.

> [A]ll the *epoch-forming* Revolutions of the Christian world, the revolutions of religion and with them the civil, social, and domestic habits of the nations concerned, have coincided with the rise and fall of metaphysical systems. So few are the minds that really govern the machine of society.[108]

Society, Coleridge claimed, is governed by a small group of "uninterested theorists." They are the ones with the true power to ameliorate the difficult conditions of the country for all of the civil, social, and domestic habits of the nation depend on the rise and fall of metaphysical systems.

In *On the Constitution of the Church and State According to the Idea of Each*, Coleridge presented a theory of society based not on a social contract or the maximization of happiness but on what he called "the fundamental ideas of the church and state". His state is a dynamic equilibrium between the forces of permanence

---

[107]For the relationship between Coleridge's clerisy and the British scientific community, see Morrell and Thackray, *Gentlemen of Science*, pp. 17-21.

[108]S.T. Coleridge, *Lay Sermons*, Ed. by R.J. White (London: Routledge and Kegan Paul, 1972), pp. 14-15.

(represented by landed property) and the forces of progression (represented by the mercantile and professional portions of the state). This equilibrium is kept stable by the National Church or clerisy, a class of intellectuals above the state who are nevertheless responsible for its well-being. They connect the present with the past by preserving traditional culture and connect the present to the future by extending the bounds of knowledge. Thus while they are crucial to society, intellectuals (and ideas) are outside of the vagaries of history and party politics. The clerisy are mediators between the material and intellectual worlds. They are the high priests of culture. Only they have the authority to decide questions about the Church or State because only they have access to the fundamental ideas.

Such a philosophy had an obvious appeal to Hamilton. Coleridge's work had been precipitated by the question of Catholic Emancipation. Daniel O'Connell represented all that Hamilton feared —utilitarianism, democracy, and the end of the Ascendancy. Hamilton dreaded that if the educated and cultured Protestant Ascendancy were no longer allowed to govern, society would crumble. In 1834, Hamilton felt that the danger was great enough to warrant a public political appearance. On August 19, Hamilton joined the Protestant Conservative Society, of which his old tutor Charles Boyton was the leading member. In his speech, he explained,

> There may be many, like myself, whose habits and inclinations lead them rather to the study than the platform. ... [T]hey may think that even if the worst were accomplished, they could still pursue in peace their private course of study and meditation. ... But if the purpose of our enemies were fully consummated, and liberty of conscience were withdrawn, how much do they suppose would remain of any other liberty? How long would we retain any semblance of liberty of thought? If ever the

papal despotism should be complete... let us not think
that the seclusion of our libraries shall long afford any
effectual protection. We may then have cause to tremble, even in that seclusion, lest the step of the inquisi-
tor be upon the stairs—lest the provisions of the *Index
Expurgatorius* should not have been sufficiently complied
with—lest our shelves perchance should be tainted by the
presence of any heretical book, even though that book
should be the heretical Principia of Newton, which dared
to stand calmly up against the thunders of the Vatican,
and, in defiance of the infallible dogmas, taught that the
earth moves round the sun.[109]

Hamilton's fear of a Catholic Ascendancy was not simply an attempt to protect Protestant privilege; it was a fear of arbitrary
intellectual authority. Like Coleridge, Hamilton saw the great revolutions of the age tied to the rise and fall of metaphysical systems.
The greatest danger, therefore, is the restriction of the freedom of
thought. Coleridgian social theory was based on the idea that
only a small intellectual minority separate from the corruption of
politics possesses the ideas necessary to govern the majority. In
Hamilton's nightmare world of democracy, popular education, and
Papal despotism, that elite could no longer govern and all society
would be destroyed. As Boyton explained in a sermon, "our concern
is—not whether it is popular, but whether it is true."[110]

Hamilton's public appearance also illustrated the tensions involved in the politics of idealism. Hamilton contrasted the contexts
of the platform and the study, yet his point is that it is impossible
to escape politics even in pure thought. This paradox is further

---

[109] *The Dublin Evening Mail* (August 20, 1834).

[110] Charles Boyton, *Observations on the Church;- Its Relation to the State;- The Author-
ity of Its Governors;- and the Duties and Obligations of Its Ministers: Being A Sermon,
Preached in the Cathedral of Derry, Before the Lord Primate...On the Occasion of the
Triennial Visitation, September 14, 1838* (Dublin, 1838), p. 16.

brought out in the press coverage of Hamilton's appearance. *The Warder* reported,

> Our Readers may rest assured that such a man would not, from any light or vain impulse, step forth into public life, from the quiet, secluded, and high studies to which he is so intensely devoted. He must have felt that the crisis is of the most awful nature; the overwhelming destruction not ideal but at hand, or he would never have shown himself in 'the assemblies of the people,' or mixed with the troubled elements of political contention.[111]

Idealism and mathematics were seen as non-political and non-partisan and to employ them in support of the Ascendancy had a tremendous impact on his audience. Hamilton's engagement with politics proved that the danger was 'not ideal but at hand.' Yet as Coleridge's philosophy illustrated, metaphysics was the most profound form of politics. Hamilton's claim to be above politics was itself a political position.

Despite claims that science is above politics and independent of religion, the special status of scientific truth in nineteenth-century Britain and Ireland meant that even those who promoted an idealized science believed that it had political relevance. Hamilton and others believed that the prediction of conical refraction demonstrated the priority of ideas over facts. In nineteenth-century Britain, this had important implications. Utilitarians like John Stuart Mill grounded their arguments for political reform on an empirical philosophy of science. Like the French philosophes, they attacked the religious and political establishments as founded on meaningless abstractions and claimed that all ideas must represent sensations. Idealists, like Hamilton's friends Whewell, Wordsworth, and Coleridge, on the other hand, believed in the existence of pure

---

[111] *The Warder* (August 23, 1834).

ideas that transcend the physical world. Not only do abstract laws
that can be intuited by the human mind govern nature, they believed, but society itself is based on fundamental ideas that cannot
be changed regardless of popular opinion. Coleridge used such
an argument to oppose the movement for Catholic emancipation,
claiming that the exclusion of Catholics is essential to the idea of
the British state. Idealism, for Hamilton, represented not only the
best kind of science but also the proper foundation for society.

Chapter 6

Imagining Quaternions

The Calculating Boy

In 1818, Zerah Colburn, the famous American Calculating Boy, arrived in Dublin. The fourteen year old boy from rural Vermont with an amazing capacity for performing complicated calculations in his head was on a tour of European capitals. He exhibited his stunning ability for lords, ladies, and heads of state, all while his father frantically tried to raise enough money to get to the next city.[1] Colburn was challenged with questions such as "What sum multiplied by itself will produce 998,001?" or "How many seconds in eleven years?" He answered each in less than four seconds. His mathematical faculties were so extreme that he was considered a freak of nature, and he was exhibited in Dublin along with an eight foot tall man, a woman who could make paper cut outs with her toes, and an albino brother and sister. It may also have helped that Colburn had an extra digit on each hand and foot (helped the publicity, that is, it does not seem to have figured into his calculating skills).

---

[1] Zerah Colburn, *Memoirs of Zerah Colburn by Himself* (Springfield: G. and C. Merriam, 1833).

Figure 6.1: Zerah Colburn (1804–1839), the American Calculating Boy.
Portrait by Henry Hoppner Meyer ©Science Museum, London

The Irish produced their own child prodigy for the occasion—
William Rowan Hamilton.[2] By age ten, Hamilton was said to
have mastered Hebrew, Persian, Arabic, Sanskrit, Chaldee, Syr-
iac, Hindi, Malay, Mahratta, and Bengali in addition to Latin,
Greek, and the modern European languages.[3] Hamilton had also
shown a strong facility for mathematics, but he was no match for
Colburn. While Colburn won the battle, however, Hamilton won
the war. Colburn fell victim to the curse of the child star. The
novelty of his ability eventually wore thin; and after a failed attempt at an acting career, Colburn had to borrow money to return
to his family's farm in Vermont. He did, however, return to Dublin
in 1820 before he left Europe and shared many of his calculating
secrets with Hamilton. (He also pressed him for a subscription to

    [2]Thomas Hankins, *Sir William Rowan Hamilton* (Baltimore: Johns Hopkins University
Press, 1980), p. 15.
    [3]Hankins, *Sir William Rowan Hamilton*, p. 13.

his forthcoming autobiography). Hamilton would later date his interest in mathematics to this encounter. "I believe it was seeing Zerah Colburn that first gave me an interest in those things," he wrote to his cousin Arthur. "For a long time afterwards, I liked to perform long operations in Arithmetic in my mind; extracting the square and cube root, and everything that related to the properties of numbers."[4]

Colburn's fame demonstrated the ambiguous role of calculation in nineteenth-century Europe. His abilities were superhuman, and yet, at the same time, there was something inhuman, even mechanical, about them. Colburn himself did not understand how he did some of his calculations, and, of course, at this very time, Charles Babbage was designing a machine that could perform similarly complicated calculations—an early version of the computer. Was this kind of calculation the height of human genius, or was it a parlor trick?

After his encounter with Colburn, Hamilton took a very different approach. He quickly became more interested in why Colburn's methods worked than in outperforming other calculating prodigies. Hamilton eventually began to explore the philosophical foundations of mathematics. In doing so, he had to confront the issues that Berkeley had raised in the eighteenth century: how is it possible that we can calculate with symbols whose meaning we do not understand? Berkeley had attacked the infinitesimal quantities at the foundation of the calculus, but other mathematicians had expressed concerns about imaginary numbers ($\sqrt{-1}$) and even negative numbers. Hamilton desperately wanted to show that analytical mathematics (the powerful set of tools based on the calculus that had transformed physics) was just as rigorously true as Euclidean geometry.

---

[4]W.R. Hamilton to Arthur Hamilton (September 4, 1822) Graves 1: 111.

Ironically, Hamilton's quaternions would lead to the radical idea that mathematics is not simply the science of calculation. Hamilton's quaternions marked a turning point in the history of mathematics.[5] They do not follow the commutative law of algebra. That is, for quaternions, $A \times B$ does not equal $B \times A$; the order of multiplication makes a difference. In a sense, quaternions were the exception that proves the rule, bringing such abstract laws of algebra to the attention of mathematicians for the first time. They led mathematicians to ask themselves about the rules that any algebraic system must follow, and they demonstrated that it is possible to invent new forms of algebra with different rules. In the process they transformed the nature of algebra from the symbolic representation of the familiar laws of arithmetic to the study of any self-consistent system of arbitrary rules. After quaternions, algebra was no longer fundamentally about counting; it had become the study of formal relationships. Mathematics, therefore, could describe logic as well as physics; and George Boole, working in Cork, created the first mathematical system of logic, the origin, in a sense, of both analytic philosophy and computer science.

Hamilton, however, saw quaternions as the ultimate expression of his idealist philosophy, the "curious offspring of a quaternion of parents, say of geometry, algebra, metaphysics and poetry." In fact, Hamilton was able to abandon the standard laws of algebra because he based his quaternions on a more fundamental geometrical interpretation. He demonstrated that they could be used to represent lines in three-dimensional space and hoped that quaternions would become the new language of geometry and physics. For Hamilton, however, more important than any application of the quaternions

---

[5]For a few historical accounts of the creation of quaternions, see E.T. Whittaker, "The Sequence of Ideas in the Discovery of Quaternions," *Proceedings of the Royal Irish Academy* (1944) 50A: 93-98' Hankins, Sir William Rowan Hamilton, pp. 245-75; T. Koetsier, "Explanation in the Historiography of Mathematics: The Case of Hamilton's Quaternions," *Studies in History and Philosophy of Science* (1995) 26: 593-616.

was the fact that they represented the culmination of his idealist philosophy.

Calculation in the nineteenth century had much broader ramifications than simply a philosophy of mathematics. While political economists began to present human beings as mechanical calculators of self-interest, Hamilton's quaternions represented man's ability to know truth intuitively. He created his quaternions within a value system that stressed the importance of beauty, imagination, and transcendent truth. Like conical refraction, quaternions represented the power of fundamental ideas derived from the mind rather than the visible world, and their practical importance paled in comparison to their metaphysical significance.

6.1 The Problem of Impossible Numbers

Hamilton's quaternions had such broad implications because of the central role of mathematics in nineteenth-century British culture. Mathematics was the standard of objective and unquestionable truth in Victorian Britain and Ireland, and its privileged epistemological status justified its predominance in the curriculum at Cambridge and at Trinity College Dublin. Beneath the public rhetoric, however, mathematicians had doubts about the rigor of certain areas of mathematics. They were at odds, in particular, over the relative value of geometry and analysis. Traditionally, Euclid's geometry had been the foundation of mathematical education.[6] Mathematicians interpreted the axioms of Euclidean geometry as descriptions of fundamental properties of reality, and educators argued that geometry offers students clear accounts of intuitive reasoning. The fact that it could be taught using Euclid's original text seemed to demonstrate its timeless nature.

---

[6]See Joan L. Richards, *Mathematical Visions: The Pursuit of Geometry in Victorian England* (London: Academic Press, 1988).

While the geometers admitted that analytical methods such as the differential and integral calculus were better problem-solving tools, they charged that analysis teaches students to follow rules rather than how to reason. Analysis, of course, had long faced questions about whether it was sufficiently rigorous. George Berkeley, for example, had attacked the philosophical foundations of the calculus in its earliest days (see chapter 2). The calculus was also seen as rather too utilitarian for the uses of gentlemen. Students, most educators agreed, need mental discipline, not technical tools. All of the latest continental research used these methods, however, and the Cambridge analysts argued that the British would have to learn these new methods if they wanted to compete scientifically. The rise of analytical mathematics in the curriculum and the growing importance of mathematical theories like the wave theory of light, therefore, left mathematicians in a quandary. Their most powerful problem solving techniques failed to live up to the very standards that gave mathematics its pride of place. Hamilton, for his part, was dedicated to the discovery of new mathematical truths and became one of the foremost analysts of his age. At the same time, he was concerned that the foundations of analysis could not match the logical standards of geometry. Quaternions arose from his attempt to solve this problem.

Most troubling to those concerned with the foundations of analysis were certain symbols that appeared in algebra that seemed to have no meaning. Algebra had long been defined as the science of quantity, and numbers were conceived of as answers to the question how much or how many. In the course of solving algebraic equations, however, new symbols arose that did not appear to fit this conception of mathematics. Given the equation $x + 6 = 3$, for example, one finds that $x = -3$. Yet -3 is not a quantity in the traditional sense. It is impossible to have -3 apples or -3 feet of cloth. But algebra would be severely restricted in its usefulness if negative numbers were not allowed.

Even more perplexing were so-called imaginary numbers, multiples of $\sqrt{-1}$ (symbolized by the letter i). Because both the square of a negative number and the square of a positive number are always positive, it is impossible to have a negative square or the square root of a negative number. Yet certain equations could not be solved without them. The solution to the equation $x^2 + 4 = 0$, for example, is $2i$ or $-2i$. Yet neither answer corresponds to a quantity. The analysts of the eighteenth century like Euler and the Bernoullis developed powerful methods using imaginary numbers, but in the late eighteenth and early nineteenth century British mathematicians began to worry about the logical foundations of such methods.[7]

William Rowan Hamilton put the problem as follows,

> [I]t has not fared with the principles of Algebra as with the principles of Geometry. No candid and intelligent person can doubt the truth of the chief properties of *Parallel Lines*, as set forth by Euclid in his Elements, two thousand years ago; though he may well desire to see them treated in a clearer and better method. The doctrine involves no obscurity nor confusion of thought, and leaves in the mind no reasonable ground for doubt.... But it requires no peculiar scepticism to doubt, or even to disbelieve, the doctrine of Negatives and Imaginaries.

Geometry, in other words, serves as a model for rational deduction, a model which algebra had failed to live up to. Hamilton concluded, "It must be hard to found a SCIENCE on grounds such

---

[7]See Ernest Nagel, "Impossible Numbers: A Chapter in the History of Modern Logic," *Studies in the History of Ideas* (1935) 3: 429-74, David Sherry, "The Logic of Impossible Quantities," *Studies in the History and Philosophy of Science* (1991) 22: 37-62, Paul J. Nahin, *An Imaginary Tale: The Story of $\sqrt{-1}$* (Princeton: Princeton University Press, 1998).

as these...."[8] Algebra worked, this no one doubted, but it did so on the basis of apparently absurd claims.

Scottish mathematician John Playfair, himself a devoted analyst, felt torn between the mathematical usefulness of impossible quantities and their troubling lack of rigor. For him they posed a profound question about the nature of mathematical reasoning. Playfair asked,

> If the operations of this imaginary arithmetic are unintelligible, why are they not also useless? Is investigation an art so mechanical, that it may be conducted by certain manual operations? Or is truth so easily discovered, that intelligence is not necessary to give success to our researches?[9]

Reasoning with mathematical symbols, he believed (along with most others), depends on knowing the meaning of the symbols, just as reasoning with words depends on understanding what each word signifies. Yet reasoning with impossible quantities appears to work even though the symbols are meaningless. In the end, Playfair allowed the use of imaginary numbers but denied that calculating with them could be called reasoning. In this way, his fears about impossible numbers resembled Berkeley's fears about the use of infinitesimals. They are mathematically useful but not logically well founded; that is, they are not attached to any sensible ideas. Berkeley, like Playfair, solved the problem by distinguishing the strict demonstrative reasoning of geometry from the useful but artificial manipulation of symbols in the calculus. Yet the price of such a tool was the knowledge that analytic mathematics failed to

---

[8]Hamilton, "Theory of Conjugate Functions, or Algebraic Couples; With a Preliminary and Elementary Essay on Algebra as the Science of Pure Time [1837]," pp. 3-96 in H. Halberstam and R.E. Ingram (eds.), *The Mathematical Papers of Sir William Rowan Hamilton*, Vol. 3: Algebra (Cambridge: Cambridge University Press, 1967), p. 4.

[9]John Playfair, "On the Arithmetic of Impossible Quantities," *Philosophical Transactions* (1778) 68: 318-43, p. 321.

live up to the ideal of mathematics as obvious, certain, and intuitive. At the heart of mathematics lay a mystery that threatened to sabotage not only the epistemological status of mathematics but also the entire educational structure of Cambridge and Trinity College Dublin.

Of course, not all mathematicians were troubled by impossible quantities. Some simply ignored any philosophical problems and kept right on calculating. Others tried to solve the problem by giving the strange symbols a real meaning. A number of mathematicians, for example, offered a geometrical interpretation for impossible quantities. If one imagines numbers placed along a line beginning with zero and increasing to the right, then negative numbers can easily be interpreted as laying along another line to the left of zero. The negative sign would then signify not a quantity less than nothing but a length in the opposite direction. Imaginary numbers could similarly be interpreted as another line perpendicular to the real numbers, creating a two-dimensional plane. Complex numbers (a real number plus an imaginary number) would then represent line segments in the plane. Such diagrams are now familiar from high school algebra, but from the 1790s to the 1820s, they represented important innovations. They seemed to bring some of the certainty of geometry to algebra and to give meaning to the symbols used in mathematical proofs. Underlying this approach was the idea that the validity of a mathematical demonstration depends on the specific nature of the subject being investigated. Mathematics, in other words, must describe some phenomena. If it does not describe the addition of apples or oranges, perhaps it describes the addition of lines in a plane. Such a method was later central to Hamilton's development of quaternions.

6.2   Formalism in Mathematics

The geometrical interpretations of negative and imaginary numbers, however, represented a compromise few were willing to make. Geometers felt that they still lacked the rigor of Euclidean geometry, while the Cambridge analysts sought to promote the autonomy of analysis as a science independent of geometry. In order to do so, the analysts found it necessary to redefine the very nature of mathematical reasoning.[10] Analytical reasoning, they argued, should focus on the rules for combining symbols, not on the meaning of the symbols themselves. This was a radical step. While geometry had traditionally been the science of space and arithmetic the science of magnitude, both would eventually become mere systems of rules for manipulating symbols. Historian Joan Richards explains,

> One of the major developments of 19th-century mathematics is that the concept of mathematical truth evolved from being one of an objective truth about an external subject matter to one of a purely abstract truth based merely on the logically consistent development of an arbitrary axiom system.[11]

The very nature of mathematics and the foundations of rational thought were being transformed.

The new view of mathematics had a number of sources. One was the attempt to understand how true results could be derived using meaningless symbols like imaginary numbers. Another was a result of the way in which the calculus developed in Britain and Ireland in the early nineteenth century. When the members of the Analytical Society and the analysts of Dublin rebelled against Newton's fluxional notation in favor of the continental differential notation,

---

[10]See Sherry, "The Logic of Impossible Quantities."

[11]Joan L. Richards, "The Art and the Science of British Algebra: A Study in the Perception of Mathematical Truth," *Historia Mathematica* (1980) 7: 343-365, p. 344.

they were also criticizing a specific view of the foundations of the calculus. Newton's fluxions represented physical quantities changing or 'flowing' in time. He had developed them in the course of his physical investigations, and it was literally impossible to separate the symbols from the ideas behind them (the fluent of x is $\dot{x}$). Of course this sat well with the belief that reasoning must involve symbols with clear (generally physical) referents. The members of the Analytical Society, however, not only adopted the continental differential notation ($\frac{dx}{dy}$), they also promoted Lagrange's algebraic definition of the derivative. Lagrange simply defined the derivative of a function to be the first coefficient of its Taylor expansion—a purely analytic definition with no underlying meaning.

The differential notation also had another advantage over the fluxional notation. The d's of the differential notation could be separated from the symbols they applied to, while the dots of the fluxional notation could not be. Fluxional symbolism always referred to flowing quantities, but the differential notation symbolized operations on variables. Following the work of a number of French mathematicians, the members of the Analytical Society, Charles Babbage, John Herschel, and George Peacock, began to study such operations independently of the functions they operated on.[12] That is, they operated on the symbol for the derivative as if it were itself a quantity, and in the process they found powerful and general methods for solving a wide range of equations. By the 1830s and 1840s, most of the major mathematicians of Cambridge and Dublin were investigating this calculus of operations.[13]

When the Cambridge mathematicians started their own journal, the *Cambridge Mathematical Journal*, its pages were filled with

[12]J.M. Dubbey, "Babbage, Peacock and Modern Algebra," *Historia Mathematica* (1977) 4: 295-302.

[13]See Elaine Koppelman, "The Calculus of Operations and the Rise of Abstract Algebra," *Archive for History of Exact Sciences* (1971) 8: 155-242, and Crosbie Smith and M. Norton Wise, *Energy and Empire: A Biographical Study of Lord Kelvin* (Cambridge: Cambridge University Press, 1989).

such investigations. The Dublin mathematicians soon joined them, and the journal was renamed the *Cambridge and Dublin Mathematical Journal*. Major contributors included Cambridge men like D.F. Gregory, William Thomson, Augustus De Morgan, and Robert Murphy (an Irishman who studied mathematics at Cambridge) and Dublin men like Charles Graves, Robert Carmichael, J.H. Jellett, and William Rowan Hamilton. George Boole, a self-taught English mathematician who taught at Queen's College Cork, was one of the pioneers of the subject. The calculus of operations served to focus attention on the laws of the combination of symbols without regard to specific operations. D.F. Gregory, for example, defined symbolical algebra as "the science which treats of the combination of operations defined not by their nature, that is, by what they are or what they do, but by the laws of combination to which they are subject."[14]

While most of the analysts pioneered new methods, a few thought carefully about their philosophical implications. Robert Woodhouse was one of the first British mathematicians to use the differential notation in his 1803 textbook, *Principles of Analytical Calculation*. His primary goal was to rid the calculus of the idea of motion and to found it on algebraical principles, and he was a major inspiration for the members of the Cambridge Analytical Society.[15] Woodhouse developed a view he claimed to have found in Berkeley's mathematical works, namely that algebra is a language of signs whose manipulation depends on their rules of combination rather than on the meaning behind the symbols.[16] Unlike Playfair, Woodhouse was happy to accept that reasoning is simply the mechanical transformation of combinations of symbols. The results of

---

[14] D.F. Gregory, "On the Real Nature of Symbolical Algebra," *Philosophical Transactions of the Royal Society of Edinburgh* (1838) 14: 208.

[15] Harvey W. Becher, "Woodhouse, Babbage, Peacock, and Modern Algebra," *Historia Mathematica* (1980) 7: 389-400.

[16] Helena M. Pycior, "Internalism, Externalism, and Beyond: 19th-Century British Algebra," *Historia Mathematica* (1984) 11: 424-441, pp. 434-35.

the operations are true, he argued, independently of what the symbols refer to. Therefore we can use $\sqrt{-1}$ without knowing what it means as long as we know what rules to use when operating on it. How do we know what rules to use? Woodhouse suggested that we simply extend the rules for arithmetic, in other words, treat $\sqrt{-1}$ as if it were a real number.

George Peacock, one of the members of the Analytical Society, became the foremost spokesman for this new view of analysis in the 1830s.[17] He wrote his *Treatise on Algebra* (1830) for the purpose of "conferring upon Algebra the character of a demonstrative science, by making its first principles co-extensive with the conclusions which were founded upon them."[18] To be demonstrative, he implied, algebra must be able to prove that each step follows logically from the step before it. Before, mathematicians believed that such continuity resulted from the subject matter being investigated; that is, they thought that the meanings of the symbols made each step true or false. Peacock argued that such continuity could be imposed from outside by defining rules. In his "Report on the Recent Progress and Present State of Certain Branches of Analysis" to the British Association for the Advancement of Science in 1833, Peacock wrote,

> [I]n symbolical algebra, the rules determine the meaning
> of the operations... we might call them *arbitrary* assumptions, in as much as they are *arbitrarily* imposed upon a
> science of symbols and their combinations, which might
> be adapted to any other assumed system of consistent
> rules.[19]

---

[17]Helena M. Pycior, "George Peacock and the British Origins of Symbolical Algebra," *Historia Mathematica* (1981) 8: 23-45.

[18]George Peacock, *Treatise on Algebra* (Cambridge, 1830), p. v.

[19]George Peacock, "Report on the Recent Progress and Present State of Certain Branches of Analysis," *Report of the BAAS* (1833), pp. 200-201.

Peacock's radical claims were jarring even for sympathetic mathematicians. Augustus De Morgan responded, "At first sight it appeared to us something like symbols bewitched, and running about the world in search of a meaning."[20]

But while Peacock often spoke as though the rules of symbolic algebra were completely arbitrary, he believed that they were in fact a generalization of the rules of arithmetic. While it might be possible to invent arbitrary new rules, there is no guarantee that such rules would allow for any useful interpretations. Though symbolical algebra no longer depended on the meaning of its symbols for its truth, Peacock made the rules themselves depend on the rules of arithmetic, which could be verified based on the meaning of its symbols. Despite his revolutionary rhetoric about arbitrary rules, therefore, Peacock's algebra was actually quite conservative.[21]

6.3   The Philosophy of Pure Time

While Hamilton shared his Cambridge colleagues' passion for analysis, he was deeply troubled by their claims about the philosophy of mathematics. Just as Hamilton differed fundamentally from them on the philosophy of science (see chapter 5), he also adopted a distinct view of the metaphysical foundations of pure mathematics. Peacock's *Treatise on Algebra*, he explained, appeared

> to reduce algebra to a mere system of symbols, and *nothing more*; an affair of pothooks and hangers, of black

---

[20] De Morgan, "Review of Peacock's Treatise of Algebra," *Quarterly Journal of Education* (1835) 9: 293-311, p. 311. See Pycior, "Early Criticism of the Symbolical Approach to Algebra," *Historia Mathematica* (1982) 9: 392-412.

[21] See Richards, "Art and Science," Joan L. Richards, "Augustus de Morgan, The History of Mathematics, and the Foundations of Algebra," *Isis* (1987) 78: 7-30, Joan L. Richards, "Rigor and Clarity: Foundations of Mathematics in France and England, 1800-1840," *Science in Context* (1991) 4: 297-319.

strokes on white paper, to be made according to a fixed but arbitrary set of rules.[22]

Hamilton had devoted his life to mathematics because he saw it as a spiritual quest for transcendental truth. And Trinity College Dublin made it the centerpiece of its education for similar reasons. Peacock's approach, Hamilton felt, reduced mathematics to little more than a meaningless game. Hamilton was deeply committed to the traditional view of mathematics as a science with real referents. He believed that algebraic demonstration, like geometry, must be based on the intuition of the nature of ideal objects.

Throughout his mathematical and metaphysical work, Hamilton strove to give the foundations of algebra the same certainty as geometry. His goal, as he explained in his most famous essay on the philosophy of mathematics, was to create

> a Science of Algebra properly so called: strict, pure and independent; deduced by valid reasonings from its own intuitive principles; and thus not less an object of a priori contemplation than Geometry.[23]

While algebra certainly entails the use of symbols and rules of operation, these are, according to Hamilton, merely superficial characteristics. "I am never satisfied," he explained, "unless I think that I can look beyond or through the signs to the things signified."[24] Behind the symbols lie intuitive principles, just as real ideas lie behind geometrical diagrams. The question for Hamilton then became, To what do the symbols of algebra refer? What is the intuition on which algebra is founded?

Hamilton found his answer through idealist philosophy. Probably the most important clue came from one of his favorite philosophers, Immanuel Kant. Kant argued that geometry is based on the

[22]Hamilton to George Peacock (October 13, 1846), Graves 2: 528.

[23]Hamilton, "Pure Time," p. 5.

[24]Hamilton to J.T. Graves (July 11, 1835), Graves 2: 143.

intuition of space. Our mind translates the confusion of our sensory impressions into events in three-dimensional space according to certain innate rules, he explained. It is for this reason that mathematicians can make discoveries in geometry without observing the empirical world. They are, in a sense, merely discovering the laws of the mind. However, while Kant understood the applicability of mathematics to nature as essentially the projection of the structure of the mind onto experience, Hamilton preferred to interpret this (as did Coleridge) as evidence of God's intervention. Nevertheless, Kant's attempt to ground geometry in the human mind rather than in the empirical world had a tremendous appeal to Hamilton. The symbols of geometry, in this view, had real referents but they were internal and ideal.

If geometry is the science of space, Hamilton believed that algebra might be the science of time.[25] The intuition of time struck Hamilton as the perfect foundation for algebra. Behind the symbols of algebra, he believed, lie real intuitions of our inner sense of time. Hamilton believed his view actually gave algebra a higher status than geometry, reversing the age-old privileging of geometry. Algebra, he explained,

> is more refined, more general, than geometry; and has its
> foundations deeper in the very nature of man; since the
> ideas of order and succession appear to be less foreign,
> less separable from us, than those of figure and extent.[26]

---

[25] John Hendry, "The Evolution of William Rowan Hamilton's View of Algebra as the Science of Pure Time," *Studies in the History and Philosophy of Science* (1984) 15: 63-81, Anthony T. Winterbourne, "Algebra and Pure Time: Hamilton's Affinity with Kant," *Historia Mathematica* (1982) 9: 195-200, Peter Øhrstrøm, "W.R. Hamilton's View of Algebra as the Science of Pure Time and His Revision of this View," *Historia Mathematica* (1985) 12: 45-55, J. Mathews, "William Rowan Hamilton's Paper of 1837 On the Arithmetization of Analysis," *Archive for History of Exact Sciences* (1978) 19: 177-200, John O'Neill, "Formalism, Hamilton and Complex Numbers," *Studies in the History and Philosophy of Science* (1986) 17: 351-372, Menachem Fisch, "'The Emergency Which Has Arrived': The Problematic History of Nineteenth-Century British Algebra-A Programmatic Outline," *British Journal for the History of Science* (1994) 27: 247-276.

[26] Hamilton, "Introductory Lecture on Astronomy, November 1832," Graves 1: 642-43.

In Hamilton's view, algebra is an expression of our innermost being.

In 1835 he presented his ideas to the Royal Irish Academy in "A Preliminary and Elementary Essay on Algebra as the Science of Pure Time." This essay was published in the *Transactions* two years later, together with his "Theory of Conjugate Functions, or Algebraic Couples." Hamilton described the combined essay as "the first installment of my long-aspired-to work on the union of Mathematics and Metaphysics."[27] In the two essays, Hamilton not only constructed the laws of arithmetic from the intuition of pure time, he also obtained the algebra of imaginary numbers, thus solving (in his own mind at least) the problem of the shaky foundations of symbolic algebra. Hamilton's philosophy was not merely an abstract doctrine. He believed that he could actually construct the real and imaginary numbers from the intuition of time; that is, he thought that he could prove that the relationships between operations on time steps are the same as the operations of algebra.

He began with the simplest thought, a moment in time. Comparing two different moments, we construct the idea of a time step. A time step can be either forward or backward, and it can be greater than, equal to, or less than another time step. Numbers are then considered as a sequence of equal time steps from an arbitrary beginning point. Hamilton solved the problem of negative numbers by interpreting them as steps backward in time. He could then define addition, subtraction, multiplication, and division in terms of operations on time steps.

Hamilton then went on to interpret complex numbers as couples of moments. Rather than consider the complex number $a + bi$, he wrote a pair of time steps (a, b) representing a step from one pair of moments to another pair of moments. Hamilton found the laws of addition and subtraction for such number couples and then constructed a rule for multiplication equivalent to the multiplication

---

[27]Hamilton to Aubrey de Vere (October 4, 1835), Graves 2: 164.

of complex numbers.

$$(a, b) \times (c, d) = (ac - bd, ad + bc)$$

Hamilton did not simply assume that the law for multiplication of couples is identical to the law for complex numbers. Instead he attempted to derive a multiplication law from his philosophy of pure time. He did this by using his definition of the addition of couples and by limiting the possible laws of multiplication using two principles. He required that the distributive law must hold,

$$(a, b)[(c, d) + (e, f)] = (a, b)(c, d) + (a, b)(e, f)$$

and he required that division should be determinate. While these conditions did not entirely determine Hamilton's definition for couple multiplication, they limited the possible choices and allowed him to present his law as non-arbitrary. Hamilton simply discarded the impossible symbol $\sqrt{-1}$ and demonstrated that the algebra of ordered pairs derived from the intuition of pure time is identical to the algebra of complex numbers. The theory, Hamilton claimed, "gives reality and meaning to conceptions that were before Imaginary, Impossible, or Contradictory."[28]

In order to demonstrate that his algebra of pure time is identical to the standard symbolic algebra, Hamilton presented one of the earliest attempts to list systematically the properties of the real number system. He demonstrated, for example, that his time steps and couples exhibited the commutative property and the distributive property:

$$A \times B = B \times A$$

$$A \times (B + C) = (A \times B) + (A \times C)$$

He also provided a definition of zero, demonstrated the existence of the additive and multiplicative inverse, and proved the law of

---

[28]Hamilton, "Pure Time," p. 7.

closure. To prove that mathematics is really the science of pure time, he had to show that operations based on the intuition of time match all of the possible procedures of algebra. In doing so, he gave one of the earliest accounts of the abstract rules of algebra. For this reason, Hamilton's system in some ways looked even more formal than the systems of the formalists who had no need to define such abstract laws because they assumed the laws of arithmetic could be generalized. The difference is that Hamilton justified all of his rules by reference to time steps. He carefully point out "these definitions are really *not arbitrarily chosen*; and... though others might have been chosen no others would be equally proper."[29] For Hamilton, algebra was unique, not arbitrary.

6.4 Imagining Quaternions

At the end of his essay on algebra as the science of pure time, Hamilton promised to extend his theory of couples to triplets. If a coherent theory of couples was possible, he supposed, why not triples or higher order numbers? Triplets also appealed to Hamilton for a number of other reasons. Hamilton (like Coleridge) had a philosophical fascination with triplets.[30] Following a number of German philosophers, Coleridge saw the philosophical triad of thesis, antithesis, and synthesis as the key to all philosophy, and both Hamilton and Coleridge appreciated the connection between philosophical triads and the Holy Trinity. Moreover, from a physical standpoint, triplets promised to be of great use. If couples could represent lines in two-dimensional space, triplets, Hamilton hoped, could represent lines in three-dimensional space. Geometers and

---

[29]Hamilton, "Pure Time," p. 83. For a detailed discussion of Hamilton's procedure, see Helena Pycior, *The Role of Sir William Rowan Hamilton in the Development of British Modern Algebra* (PhD Thesis: Cornell University, 1976), pp. 102-6.

[30]Thomas L. Hankins, "Triplets and Triads: Sir William Rowan Hamilton on the Metaphysics of Mathematics," *Isis* (1977) 68: 175-193.

physicists would then be able to operate on lines in space through pure analysis.

Hamilton's search for triplets, however, was a failure. It was not that Hamilton could not create a system of triplets (he created many). Rather, none of his triplet systems followed the rules that he deemed essential for any proper algebraic system. The steps in Hamilton's discovery illustrate the crucial role of choice in mathematics. Hamilton had to decide which rules the new algebra should follow. After he decided that preserving a relationship to geometry was essential, he found that he had to abandon other rules, like the commutative law. To preserve the geometrical interpretation, Hamilton even had to create new imaginary numbers. In the end he extended his view from three-component numbers (triplets) to four-component numbers (quaternions). It was an act of pure imagination, and Hamilton believed his idealist philosophy guaranteed its legitimacy. The process through which Hamilton discovered quaternions, therefore, provides an intriguing example of the workings of mathematical creativity.[31]

In contrast to his earlier algebraic work on number couples, the geometric interpretation played a crucial role in Hamilton's triplets. The reason is that, like Peacock, Hamilton wanted to guarantee that his algebra would be useful and interpretable. One could create any number of triplet systems with arbitrary rules, but Hamilton wanted to find one in which operations on triplets could be interpreted as operations on lines in three-dimensional space. His starting point, therefore, was Warren's work on complex num-

---

[31]See Hankins, *Sir William Rowan Hamilton*, pp. 281-325, E.T. Whittaker, "The Sequence of Ideas in the Discovery of Quaternions," *Proceedings of the Royal Irish Academy* (1944) 50A: 93-98, M.J. Crowe, *A History of Vector Analysis* (South Bend: Notre Dame University Press, 1967), pp. 17-46, Koppelman, "Calculus of Operations," pp. 222-31, Pycior, *The Role of Sir William Rowan Hamilton*, Pickering and Stephanides, "Constructing Quaternions," Pickering, "Concepts and the Mangle of Practice," Teun Koetsier, "Explanation in the Historiography of Mathematics: The Case of Hamilton's Quaternions," *Studies in History and Philosophy of Science* (1995) 26: 593-616.

bers.[32] Warren interpreted the complex number $x + iy$ as a line directed from the origin to the point (x, y). He then defined the addition, subtraction, multiplication, and division of line segments. Consider the lines $A = a + bi$ and $B = c + di$. Warren showed that the sum of the lines,

$$A + B = (a + bi) + (c + di) = (a + c) + (b + d)i$$

can be interpreted geometrically as a new line formed by putting the lines A and B end to end. He also showed that the product,

$$C = A \times B = (a + bi)(c + di) = (ac - bd) + (ad + bc)i$$

has a geometrical interpretation. The angle that C makes with the x-axis is equal to the sum of the angles made by A and B with the x-axis. And the length of C equals the product of the lengths of A and B. Another way to say this is that the law of the modulus holds, namely,

$$\text{Length of } C = (\text{Length of } A) \times (\text{Length of } B)$$

$$\sqrt{[(ac - bd)^2 + (ad + bc)^2]} = \sqrt{(a^2 + b^2)} \times \sqrt{(c^2 + d^2)}$$

The law of the modulus guarantees that the product is well defined and therefore implies that it is also possible to divide complex numbers.

Hamilton hoped that he could extend this geometrical interpretation to three dimensions. One of the most interesting accounts of Hamilton's final attempt comes from his notebook for October 16, 1843, the day he discovered quaternions.[33] He begins,

---

[32] John Warren, *A Treatise on the Geometrical Representation of the Square Roots of Negative Quantities* (Cambridge, 1823).

[33] "Quaternions: Note-book 24.5, entry for 16 October 1843," *Mathematical Papers* 3: 103-105, "Letter to Graves on Quaternions; or on a New System of Imaginaries," *Philosophical Magazine* (1844) 25: 489-95 in *Mathematical Papers* 3: 106-110, "On a New Species of Imaginary Quantities Connected with the Theory of Quaternions," *Proceedings of the Royal Irish Academy* (1844) 2: 424-34 in *Mathematical Papers* 3: 111-16. See also the preface to W.R. Hamilton, *Lectures on Quaternions* (Dublin, 1853).

> Couples being supposed known, and known to be repre-
> sentable by points in a plane, so that $\sqrt{-1}$ is perpen-
> dicular to 1, it is natural to conceive that there may be
> another sort of $\sqrt{-1}$, perpendicular to the plane itself.[34]

If i represents an axis perpendicular to the real axis then, Hamilton
supposes, perhaps j can represent a third axis. Just as $i^2 = -1$, he
assumes that $j^2 = -1$. Note that Hamilton has just created a new
number, 'another sort of $\sqrt{-1}$,' presumably no more intelligible
than the original $\sqrt{-1}$. Yet, because he had demonstrated that
imaginary numbers could be explained by his philosophy of pure
time, he seems to feel comfortable creating new ones. If $x + iy$
represents a line in two dimensions, Hamilton hypothesizes that
$x + iy + jz$ will represent a line in three dimensions.

Hamilton's task then is to try basic operations on triplets (such
as addition, subtraction, multiplication, and division) in order to
see if the geometrical analogy holds. But when he calculates the
square of a triplet, he finds that the product is not a triplet; it has
terms with coefficients ij and ji. Hamilton now has to determine
what the product ij equals. But how can he find the rules that
govern a symbol he has just created? Given that $i^2 = -1$ and
$j^2 = -1$, it is possible that $ij = -1$. However, it is also possible
that $ij = 0$ or that $ij = -ji$. Each of these rules would make the ij
and ji terms disappear and guarantee that the square of a triplet
is another triplet. Hamilton decides to try assuming $ij = -ji$.
Note that even at this stage he admits the possibility that the
commutative law might not always hold.

Hamilton then checks to see if his triplets act like Warren's two-
dimensional lines when they are in the same plane. He wants to
show that the product of the two lines fulfills the necessary geometric conditions. Most importantly, he wants to show that the
law of the modulus holds. He finds that this condition holds on the

---

[34]"Notebook Entry for 16 October 1843," p. 103.

assumption that $ij = -ji$. The correspondence with Warren's geometrical interpretation, therefore, convinces him that his new rule
is not just an arbitrary choice. He then decides to try multiplying
two triplets. Once again, he fails to obtain a triplet. The ij term
is still there. Perhaps, Hamilton thinks, $ij = 0$. But while $ij = 0$
would make the general product a triplet, Hamilton soon realizes
that if this were true the law of the modulus would fail. When
Hamilton performs the calculation, he finds once again an extra
term. But the extra term is identical to the troublesome ij term in
the general product. In other words, if Hamilton sets $ij = -ji$, the
product of two triplets is not a triplet, and if he sets $ij = 0$, the law
of the modulus does not hold and the geometrical interpretation
is impossible. The situation, Hamilton explained, "*forced* on me
the non-neglect of ij; and *suggested* that it might be equal to k,
a new imaginary."[35] Hamilton found it impossible to create a system of triplets that followed all the rules he required, but he began
to suspect that it might be possible to satisfy the rules with four
component numbers, the quaternions. When he adds the new component k, he finds that all of the equations work. It was essential
for Hamilton that his algebraic system have a clear and consistent
geometrical interpretation. It had to be possible to multiply and
divide and, most importantly, it had to follow the geometrical law
of the modulus, "without which consistency being verified," he later
explained, "I should have regarded the whole speculation as a failure."[36] Hamilton also assumed that his algebra would follow the
commutative law, but he was willing to give this up in order to
preserve the geometrical interpretation.

Hamilton's quaternions represent the point (x, y, z) with the
symbol,

$$w + ix + jy + kz$$

[35] Hamilton, "Preface to *Lectures on Quaternions*," p. 104, n. 32.
[36] Hamilton, "Letter to Graves on Quaternions," p. 108.

Operations are governed by the "multiplication assumptions, or definitions,"

$$i^2 = j^2 = k^2 = 1,$$
$$ij = k, \quad jk = i, \quad ki = j$$
$$ji = -k, \quad kj = -i, \quad ik = -j$$

These assumptions might look arbitrary and even unnatural when compared to standard algebra, where $ij = ji$. The new imaginaries j and k, after all, are even more mysterious than $\sqrt{-1}$. Yet together they form an algebra that follows all of the rules for operations on lines in space. Hamilton was convinced that his rules were not arbitrary at all, but rather that he had discovered a unique mathematical language for describing three-dimensional space. He spent the rest of his life (over twenty-two years) exploring the uses and implications of the quaternions.

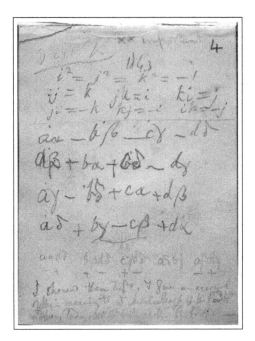

Figure 6.2: Hamilton's Notebook on Quaternions (1843).
Reprinted with permission from Trinity College Dublin

At first, other mathematicians were nervous about Hamilton's introduction of new entities that broke established rules. John Graves wrote,

> There is still something in the system which gravels me.
> I have not yet any clear view as to the extent to which
> we are at liberty arbitrarily to create imaginaries, and to
> endow them with supernatural properties.[37]

How can mathematics, the epitome of objective and universal knowledge, be the result of the imagination? Hamilton felt justified in taking such an imaginative step because he felt the geometrical analogy and his philosophy of algebra as pure time kept

---

[37]J.T. Graves to Hamilton (October 31, 1843), Graves 2: 443.

him grounded in reality. Like the prophecy of conical refraction, the imagination of quaternions demonstrated Hamilton's ability to find truth through creative mathematics.

Other mathematicians, however, did not share Hamilton's belief in the uniqueness of his system. Ironically, Hamilton's intuitive approach to algebra led him to what other mathematicians saw as the first purely formal or arbitrary system of algebra. Historian of mathematics Ernest Nagel explains,

> For the first time an algebra was discovered with an indisputably useful geometrical interpretation, which illustrated a set of principles of operation radically different, even formally, from the operations of arithmetic and the algebras suggested by arithmetic. Mathematics had really come into its maturity as the science of the abstract structure of anything whatsoever.[38]

For Hamilton, the correspondence between his algebraic system and geometry gave him the courage to break away from the rules of ordinary arithmetic. Soon other mathematicians would even abandon Hamilton's geometrical strictures to create an enormous number of new algebras with no clear reference to any underlying system. In one of his papers De Morgan explained, "These imaginaries are not deductions but inventions: *their laws of action on each other are assigned.* This idea Mr. De Morgan desires to acknowledge was entirely borrowed from Sir William Rowan Hamilton."[39] After the quaternions, numerous mathematicians continued to develop new systems of triplets. They merely chose different multiplication assumptions for i, j, and k and abandoned Hamilton's insistence that the geometrical interpretation hold.

Hamilton's emphasis on the geometrical interpretation of quaternions bore fruit in the second half of the nineteenth century,

---

[38] Nagel, "Impossible Quantities," p. 464.

[39] De Morgan, "Abstract of a Memoir on Triple Algebra (1844)," Graves 3: 251.

as quaternions and their successor—the vector calculus—became
the fundamental language of theoretical physics. Hamilton spent
the rest of his life trying to convince physicists to use his quaternions rather than Cartesian coordinates,[40] and after his death in
1865, Peter Guthrie Tait, a Scot, continued the crusade.[41] Tait
is most famous for co-authoring *A Treatise on Natural Philosophy*
(1867) with William Thomson, Lord Kelvin. Thomson and Tait's
Treatise presented physics in terms of Hamiltonian dynamics, and
Tait hoped that it might also present quaternions as the language of
physics. Thomson, however, would have none of it. The metaphysical resonances that attracted Hamilton and others to quaternions
annoyed the practical and industrially minded Thomson.[42]

While Tait failed to convince Thomson to take up the quaternions, he succeeded in convincing James Clerk Maxwell.[43] Maxwell
found in quaternions a way to visualize the physical processes central to electric and magnetism. As Maxwell gave mathematical
form to Faraday's concept of the electrical and magnetic fields, he
found the concepts of vector and scalar (the imaginary and real
parts of the quaternion) essential. And the mathematical operations that Hamilton and Tait had developed that would later be
known as divergence, gradient, and curl were easily interpretable
in terms of the lines of force that Maxwell imagined around all
bodies. Maxwell thought it essential to see behind the symbols
to the phenomena they represent, and he believed that quaternions achieved this much better than the standard Cartesian coordinates. To prove his point Maxwell presented all the equations in
his epochal *Treatise on Electricity and Magnetism* (1873) in both

---

[40]Luc Sinègre, "Les Quaternions et le Mouvement du Solide autour d'un Point Fixe Chez
Hamilton," *Revue d'histoire des mathématiques* (1995) 1: 83-109.

[41]See David R. Wilkins (ed.), *Perplexingly Easy: Selected Correspondence between
William Rowan Hamilton and Peter Guthrie Tait* (Dublin: Trinity College Dublin Press,
2005), Crowe, *Vector Analysis*, pp. 117-25.

[42]See Smith and Wise, *Energy and Empire*, pp. 363-65.

[43]See Crowe, *Vector Analysis*, pp. 127-39, P.M. Harman, *The Natural Philosophy of
James Clerk Maxwell* (Cambridge: Cambridge University Press, 1998), pp. 145-54.

Cartesian and quaternion notation. This 'bilingual' approach, he believed, would demonstrate the superiority of quaternions.

619.] QUATERNION EXPRESSIONS. 237

	Symbol of Vector.	Constituents.
The electromotive force	$\mathfrak{E}$	$P\ Q\ R$
The mechanical force	$\mathfrak{F}$	$X\ Y\ Z$
The velocity of a point	$\mathfrak{G}$ or $\dot{\rho}$	$\dot{x}\ \dot{y}\ \dot{z}$
The magnetic force	$\mathfrak{H}$	$\alpha\ \beta\ \gamma$
The intensity of magnetization	$\mathfrak{J}$	$A\ B\ C$
The current of conduction	$\mathfrak{K}$	$p\ q\ r$

We have also the following scalar functions :—

The electric potential Ψ.

The magnetic potential (where it exists) Ω.

The electric density e.

The density of magnetic 'matter' m.

Besides these we have the following quantities, indicating physical properties of the medium at each point :—

C, the conductivity for electric currents.

K, the dielectric inductive capacity.

μ, the magnetic inductive capacity.

These quantities are, in isotropic media, mere scalar functions of ρ, but in general they are linear and vector operators on the vector functions to which they are applied. K and μ are certainly always self-conjugate, and C is probably so also.

619.] The equations (A) of magnetic induction, of which the first is,

$$a = \frac{dH}{dy} - \frac{dG}{dz},$$

may now be written $$\mathfrak{B} = V \nabla \mathfrak{A},$$

where ∇ is the operator

$$i\frac{d}{dx} + j\frac{d}{dy} + k\frac{d}{dz},$$

and V indicates that the vector part of the result of this operation is to be taken.

Since $\mathfrak{A}$ is subject to the condition $S\nabla\mathfrak{A} = 0$, $\nabla\mathfrak{A}$ is a pure vector, and the symbol V is unnecessary.

The equations (B) of electromotive force, of which the first is

$$P = c\dot{y} - b\dot{z} - \frac{dF}{dt} - \frac{d\Psi}{dx},$$

become $$\mathfrak{E} = V\mathfrak{G}\mathfrak{B} - \dot{\mathfrak{A}} - \nabla\Psi.$$

The equations (C) of mechanical force, of which the first is

$$X = cv - bw - e\frac{d\Psi}{dx} - m\frac{d\Omega}{dx},$$

become $$\mathfrak{F} = V\mathfrak{C}\mathfrak{B} - e\nabla\Psi - m\nabla\Omega.$$

Figure 6.3: Maxwell expresses his famous electromagnetic equations in quaternion form (1873). James Clerk Maxwell, *A Treatise on Electricity and Magnetism* (1873), Vol. II, p. 237.

Few of Maxwell's readers agreed, however. The popularity of Maxwell's *Treatise* introduced Hamilton's quaternions to a large audience of physicists, but many found them difficult to use. The American Josiah Willard Gibbs and the Englishman Oliver Heaviside, who both learned quaternions from Maxwell's *Treatise*, created vectors as a simpler, more intuitive method in the 1880s. By the 1890s, the vector calculus challenged quaternions for predominance.[44] The debate essentially involved different notions of the proper language for mathematical physics. Quaternionists tended to put more emphasis on mathematical elegance and algebraic simplicity, while vectorialists privileged ease of understanding and physical expressiveness. Hamilton felt the algebraically rigorous foundations of the quaternions guaranteed their ability to produce true results, but for physicists like Gibbs and Heaviside it was more important to be able to visualize physical situations. The needs of mathematicians and the needs of mathematical physicists had begun to diverge.

6.5   The Laws of Thought

While physicists took quaternions in the direction of more concrete applications, algebraists looked for increasingly abstract systems. Hamilton's breakthrough not only seemed to demonstrate that any number of arbitrary algebraic systems could be invented, but it also proved that mathematics need not be the science of quantity. While Hamilton argued that it is actually the science of time, other mathematicians, following the formalist claims of Woodhouse and Peacock, saw mathematics as an abstract science with no particular referent. When algebra was loosed from any concrete foundation in this way, it could be applied to other, previously non-mathematical fields, like logic. If algebra is merely a formal language with no nec-

---

[44]See Crowe, *Vector Analysis*, pp. 109-246.

essary connection to magnitude, why could not the formal system of logic be expressed in similarly mathematical terms. There was already a long tradition linking logical and mathematical analysis and both with calculating machines, but Hamilton's quaternions opened up a new connection.

Augustus De Morgan, for example, published a series of essays in which he abstracted general rules from algebra, such as the commutative and distributive properties. In 1842, even before Hamilton's quaternions, De Morgan had proposed a set of formal axioms for algebra. At the same time, he made a study of logic and was struck by the similarities between the formal approaches to algebra and logic.[45] De Morgan approached logic as a similarly formal system in which reasoning can be performed without attention to the meaning of the symbols. While he advocated uniting mathematics and logic in principle, however, in practice he brought few mathematical techniques to the study of logic. De Morgan's friend, George Boole, however, revolutionized the field of logic by presenting it as essentially a branch of analysis. In *Mathematical Analysis of Logic* (1847) and *An Investigation of the Laws of Thought* (1853), Boole demonstrated that logic can be treated as a branch of mathematics.[46] "[W]e ought no longer to associate Logic and Metaphysics," he argued, "but Logic and Mathematics."[47]

---

[45] Helena M. Pycior, "Augustus De Morgan's Algebraic Work: The Three Stages," *Isis* (1983) 74: 211-266, Luis M. Laita, "Influences on Boole's Logic: The Controversy Between William Hamilton and Augustus De Morgan," *Annals of Science* (1979) 36: 45-65, N.I. Styazhkin, *History of Mathematical Logic from Leibniz to Peano* (Cambridge: MIT Press, 1969), pp. 161-69, Daniel D. Merrill, *Augustus de Morgan and the Logic of Relations* (Dordrecht: Kluwer, 1990).

[46] John Richards, "Boole and Mill: Differing Perspectives on Logical Psychologism," *History and Philosophy of Logic* (1980) 1: 19-36, Nicla Vasallo, "Analysis versus Laws: Boole's Explanatory Psychologism versus His Explanatory Anti-Psychologism," *History and Philosophy of Logic* (1997) 18: 151-63, Luis M. Laita, "The Influence of Boole's Search for a Universal Method in Analysis on the Creation of His Logic," *Annals of Science* (1977) 34: 163-76, Styazhkin, *Mathematical Logic*, pp. 170-202, Desmond MacHale, *George Boole: His Life and Work* (Dublin: Boole Press, 1985).

[47] George Boole, *The Mathematical Analysis of Logic, Being an Essay Towards a Calculus of Deductive Reasoning* (Cambridge, 1847), p. 13.

Boole's early work had been in the calculus of operations, and he was greatly impressed by the formal view of mathematics. He explained,

> Those who are acquainted with the present state of the theory of Symbolical Algebra are aware that the validity of the processes of analysis does not depend upon the interpretation of the symbols which are employed, but solely upon the laws of their combination.[48]

Mathematics, therefore, can be interpreted in terms of space, quantity, mechanics, optics, or even logic. Boole stated, "it is not of the essence of mathematics to be conversant with the ideas of number and quantity."[49] Mathematics is a purely abstract system, he believed, that can be interpreted in any number of ways.

Figure 6.4: George Boole (1815–1864), Professor at Queen's College Cork and one of the founders of mathematical logic. *Popular Science Monthly* (1880), Vol. 17.

---

[48] Boole, *Mathematical Analysis of Logic*, p. 3.

[49] George Boole, *An Investigation of the Laws of Thought, On Which Are Founded the Mathematical Theories of Logic and Probabilities* (London, 1854), p. 12.

Boole's goal was "to investigate the fundamental laws of those operations of the human mind by which reasoning is performed."[50] He began, therefore, with the assumption that the human mind operates according to definite laws. Logic, in Boole's view, expresses the rules of mental operations just as the calculus of operations expresses the rules of mathematical operations. In fact, according to Boole, both sets of operations follow the same rules. For Boole, the basic mental operation is the act of selecting an object from a class. The concept 'the gray cat,' for example, selects from the class of all cats those that are gray. The selection operation, he noted, does not depend on the order of operations. In other words, the class of all cats that are gray is the same as the class of all gray objects that are cats. If the letter x represents the class of cats, and y represents the class of things that are gray, the law can be written as,

$$xy = yx$$

Of course this is identical to the commutative law in algebra. Boole noted that logical operations also follow the distributive rule. If z is the class of small objects, then,

$$x(y + z) = xy + xz$$

states that the class of cats that are gray or small is equal to the class of cats that are gray or cats that are small. Boole concluded,

> There is not only a close analogy between the operations of the mind in general reasoning and its operations in the particular science of Algebra, but there is to a considerable extent an exact agreement in the laws by which the two classes of operations are conducted.[51]

The laws of logic, he discovered, are identical to the laws of algebra. There was one important exception, however. Repeating a selection

---

[50] Boole, *Laws of Thought*, p. 3.
[51] Boole, *Laws of Thought*, p. 6.

is identical to making the selection once. In other words, if we chose all gray objects from the class of cats and then chose all gray objects from the resulting class of gray cats we are once again left with the class of gray cats. Boole wrote this as,

$$xx = x,$$

or

$$x^2 = x$$

This statement does not hold in general for algebra. In fact, it only holds for the numbers 0 and 1. Boole explained,

> Let us conceive, then, of an Algebra in which the symbols x, y, z, &c. admit indifferently of the values 0 and 1, and of these values alone. The laws, the axioms, and the processes, of such an Algebra will be identical in their whole extent with the laws, the axioms, and the processes of an Algebra of Logic.[52]

In other words, Boole found that all of logic can be expressed as an algebra of 1 and 0. On this basis Boole could write syllogisms and other logical statements as equations. He could then transform them according the rules of algebra and then translate the solutions back into logical statements. He noted,

> We may in fact lay aside the logical interpretation of the symbols in the given equation; convert them into quantitative symbols, susceptible only of the values 0 and 1; perform upon them as such all the requisite processes of solution; and finally restore to them their logical interpretation.[53]

Logic, on Boole's view, is simply a branch of a more universal mathematical calculus, defined by the same laws of operation but

---

[52] Boole, *Laws of Thought*, pp. 37-38.
[53] Boole, *Laws of Thought*, p. 70.

with a different interpretation. This was a key moment in the creation of the distinction between pure and applied mathematics.
On Boole's view, pure mathematics is a formal, rule-based system
with no concrete reference, while applied mathematics is a particular interpretation of such a formal system. Bertrand Russell, in
fact, claimed, "Pure mathematics was discovered by Boole."[54]

On the surface, Boole's approach looks rather similar to Hamilton's. He discusses the laws of thought and emphasizes the purity of
mathematics. But while Hamilton found purity by founding mathematics on the intuition of pure time, Boole found it by freeing
mathematics of any external referent. Mathematics, for Boole, is
pure because it is empty. Moreover, while both Boole and Hamilton
lived and worked in Ireland, they came from divergent backgrounds
and differed greatly in their views on religion and society. Boole
was born in England, the son of a cobbler, and taught himself
mathematics while he worked as a teacher in various Dissenting
Academies and mechanics institutes.[55] He eventually befriended
a number of Cambridge mathematicians and published important
work, but he never attended university. In 1849 he was appointed
professor of mathematics at the new Queen's College Cork, one
of the new secular colleges established in Ireland by the British
government in an attempt to solve the problem of Catholic education. Boole was disgusted by the sectarian squabbles that marred
Ireland. He wrote,

> [T]he Roman Catholic Priesthood seem to have been do-
> ing all they can to preach disloyalty. Between them and
> a bigoted Calvinistic Protestant population this is a coun-

---

[54] Bertrand Russell, *Mysticism and Logic* (London, 1963), p. 59.
[55] See MacHale, *George Boole.*

try which does not on the whole present the most
favourable picture of Christianity.[56]

In fact, Boole believed that his mathematics could help. His wife
explained,

> The hope in his heart had been to work in the cause of
> true religion. Mathematics had never been more than a
> secondary interest for him; and even logic he cared for
> chiefly as a means of clearing the grounds of doctrines
> imagined to be proved, by showing that the evidence on
> which they were supposed to rest had no tendency to
> prove them.[57]

Boole, like De Morgan, tended towards the rationalistic faith of
the Unitarians that downplayed the differences between Christian
sects. Boole believed that a clear knowledge of the laws of thought
would lead to fewer religious conflicts and a more universal faith.
He even devoted a chapter of his *Laws of Thought* to the logical
analysis of the religious arguments of Samuel Clarke and Baruch
Spinoza. He concluded,

> It is not possible I think to rise from the perusal of the
> arguments of Clarke and Spinoza without a deep convic-
> tion of the futility of all endeavours to establish *a priori*
> the existence of an Infinite being, His attributes and His
> relation to the Universe.[58]

While Hamilton saw his mathematics as a support for established
religion, Boole saw it as a weapon to explode the dangerous claims
of Christian dogmatists. Like the analytical arguments of the

---

[56]George Boole to A.T. Taylor (January 26, 1860), Boole Papers, quoted in Daniel J.
Cohen, *Symbols of Heaven, Symbols of Man: Pure Mathematics and Victorian Religion*,
(PhD Thesis: Yale University, 1999), p. 129.

[57]Mary Everest Boole, quoted in MacHale, *George Boole*, p. 196.

[58]Quoted in MacHale, *George Boole*, pp. 199-200.

philosophes, Boole's mathematical logic developed as a critique of the claims of established religion. Despite his respect for Hamilton, De Morgan saw the idealist philosophy of mathematics as dogmatic. He explained,

> [A] symbol is not the representation of an external object absolutely, but of a state of mind in regard to that object; of a conception formed, for the formation of which the mind knows that it is indebted to the presence, bodily or ideal, of the object. Those who do not remember this, the real use of a symbol, are apt to dogmatize, declaring one or another impression produced on their own minds to be real, true, natural or necessary: it being neither one nor the other, except with reference to the particular mind in question.[59]

As Daniel Cohen has argued, both Boole and De Morgan (also a Dissenter and outspoken critic of the establishment) developed mathematical logic in an attempt to create an independent and professional mathematics insulated from traditional concerns of metaphysics and religion. In the late nineteenth century, according to Cohen, mathematics became less transcendental, more autonomous and professional. He explains,

> Rather than claiming that mathematics transcribed the mind of God, late nineteenth-century mathematicians proffered a baser and more pragmatic vision: mathematics was a set of laws and a system of notation created in the *human* mind.[60]

Both Boole and De Morgan hoped that mathematical logic would offer a tool to bring clarity to the religious disputes of the age while

---

[59] Quoted in Pycior, *The Role of Sir William Rowan Hamilton*, p. 132.
[60] Cohen, *Symbols of Heaven*, p. 17.

remaining above the fray itself. They fought against dogmatism in religion and politics as well as in mathematics and logic.

6.6 The Calculating Man

The new formalism in algebra and logic represented more than just a philosophy of mathematics. What drew Hamilton, De Morgan, Boole and others to algebra was the hope that they might come to understand something fundamental about rationality and man's relationship to truth. Such profound issues involved philosophy and religion as well as mathematics. For Hamilton, the question of whether mathematics is based on the intuition of pure ideas or simply the mechanical manipulation of symbols went to the very heart of his philosophical, religious, and social views. Formalism was closely associated with a mechanical view of thought. De Morgan described mathematical thinking as "watch[ing] the machine in operation without attending to the matter operated on,"[61] while Boole noted, "the process of inference is conducted with a precision which might almost be termed mechanical."[62] Charles Babbage, of course, created the most powerful symbol of this view with his analytical engine. If mathematics is simply the application of rules to meaningless symbols it can be done by a machine as well as by a human. The mechanical view had much broader implications. Babbage visited factories, promoted political reform, and constructed "a peculiarly industrial notion of intelligence" that emphasized the values of efficiency, power, and production.[63] Babbage wrote on factory management as well as analysis and explained miracles in

---

[61] Augustus De Morgan, "On Syllogisms: III and on Logic in General (1858)," pp. 74-146 in Heath (ed.) *On the Syllogism*, p. 75.

[62] Boole, *Laws of Thought*, p. 186.

[63] William J. Ashworth, "Memory, Efficiency and Symbolic Analysis: Charles Babbage, John Herschel and the Industrial Mind," *Isis* (1996) 87: 629-653, p. 632; Simon Schaffer, "Babbage's Intelligence: Calculating Engines and the Factory System," *Critical Inquiry* (1994) 21: 203-227.

analogy to the mechanical functioning of his programmed calculating machine.

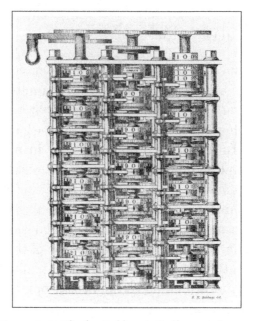

Figure 6.5: Charles Babbage's Difference Engine,
No. 1 (1830s), an early calculating machine.
©Science Museum, London

Babbage represented a larger trend in nineteenth-century Britain and Ireland embodied in the wide-ranging connotations of the term 'calculation'. Calculation was seen by many as mechanical and amoral, disrespectful of tradition and interested only in personal gain. Edmund Burke had proclaimed, "the age of chivalry is gone.— That of sophisters, oeconomists, and calculators, has succeeded; and the glory of Europe is extinguished forever."[64] And the poet Shelley similarly saw in his age "an excess of the selfish and cal-

---

[64] Edmund Burke, *Reflections on the Revolution in France* [1795] (New York: Anchor Books, 1973), p. 89.

culating principle."[65] Thomas Carlyle's essay "Signs of the Times" was an even more elaborate investigation of the growing mechanization of man and society and their connection to calculation. Carlyle proclaimed,

> It is the Age of Machinery, in every outward and inward sense of that word; the age which, with its whole and undivided might, forwards, teaches and practises the great art of adapting means to ends. Nothing is now done directly, or by hand; all is by rule and calculated contrivance.[66]

Carlyle was not simply bemoaning the coming of industrialization. He indicted mechanism in politics, religion, and education. His greatest fear was that man himself was becoming mechanical as a result of such social changes. He explained, "the same habit regulates not our modes of action alone, but our modes of thought and feeling. Men are grown mechanical in head and in heart, as well as in hand."[67] Even mathematics had become tainted. "Our favourite Mathematics," he lamented,

> the highly prized exponent of all these other sciences, has also become more and more mechanical. Excellence in what is called its higher departments depends less on natural genius than on acquired expertness in wielding its machinery. ... [The] calculus, differential and integral, is little else than a more cunningly-constructed arithmetical mill; where the factors being put in, are, as it were, ground into the true product, under cover, and without

---

[65]Percy B. Shelley, "A Defence of Poetry [1821]," pp. 478-508 in Donald H. Reiman and Sharon B. Powers (eds.) *Shelley's Poetry and Prose* (New York: W.W. Norton & Company, 1977), p. 503.

[66]Thomas Carlyle, "Signs of the Times [1829]," pp. 31-54 in G.B. Tennyson (ed.), *A Carlyle Reader* (Cambridge: Cambridge University Press, 1984), p. 34.

[67] Carlyle, "Signs of the Times," p. 37.

other effort on our part than steady turning of the handle.[68]

Social reform, industrialization, and the calculus all indicated to Carlyle a frightening shift in the social, moral, and intellectual foundations of British society.

What Carlyle feared, the utilitarians welcomed. Bentham had made the calculating man the foundation of the utilitarian philosophy, and in the process he made explicit all of the implications feared by the critics of calculation.[69] Most famously, he rejected traditional morality based on religious injunctions and man's inherent moral sense and attempted to replace it with a 'felicific calculus' that recast ethics as the calculation of the total quantity of happiness that would result from a given action. Adam Sedgwick, fellow at Cambridge University, scoffed,

> The utilitarian scheme starts... with an abrogation of the authority of conscience—a rejection of the moral feelings as the test of right and wrong. ... Virtue becomes a question of calculation—a matter of profit or loss; and if man gain heaven at all on such a system, it must be by arithmetical details—the computation of his daily work—the balance of his moral ledger.[70]

Bentham extolled the rational pursuit of self-interest and his followers helped to erect this principle as the foundation of classical economics. Human behavior, they found, could be described on the model of a simple machine calculating the most profitable of a range of options.

Such were the implications of the formalist view of mathematics and the reason that Hamilton worked so hard to develop an ide-

---

[68] Carlyle, "Signs of the Times," p. 38.

[69] See Elie Halévy, *The Growth of Philosophic Radicalism* [1928] Trans. Mary Morris (London: Faber and Faber, 1972).

[70] Adam Sedgwick, *A Discourse on the Studies of the University of Cambridge*, 5th Ed. (London, 1850), p. 65.

alist alternative. Analytical mathematics, he tried to show, need not lead to mechanism and the denial of sympathy and tradition. Hamilton wanted to retain the powerful methods of analysis while at the same time giving algebra the cultural status of geometry. The creation of quaternions can be read as the search for a solution to this problem. Hamilton wanted to develop a form of analysis with powerful applications in physics and geometry while demonstrating at each step the ideas behind the symbols.

Hamilton could never accept that the human mind works mechanically or that truth could be found through the manipulation of meaningless symbols. Mathematics meant something very different to him. He once explained,

> [Mathematics] presents to its votaries some of sublimest objects of human contemplation... its results are eternal and immutable verities... it seems to penetrate the counsels of Creation, and soar above the weakness of humanity. For it sits enthroned in its sphere of isolated intellect, undisturbed by passion, unclouded by doubt."[71]

Quaternions for him demonstrated the possibility of transcending the mechanical rules of calculation through pure thought. They even represented the role that imagination plays in science. For Hamilton, doing mathematics is a creative expression of man's relationship to the divine explored through the process of introspection.[72] Despite Hamilton's intentions, however, other mathematicians refused to read them as he did. The physicists stripped off what Hamilton believed was most essential in order to forge a more useful tool, while the analysts read them as proof that mathematics is essentially contentless, the mere play of formal systems. Quaternions live on today in algorithms for modeling three dimensional

---

[71]Graves 1: 194.

[72]See Joan L. Richards, "God, Truth and Mathematics in Nineteenth-Century England," pp. 51-78 in M. J. Nye (ed.), *The Invention of the Physical Sciences* (Dordrecht: Kluwer, 1992).

rotations for satellite navigation and video game design—hardly
the uses to which Hamilton aspired.

Chapter 7

Engineering the Empire

News by Ether

July 20, 1898. It was the first day of the Queen's Cup Regatta at the Royal St. George Yachting Club in Kingstown (now called Dún Laoghaire) just outside of Dublin. Hundreds of spectators were gathered in their most elegant summer clothing. Most had brought binoculars to observe the ships at sea. Unfortunately, the weather was not cooperating. A thick haze made it impossible for those on the shore to see anything in the race. And yet for the first time in history, the assembled crowd was receiving minute-by-minute updates on the progress of the race.

A twenty-three year-old Italian inventor, Guglielmo Marconi was observing the race from "The Flying Huntress," a tug boat loaded with electrical equipment, including an 80 foot wire hung from the mast. The wire served as an antenna for sending Morse code messages wirelessly to a 110 foot high receiving mast at the Harbor Master's house in Kingstown. The messages came through clearly despite the fog, even when Marconi's tug was 25 miles off shore. The station in Kingstown then relayed the messages by telephone to the *Daily Express* in Dublin, which had commissioned Marconi especially for this event. The *Express* was able to print full accounts of the races even before they were over and even when the yachts were out of range of any telescope. It was the first ever commercial use of wireless communication. At least one observer realized the importance of the moment: "This great achievement, this wireless telegraphy will produce a revolution in the conditions of life." [1]

Figure 7.1: Guglielmo Marconi covering the Kingstown Regatta (1898). A line drawing created for the booklet that accompanied the Kingstown Regatta.

The possibility of wireless communication arose from Maxwell's theory of electromagnetism. In creating his mathematical theory, the Scottish physicist James Clerk Maxwell had come to believe that light itself is an electromagnetic wave, and he predicted that other electromagnetic waves of different frequencies should exist as well. In 1888, the German scientist Heinrich Hertz succeeded in

---

[1] Elettra Marconi, "The Queen's Cup Regatta Kingstown 1898" in Maria Cristina Marconi and Elettra Marconi, *Marconi My Beloved* (Boston: Dante University of America Press, 1999), p. 82.

generating and detecting these waves just as Maxwell's theory had predicted. The press coverage of Hertz's death in 1894 spurred the twenty-one-year-old Marconi to start experimenting with the so-called 'Hertzian waves' on his father's estate in Bologna in 1895. By tinkering with various configurations of the equipment invented by university professors, Marconi was soon able to transmit messages over a mile and half, much farther than any academic scientist had attempted. Marconi knew that he was on to something, and, luckily, he had family connections to help him with his invention. Marconi's mother was Annie Jameson, granddaughter of John Jameson, founder of the Irish whiskey distillers Jameson & Sons. The Jameson family supported Marconi to come to England in 1896 to demonstrate his invention; and when Marconi set up a company to commercialize his invention the following year, one of the Jameson's became the first managing director.

The Kingstown Regatta was one of the first paying jobs for the new company, and the publicity it generated rapidly led to more. Just a few days later, Marconi was asked to set up wireless communication between Osborne House on the Isle of Wight and the royal yacht so that the Queen could get frequent updates about the Prince's injured knee. The following October, Marconi covered the America's Cup for the *New York Herald* and established a Marconi Company in America. By 1901 (just three years after the Kingstown Regatta), Marconi succeeded in sending messages all the way across the Atlantic from Cornwall to Newfoundland. Ireland, however, continued to play an important role in his work. He returned in 1905 to set up a high power transmitting station in Clifden, County Galway to handle transatlantic commercial traffic, and in 1919 he transmitted the first transatlantic telephone message from Ballybunion, County Kerry. He even married an Irish woman.[2]

---

[2] Noel Barry, "Irish Contributions to Wireless Communications," *Technology Ireland* (May 1994), pp. 23-26.

Two Irish university professors observed the trial at Kingstown: George Francis Fitzgerald from Trinity College Dublin and Gerald Molloy from University College Dublin. In fact, Marconi borrowed equipment from Fitzgerald's laboratory for the event. Fitzgerald wrote a short piece for the *Daily Telegraph* the following day on "The Meaning and Possibilities of Wireless Telegraphy."[3] This was the first time in history, he explained, that wireless was used to convey useful information; and he looked forward to the day in the near future when it would be used for doing even more "than giving information as to what the odds should be in betting transactions— though, from the amount of space and interest devoted thereto, one would imagine that nothing was so important as this latter."[4]

For Fitzgerald, the reality of "news conveyed by the ether" demonstrated conclusively the importance of mathematical theories like Maxwell's to whose development Fitzgerald had devoted most of his career. Marconi's chief assistant George Kemp, however, had a very different perspective. Kemp had served in the Royal Navy as an electrical and torpedo instructor and then worked at the English post office for its Engineer-in-Chief, William Preece. Preece and Kemp were practical men with little respect for university scientists. Kemp drew a particularly harsh conclusion from the successful experiment at Kingstown: "The only thing to do if you expect to find out anything about electricity is to work, for you can do nothing with theories. Signor Marconi's discoveries prove that the professors are all wrong and now they will have to go and burn their books."[5]

The four decades from 1870 to 1910 witnessed an explosion of scientific inventions that transformed everyday life—the telephone, the internal combustion engine, the phonograph, moving

---

[3] George Francis Fitzgerald, "The Meaning and Possibilities of Wireless Telegraphy," *Daily Express* (July 21, 1898).

[4] Fitzgerald, "The Meaning and Possibilities of Wireless Telegraphy."

[5] Marconi, "The Queen's Cup Regatta," p. 82.

pictures, radio, the electric light bulb, the steam turbine, the automobile, and the airplane. These were all science-based innovations, and yet for the most part their inventors were not university-trained scientists or engineers. Thomas Edison, Henry Ford, Nikola Tesla, Alexander Graham Bell, Orville and Wilbur Wright, and Guglielmo Marconi were largely self-taught tinkerers and businessmen. Fitzgerald and his academic colleagues struggled to make the case that university-based research was essential for technical and industrial progress. Fitzgerald himself was fascinated by the new technologies and experimented with most of the innovations of his time—x-rays, dynamos, electric trams, the electric arc, and gliders, to name only a few. Fitzgerald was also responsible for bringing electric lights to Trinity in 1889.[6] Fitzgerald believed that technology is the application of scientific theory, and he argued that engineers should be trained in engineering schools like the one at Trinity College Dublin rather than through apprenticeship. Only in a university, Fitzgerald explained, can engineers learn the principles of science that lie behind modern technology. In fact, Fitzgerald claimed that all people should be given a scientific education, and he attributed Ireland's economic backwardness to the lack of science teaching in schools.

For Humphrey Lloyd, who helped to found Trinity's engineering school in 1841, and for Fitzgerald who directed it at the end of the century, the connection between science and industry would justify Trinity's privileged place in society. While the world might dismiss Trinity's geometers as irrelevant in an age of industry, Trinity would train the engineers who would run the empire. They would apply their ascendancy in mathematics and science to the problems of industry and empire. In the late nineteenth century, however, the connections between academic research, technological progress, economic development, and social improvement were far

---

[6] J. Edmundson to G.F. Fitzgerald (August 7, 1889) Fitzgerald Collection, Royal Dublin Society.

from clear.[7] In Ireland in particular, science failed to lead to industry, and technical education had little or no effect on the Irish economy.[8] At the end of the century, Fitzgerald's vision of a prosperous and technologically advanced Ireland collapsed along with the Ascendancy that promoted it.

7.1   Trinity in the Service of the Empire

By the mid-nineteenth century, Trinity had built a reputation for high standards in mathematics and mathematical physics. Its professors were famous throughout Britain and the continent. They not only dominated Irish scientific institutions but also played prominent roles in the Royal Society of London, the Royal Astronomical Society, and the British Association for the Advancement of Science. But the first half of the nineteenth century had transformed Britain. Dublin, once the second largest city in the empire, now paled in comparison to the industrial towns of Manchester, Glasgow, and even Belfast. Factories, canals, railroads, and telegraph lines covered England and Scotland and were even slowly making inroads into Ireland. Moreover, Britain now had a global

[7] Bruce J. Hunt, "'Practice vs. Theory': The British Electrical Debate, 1888-1891," *Isis* (1983) 74: 341-55, Bruce J. Hunt, *The Maxwellians* (Ithaca: Cornell University Press, 1990), Sungook Hong, "Marconi and the Maxwellians: The Origins of Wireless Telegraphy Revisited," *Technology and Culture* (1994) 35: 717-49.

[8] J.H. Weiss, *The Making of Technological Man: The Social Origins of French Engineering Education* (Cambridge: Harvard University Press, 1982), Robert Fox and Anna Guagnini (eds.), *Education, Technology and Industrial Performance in Europe, 1850-1939* (Cambridge: Cambridge University Press, 1993); Richard A. Jarrell, "The Department of Science and Art and the Control of Irish Science, 1853-1905," *Irish Historical Studies* (1983) 23: 330-47, W. Garrett Scaife, "Technical Education and the Application of Technology in Ireland, 1800-1950," pp. 85-100 in P. J. Bowler and N. Whyte (eds.), *Science and Society in Ireland: The Social Context of Science and Technology in Ireland, 1800-1950* (Belfast: The Institute of Irish Studies-Queen's University of Belfast, 1997), Richard A. Jarrell, "Some Aspects of the Evolution of Agricultural and Technical Education in Ireland, 1800-1950," pp. 101-17 in Bowler and Whyte (eds.), *Science and Society in Ireland*, Richard Jarrell, "Technical Education and Colonialism in 19th Century Ireland," pp. 170-87 in N. McMillan (ed.) *Prometheus's Fire: A History of Scientific and Technological Education in Ireland* (Carlow: Tyndall Press, 2000).

empire linked by the new technologies. In this context few could sustain William Rowan Hamilton's proud disdain for technology and his belief in metaphysics as the salvation of humanity. The members of Trinity College Dublin believed it was their duty and their privilege to take part in the imperial project, and they sought ways to leverage their reputation for scientific research to extend their influence.

Trinity's first attempt to bring science to the service of empire involved the so-called Magnetic Crusade, a global project to map the earth's magnetic field in the 1830s and 1840s.[9] Humphrey Lloyd, Professor of Natural and Experimental Philosophy, played a crucial role in the project, as did Edward Sabine and Francis Beaufort, two Irish military men. They brought together scientists, the military, and the British government in the largest and most expensive scientific endeavor of its time.

Since the sixteenth century, navigators had known that magnetic north is not identical to true north. In fact, the position of magnetic north differs from place to place and even changes over time. Clearly, such changes in magnetic declination were of interest to navigators dependent on magnetic compasses. Some even thought that a better understanding of declination might lead to a method to determine longitude at sea. In 1829 Alexander von Humboldt established a Magnetic Observatory in Berlin to make geomagnetic measurements. Gauss and Weber began similar work at Göttingen in the late 1830s and soon established a global network of magnetic observatories.[10] Lloyd first became involved in 1834 when he began a series of magnetic surveys of Ireland, Scotland, England, and Wales.[11] Then in 1837 Lloyd proposed the construction of a per-

---

[9]Christopher Carter, "Magnetic Fever: Global Imperialism and Empiricism in the Nineteenth Century," *Transactions of the American Philosophical Society* (2009) 99: i-168.

[10]John Cawood, "Terrestrial Magnetism and the Development of International Collaboration in the Early Nineteenth Century," *Annals of Science* (1977) 34: 551-87.

[11]James G. O'Hara, *Humphrey Lloyd (1800-1881) and the Dublin Mathematical School of the Nineteenth Century* (PhD Thesis: University of Manchester, 1979), James G. O'Hara,

manent magnetic observatory at Trinity College to make precision geomagnetic measurements.[12]

Terrestrial magnetism combined the interests of scientists who wanted to better understand the composition of the earth and the military who were concerned with improvements to navigation. The British Association and the Royal Society both lobbied the government to support an ambitious scheme of observatories. Lloyd's primary collaborator was Edward Sabine, an officer in the Royal Artillery with an interest in science. Sabine was also a fellow of the Royal Society and would later become its president.[13] Another critical player was Francis Beaufort, hydrographer to the Royal Navy and so responsible for mapping the oceans.[14] Born in Ireland, Beaufort had spent five months studying with Brinkley at Dunsink Observatory before joining the Navy. His interest in meteorology and surveying led him to join the Royal Society, the Royal Astronomical Society, and the Royal Geographical Society, though he is perhaps most famous for recommending Charles Darwin for the position of naturalist on the *Beagle*. Beaufort married a sister of Maria Edgeworth, and his son Francis Beaufort Edgeworth was a close friend of William Rowan Hamilton. These three Irishmen, Lloyd, Sabine, and Beaufort, spearheaded the project that came to be known as the Magnetic Crusade because of its enormous scope. In 1839 the government agreed to establish four observatories in Toronto, St. Helena, the Cape of Good Hope, and Van Diemen's Land, while the East India Company established ob-

---

"Humphrey Lloyd: Ambassador of Irish Science and Technology," pp. 124-39 in J. Nudds, *Science in Ireland*, T.D. Spearman, "Humphrey Lloyd, 1800-1881," *Hermathena* (1981) 130: 37-52.

[12]H. Lloyd, *Account of the Magnetical Observatory of Dublin and of the Instruments and Methods of Observation Employed There* (Dublin, 1842).

[13]Paul Hackney, "Edward Sabine," pp. 331-36 in J. W. Foster and H. C. G. Chesney (eds.), *Nature in Ireland: A Scientific and Cultural History* (Dublin: Lilliput, 1997).

[14]Sheila Landy, "Francis Beaufort," pp. 327-30 in Foster and Chesney (eds.), *Nature in Ireland*.

servatories in Simla, Madras, Singapore, and Bombay.[15] All were
outfitted with instruments designed by Lloyd, and Lloyd trained
their directors (mostly naval officers) at the magnetic observatory
in Dublin. Lloyd's work literally put Dublin on the map, making
it an important site in a global network of observatories designed
to support Britain's sea power.

Lloyd's involvement with the Magnetic Crusade demonstrated
the new possibilities for science in an age of empire. Such considerations may well have been on his mind when in April of 1841 he
proposed the creation of an engineering school at Trinity. While
the education of engineers in a university may seem natural today,
it was a radical proposal in Ireland in 1841. British engineers were
trained through apprenticeship, while university students went into
traditional professions such as the church, medicine, and law. Most
saw the two as antithetical. Engineers were not the type to study
the classics, and university men were not the type to get their hands
dirty working with machines. Trinity (like Oxford and Cambridge)
had built its identity around the promotion of pure truth without regard to material gain. Even the proposal that the university
might teach experimental science would have been considered radical at the time. Only mathematical science reached the standards
of proof the College believed essential for the teaching of students.

Engineering, however, was slowly and fitfully making its way
into the new English universities. In 1827 the University of London
created a chair of engineering, though the first incumbent soon left
for more lucrative work and was not replaced until 1841. In 1838
King's College London and the University of Durham set up engineering courses, and in 1840 Glasgow University received a Regius
Chair of Civil Engineering. It was the new universities, located in
industrial centers and catering to middle class and professional students rather than to the traditional liberal arts students, that took

---

[15]John Cawood, "The Magnetic Crusade," *Isis* (1979) 70: 493-518.

the first steps to incorporate engineering. As Lloyd noted, however, on the continent the situation was quite different. In France and Germany, for example, the government established technical schools, fearing that they had already fallen behind England's industrial progress.

But while the older English universities ignored the new trend, Trinity saw an opportunity. A certain amount of political pressure was involved as well. In 1838 a Report of the Select Committee of the House of Commons on Education in Ireland recommended the creation of four non-denominational provincial colleges (the Queen's Colleges in Belfast, Cork, and Galway that would be established in 1845) as well as a Central Polytechnic Institute in Dublin. For the first time in its history Trinity was about to lose its monopoly on education in Ireland. It needed to find a new way to maintain its special status. Moreover, by the 1840s engineering was on its way to becoming a respectable profession. As the church declined as an outlet for students, new careers needed to be found.

Trinity College Dublin was the first of the older universities to attempt to combine engineering education with liberal education. Lloyd and his colleagues had to forge new arguments to reconcile these two divergent traditions. In their proposal they argued for the importance of training engineers in universities, "on account of the close connection of the knowledge which they require with the science already taught within the walls of these institutions."[16] Engineering, they claimed, requires science, and as the home of science, the university is the best place to train engineers. "The subjects requisite to be taught in such a school," the proposal continued, "are the *principles* of Mathematics, Mechanics, Chemistry and Geology and the *application* of these principles to the Arts of Construction, practical Engineering and Architecture."[17] While most British engineers believed that they applied their experience,

---

[16]R.C. Cox, *Engineering at Trinity* (Dublin: School of Engineering, 1993), p. 124.
[17]Cox, *Engineering at Trinity*, p. 124.

the Trinity fellows argued that engineers in fact apply scientific principles. Trinity could therefore staff an engineering school with its current professors of natural philosophy and mathematics. They required only two new professors, one in chemistry and geology and another in civil engineering. Trinity's engineering students would have to attend the first two years of the arts course and were encouraged to complete the entire course. Mathematics, logic, Greek, and Latin were held to be just as essential to engineers as to clergymen.

While controversial in the broader context, the proposal appealed greatly to the Board of Trinity College. In less than three months, Trinity was advertising for the new professors, and Lloyd was planning a trip to France to investigate French engineering education. The school opened in November with an address by Lloyd justifying the role of engineering in a university.[18] While Lloyd pointed to the precedent of the continental engineering schools and the new courses at King's College and Durham, he could not disguise the fact that Trinity was embarking on a radical new departure. Recall that Lloyd's own father had warned, "We do not wish the great University of Dublin, founded chiefly to promote the all-important study of Theology, to degenerate into a mere École Polytechnique...."[19]

Lloyd's father had done more than anyone in the history of the College to prove that Trinity could promote the progress of science. He had forged a new role for Trinity in the 1830s and in the process saved it from intrusive attempts at reform. Now Humphrey Lloyd charted a new course for the 1840s. It was no longer enough for Lloyd to gesture towards the university's ability to promote science,

---

[18] H. Lloyd, *Praelection on the Studies Connected with the School of Engineering. Delivered on the Occasion of the Opening of the School, November 15, 1841* (Dublin, 1841).

[19] Charles Parsons Reichel and William Anderson, *Trinity College Dublin and University Reform* (Dublin 1858), p. 38.

he now argued that the university could also promote "applications of science to the wants and to the uses of man."[20]

In his speech, Lloyd extolled the practical benefits of science, but at the same time he maintained the college's traditional belief in its intellectual and moral benefits. He cautioned students that the introduction of an engineering school was not intended

> as a concession to that unphilosophical, but unhappily too popular feeling, which decries as worthless every intellectual pursuit, if unaccompanied by results of a practical nature,—which can discern no value in speculation, unless it conducts us to conclusions bearing immediately upon the uses and the necessities of daily life.[21]

The practical application of science, he explained, does not mean that science exists merely to be applied. Lloyd reiterated the university's traditional justifications for science, praising "the high and disinterested pleasures that attend the discovery of the links in the chain of *a priori* knowledge."[22] It was crucial for Lloyd to demonstrate the harmony and interconnectedness of these two divergent views of science—the ultimate agreement between Hamilton's vision of science as the pursuit of truth and beauty with Petty's view of science as the pursuit of wealth and power. Lloyd explained,

> I think there can hardly be a doubt, that where the practical *applications* of science are concerned, the *sciences* themselves must be systematically taught. ... I am sure, above all, that it is in the universities, the established schools of science, that such applications may be best unfolded.[23]

---

[20] Lloyd, *Praelection*, p. 5.
[21] Lloyd, *Praelection*, p. 6.
[22] Lloyd, *Praelection*, pp. 6-7.
[23] Lloyd, *Praelection*, p. 22.

Engineering belongs in the university, he argued, because it is essentially the application of science.

By 1841 engineering had also come to play an important role in British society and the British Empire. Britain's preeminent position in the world, for example, depended her mastery of the seas. Lloyd explained,

> She carries the products of her industry across the ocean, to exchange them for the riches of distant nations. She spreads the terror of her arms to the remotest regions of the globe. She peoples the most distant shores with her colonies; and gives birth to nations affiliated to her by the ties of a common blood, a common language, and a common interest. She diffuses in every quarter her science and her arts,—the gifts of civilization, and the blessings of religion.[24]

Science, in Lloyd's view, links technology, industry, imperialism, religion, and civilization itself. Applied science, Lloyd claimed, has effected a moral revolution "in delivering man from the thraldom of brute toil, and leaving him more free to cultivate the intellectual part of his nature."[25] While some critics saw industrialization and imperialism in rather darker terms, Lloyd expressed an unremitting optimism about the power of science and technology to improve the world.

By linking science, industry, and the university in this way, Lloyd could meet the common objections that abstract science is useless and that the traditional university has no role in industrial society, while at the same time he could raise the status of engineering as a profession and find a new outlet for Trinity students. Lloyd concluded, "The profession you have adopted... has now risen

---

[24]Lloyd, *Praelection*, p. 9.
[25]Lloyd, *Praelection*, p. 16.

to take its rank among the first of the liberal professions." [26] Trinity
graduates had traditionally taken responsibility for the soul of the
nation as clergymen. Now they would take responsibility for its
body as well.

The crucial role of the imperial civil service as a career for Trinity
engineers demonstrates that engineering at Trinity was at least
as much about empire as industry. The Ascendancy was deeply
committed to the Union and prided themselves on the power of the
British Empire. They also took seriously their duties as members
of that empire. James Booth, a Trinity mathematics graduate,
clergyman, and proponent of technical education, [27] explained,

> The vast regions of India and Central Asia, the trackless
> forests of Canada and the barren steppes of Australia
> are yet to be scored with railways, while the earth's globe
> itself must be interlaced with the invisible bands, and the
> network of the electric telegraph. Canals are to be dug,
> deserts to be irrigated, forests cut down, savage regions
> to be opened up and made accessible to civilization and
> Christianity. [28]

"We must supply," he concluded, "if not the manual labour, the
heads to guide, the science to suggest, and the capital to provide
for their execution." Engineering at Trinity was not about making money, nor was it primarily about improving technology. At
its foundation, engineering was an expression of imperial values.
Trinity would provide 'the heads to guide' the expansion of British
civilization and Anglican religion throughout the world.

Trinity engineers would not be mechanics. They would be civil
engineers, trained to plan and manage large public works. In fact

---

[26] Lloyd, *Praelection*, p. 32.

[27] Frank Foden, "James Booth: The Father of Technical Examinations," pp. 367-75 in N.
McMillan (ed.) *Prometheus's Fire*.

[28] James Booth, *How to Learn and What to Learn: Two Lectures Advocating the System
of Examinations Established by the Society of Arts* (London, 1856), p. 42.

few of them were ever employed in private industry; the vast majority spent their lives in the imperial civil service. Trinity engineers became Chairman of the Madras Port Trust, Chief Engineer of Sydney Harbor, City Engineer of Bangkok, Director of Works for the Egyptian Government, and Chief Engineer of Calcutta.[29] A number became professors of engineering at the new universities such as the Queen's Colleges and University College London. As engineers became powerful figures in the British Empire, Trinity claimed its right to train them.

While Trinity's new departure drew the support of the Board, support from students and the civil service was much slower in coming. Trinity was willing to train engineers according to scientific principles, but few people were willing to hire them. Graduates of the engineering course were charged the same premium to enter apprenticeships as those without an engineering diploma. Consequently student numbers remained low. Two diplomas were awarded in 1843 and in 1844, six in 1845, and two in 1846.

Trinity found it easier to convince the government to hire its graduates than industry, but even government employment was a struggle. Despite their strong success on the civil service examinations (see chapter 3), Trinity graduates faced a number of obstacles. The Royal Military Academy at Woolwich refused to accept any of the first candidates that Trinity offered, complaining that they were too inexperienced. In 1859 the Indian Civil Service threatened to require three year's apprenticeship for all of its new recruits, but after some skillful lobbying by Trinity fellows, it agreed to accept Trinity's graduates. In 1871, however, the Indian Public Works Department opened its own Engineering College at Cooper's Hill and would accept only its own graduates. Even in the 1880s George Francis Fitzgerald, as registrar of the engineering school, continued

---

[29] *School of Engineering, Trinity College, Dublin: A Record of Past and Present Students* (Dublin, 1909), John Purser, "A Note on the School of Engineering Since Its Foundation," *Hermathena* (1941) 58: 53-6, Cox, *Engineering at Trinity.*

to fight for official recognition for Trinity engineers.[30] Engineering
schools should be recognized as at least part of the training for engineering civil service appointments, he pleaded. Even by the end
of the nineteenth century, however, the belief in the essential relationship between science, technology, and engineering education
was far from widespread.

7.2   The Science of Technology

While Lloyd described technology as the application of science, he
did not mean simply the application of Trinity's traditional science. By the mid-nineteenth century, industrialization was transforming science as well as society. Newtonian mechanics might lie
behind the motions of machines in principle, but the mathematical methods of particles and forces were not particularly useful
even for idealized industrial problems. The new sciences of heat,
electricity, and magnetism rose in importance throughout the nineteenth century as steam engines and telegraphs came to dominate
the landscape. Scientists developed new concepts like energy and
fields to understand the new technologies and in the process formed
a new view of nature's most basic processes. Science and engineering came to be related because each changed to suit the other.
Throughout the industrializing world, university scientists began
to approach industrial problems and even to collaborate with industry. In Ireland, however, there was little industry, and Trinity
saw its role more in terms of running the empire than building new
technologies. Consequently, the trends that were reshaping physics
only slowly made their way into Trinity's curriculum.

The French engineers of the École Polytechnique were among
the first to develop a new form of physics suited to the description

---

[30]George Francis Fitzgerald, "Engineering Schools," pp. 224-28 in J. Larmor (ed.) *The
Scientific Writings of George Francis Fitzgerald* (Dublin: Hodges, Figgis and Co., 1902).
[Originally published in *Nature* (August 2, 1888).]

of machines. In the first decades of the nineteenth century, they found the old concept of force less useful than that of work, the exertion of a force through a given distance.[31] Work could be used to measure the output and efficiency of a machine, and it had a close connection to economic value. It measured how much a machine was worth. Calculating such equivalences became a central feature of the new physics. How much weight can a given amount of coal lift? How much heat can a weight falling through a given distance generate? How much heat can a given amount of current generate? These questions had practical importance for industry and deep theoretical importance for science. The study of such conversion processes was essential to the creation of the concept of energy.[32] Each process came to be seen as the conversion of energy from one form to another. William Thomson (later known as Lord Kelvin), an Irishman educated at the Royal Belfast Academical Institution and Cambridge University, developed the science of thermodynamics and rewrote dynamics as the science of energy through his study of the steam engine.[33] Thomson's analytical dynamics utilized the Hamiltonian function, but Hamilton himself admitted that he could not understand the new concept of work. In 1863, he wrote to Tait,

> The world of Science seems to admit that I had not only read but written to some purpose on *Dynamics* about *thirty years ago*; but a new generation has arisen. *Energy* and *work*, in the *old* English meaning, are things not

---

[31]Ivor Grattan-Guinness, "Work for the Workers: Advances in Engineering Mechanics and Instruction in France, 1800-1830," *Annals of Science* (1984) 41: 1-33.

[32]Thomas S. Kuhn, "Energy Conservation as an Example of Simultaneous Discovery," pp. 66-104 in T. S. Kuhn (ed.) *The Essential Tension: Selected Studies in Scientific Tradition and Change* (Chicago: University of Chicago Press, 1977).

[33]M. Norton Wise, "Mediating Machines," *Science in Context* (1988) 2: 81-117, Crosbie Smith and M. Norton Wise, *Energy and Empire: A Biographical Study of Lord Kelvin* (Cambridge: Cambridge University Press, 1989), Crosbie Smith, *The Science of Energy: A Cultural History of Energy Physics in Victorian Britain* (Chicago: University of Chicago Press, 1998).

unfamiliar to me. But I have only the dimmest views of
the modern meanings attached to those terms.[34]

Thomson transformed Hamilton's dynamics from an abstract
mathematical principle into a description of the flow of energy.

At Trinity College Dublin, however, science still meant mathematical science, and topics such as heat and electricity were not
considered scientific. The textbooks of the 1820s and 1830s had
taught students the calculus and its application to mechanics. Optics was introduced when it could be taught as an analytical science.
In 1899, George Salmon, one of Trinity's great mathematicians and
Provost at the end of the nineteenth century, recalled, "In my time
[science] meant pure and applied mathematics. We did not call
chemistry Science, until by many formulae they seemed to place
themselves within the pale."[35] In 1849, however, Trinity introduced
a fourth moderatorship in experimental science. (Cambridge introduced a Natural Science Tripos in 1851.[36]) Robert Vickers Dixon,
Lloyd's successor as Erasmus Smith Professor of Natural and Experimental Philosophy published *A Treatise on Heat* that same
year. In the preface he explained,

> [T]he leading principles of the more important of the
> Physical Sciences are now established by such evidence
> as entitles them to rank, in point of certainty, with the
> deductions of the Abstract Sciences, and to justify their
> being referred to as the authentic results of a correct sys-
> tem of physical investigation.[37]

Unlike earlier Trinity textbooks, Dixon's *Treatise* focused primarily
on practical experimental methods and the reduction of data. His

---

[34]W.R. Hamilton to P.G. Tait (August 28, 1863) in Robert P. Graves, *Life of Sir William
Rowan Hamilton* (Dublin: Hodges, Figgis, 1882-89), 3: 150.

[35]George Salmon to J.H. Bernard (August 8, 1899), TCDMS 2384.

[36]D.B. Wilson, "Experimentalists Among the Mathematicians: Physics in the Cambridge
Natural Sciences Tripos, 1851-1900," *Historical Studies in the Physical Sciences* (1982) 12:
325-71.

[37]Robert V. Dixon, *A Treatise on Heat* (Dublin, 1849), p. viii.

was a very different view of science than Brinkley's astronomy or Bartholomew Lloyd's analytical geometry. Even Humphrey Lloyd's experimental work in optics and geomagnetism emphasized the reduction of science to analytical theory. Despite Dixon's textbook, however, the theory of heat failed to become a research interest at Trinity, nor was it the focus of teaching (apart from the engineering students).

While the steam engine was central to the development of the science of energy, the telegraph played a crucial role in the development of the field theory of electricity and magnetism.[38] Scientists attempted to apply their theories to the practical needs of telegraphers, and in the process, they not only remade their theories but also remade the role of the scientist. Scientists constructed teaching laboratories for the first time, instructing students in practical methods of precision measurement. They began to consult for the telegraph industry and to design commercial instruments. In the process they challenged the authority of the traditionally-trained engineers and asserted the necessity of theory in engineering practice.

Academic scientists at first had little to do with the early history of the telegraph; sending signals over a wire required practical rather than theoretical knowledge. Around 1848, however, the introduction of the insulator gutta percha made underground and submarine cables possible, and in 1851 the first submarine cable was laid from Dover to Calais. Telegraphists, however, soon found that signals over such cables were distorted. A sharp signal was received as a blur, dots could not be distinguished from dashes, and it eventually became impossible to send any signal at all. Michael Faraday, the originator of field theory, and William Thomson studied the problem in the early 1850s. They found that insulated cable acts like a large capacitor, gradually accumulating a charge until

---

[38]M. Norton Wise, "Mediating Machines," Smith and Wise, *Energy and Empire*, Hunt, *The Maxwellians*.

no more signals can be sent. Thomson found mathematically that the retardation should increase with the square of the length. His calculations implied that it would be impossible to lay a workable cable across the Atlantic Ocean.

Cyrus Field, however, had already begun to promote just such a transatlantic cable in 1854. E.O. Wildman Whitehouse, official electrician to Field's Atlantic Telegraph Company, denounced Thomson's law of squares as a 'fiction of the schools.'[39] He argued on the basis of his experience that the cable should be thin, while Thomson argued on the basis of theory that it should be thick. Thomson was appointed to the Board of Directors, however, and both he and Faraday eventually acquiesced in the thin cable. The cable ran 2,350 miles from Nova Scotia to Valentia Island, off the west coast of Ireland. It was connected on August 5, 1858, and within a few weeks the enormously expensive cable had failed completely. When the British government's cable to India failed two years later, the government and the Atlantic Telegraph Company undertook a joint investigation of the failures. In the end, they endorsed Thomson's law of squares, and Thomson developed extremely sensitive signaling and receiving instruments that would allow the use of smaller currents. On his second attempt, Field followed Thomson's advice and the cable was successful. By the mid-1870s Britain had telegraph lines connecting its vast empire, stretching to India, Australia, and the Far East. Thomson was knighted, and his instruments were used on many of the cables. The Atlantic cable appeared to prove that where experience had failed, science could succeed.

---

[39]Wildman Whitehouse, "The Law of Squares—Is It Applicable or Not to the Transmission of Signals in Submarine Circuits," *British Association Report* (1856), pp. 21-23.

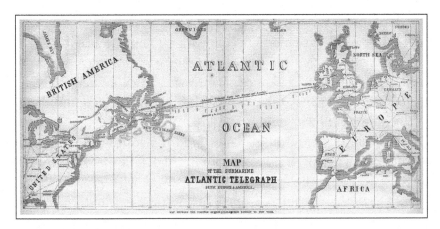

Figure 7.2: The Atlantic Telegraph (1858) stretched from Newfoundland to Ireland

As part of his cable work, Thomson started a laboratory at Glasgow University in the 1850s devoted to electrical measurements. While university professors in the sciences had long had 'physical cabinets' of instruments for lecture demonstrations and personal research, the scope of Thomson's laboratory and its use in teaching practical techniques to students was novel. Following Thomson's example, the 1860s and 1870s witnessed a 'laboratory revolution' in Britain with new laboratories constructed at University College London (1866), Oxford University (1866), Edinburgh University (1868), and Cambridge (1874). In 1885, Thomson wrote, "No University can now live unless it has a well-equipped laboratory."[40] Trinity College Dublin, however, did not receive its physical laboratory until 1905, though Fitzgerald managed to outfit a makeshift laboratory in the 1890s.

The central mission of these laboratories was precision measurement, particularly in electrical science. In fact it was due to the need for laborious observations that Thomson first invited students

[40]Cited in Graeme Gooday, "Precision Measurement and the Genesis of Physics Teaching Laboratories in Victorian Britain," *The British Journal for the History of Science* (1990) 23: 25-51, p. 42.

to work in his private laboratory, which soon expanded to become a central part of his teaching. In his work on the Atlantic telegraph, Thomson realized that the success of the telegraph depended on a standardized unit of resistance, the testing of cable materials, and sensitive signaling instruments. If the resistance per unit length of the cable is known, it is possible to locate any cable break by balancing the potential across a standard resistance against that of the unknown length of cable up to the fault. Not only is it necessary to know the standard of resistance with great accuracy, however, but the cable itself must be of standard material. Precision measurement and accurate electrical standards were crucial to the success of submarine telegraphy.

While Thomson's interest in commercial and industrial work was welcomed at Glasgow, and while the new redbrick universities were often designed with rooms for laboratory work, such industrial values were difficult to integrate into the traditional universities. In Cambridge, Maxwell had to work hard to make precision measurement look like a moral enterprise rather than a commercial one.[41] Maxwell emphasized the self-discipline required by the precision experimenter as well as values of imperialism. The integrity of the empire depended on the integrity of the telegraph network, which in turn depended on the integrity of the measurements made at the Cavendish laboratory (which Maxwell founded at Cambridge) and the integrity of the young men who made them. Electrical measurement also intersected with Maxwell's theoretical interests as the measurement of the ratio of electrostatic to electromagnetic units, crucial to telegraphy, was also crucial to his electromagnetic theory of light as he predicted that it would be equal to the speed of light.

---

[41]Simon Schaffer, "Late Victorian Metrology and Its Instrumentation: A Manufactory of Ohms," pp. 23-56 in S. Cozzens and R. Bud (eds.), *Invisible Connections* (Bellingham: SPIE Optical Engineering Press, 1992), p. 23.

A similar transformation occurred in German physics from about 1865 to 1914.[42] Beginning in the 1830s and 1840s, Weber and Gauss became interested in precision measurement through the magnetic crusade and established private laboratories. Around the same time Neumann and Jacobi created the first physics seminar in Königsberg with an emphasis on precision measurement and the theory of measurement.[43] But it was not until the 1860s and particularly after unification that the German government (now interested in the support of science-based industry) began to fund the construction of physics institutes with full laboratories and staff. It was the industrialist Siemens who provided the impetus for the establishment of the Physikalisch-Technische Reichsanstalt in 1887, the first of the national standards laboratories and the model for the National Physical Laboratory in Britain and the National Bureau of Standards in the United States.[44]

Scientists, governments, and industry all began to promote the importance of precision measurement as large technology-based companies such as General Electric, Dupont, Dow, Eastman Kodak, and Bell Telephone all established research laboratories in the first two decades of the twentieth century.[45] In less than fifty years, the science laboratory had become a key site for the application of scientific techniques to technological and industrial problems.

In Ireland, however, there were no large-scale technological industries (with the possible exception of Guinness), and Trinity

[42]David Cahan, "The Institutional Revolution in German Physics, 1865-1914," *Historical Studies in the Physical and Biological Sciences* (1985) 15: 1-65.

[43]Kathryn M. Olesko, *Physics as a Calling: Discipline and Practice in the Königsberg Seminar for Physics* (Ithaca: Cornell University Press, 1991).

[44]David Cahan, *An Institute for an Empire: The Physikalisch-Technische Reichsanstalt, 1871-1918* (Cambridge: Cambridge University Press, 1989).

[45]George Wise, "A New Role for Professional Scientists in Industry: Industrial Research at GE, 1900-16," *Technology and Culture* (1980) 21: 408-29, Lillian Hoddeson, "The Emergence of Basic Research in the Bell Telephone System, 1875-1915," *Technology and Culture* (1981) 22: 512-44, L. Reich, *The Making of American Industrial Research: Science and Business at GE and Bell, 1876-1926* (Cambridge: Cambridge University Press, 1985), Paul Israel, *From Machine Shop to Industrial Laboratory: Telegraphy and the Changing Context of American Invention, 1830-1920* (Baltimore: Johns Hopkins University Press, 1992).

College trained students for the civil service rather than industry. Trinity did not build a physical laboratory until 1905, long after most other British universities had new facilities.[46] Despite the interests of Lloyd and Fitzgerald, there was no community of industrialists with an appreciation for the role that science might play in industry. Lloyd had succeeded in establishing a magnetic observatory where students were employed to make precision measurements, but the observatory does not seem to have become a site for formal teaching. Fitzgerald was a friend of Thomson and an important promoter of the connection between Maxwell's field theory and practical electrical engineering, but he rarely worked with industry or created patentable inventions.

In 1899 Trinity's Board appointed a small committee to consider adding electrical and mechanical engineering to the engineering course, which soon led to the construction of a small teaching laboratory. While the Professor of Civil Engineering was opposed, Anthony Traill, the future provost (who had constructed the first hydroelectric passenger tramway in the British Isles in Antrim), and Fitzgerald of course, were both committed to electrical engineering. This led to another committee with a broader mandate to report on Trinity's needs in science. The new Science Committee reported in 1899 that a minimum of £70,000 for capital improvements and £5,000 annually was required to support advanced science. The College's total annual income, however, was only £80,000, and the Board asked the committee to submit a more economical report. Fitzgerald was livid and replied that at least £250,000 would be necessary to ensure that Trinity remain at the forefront of science. Fitzgerald, however, died in 1901. Ultimately, Edward Cecil Guinness (Lord Iveagh), Chairman of the Guinness

---

[46]R.B. McDowell and D.A. Webb, *Trinity College Dublin, 1592-1952: An Academic History* (Cambridge: Cambridge University Press, 1982), pp. 405-11, J.V. Luce, *Trinity College Dublin: The First 400 Years* (Dublin: Trinity College Dublin Press, 1992), pp. 126-28, Nicholas Whyte, *Science, Colonialism and Ireland* (Cork: Cork University Press, 1999), pp. 46-50.

Brewery, offered to cover the capital costs. Apart from Guinness, however, Trinity lacked the wealthy industrialists who contributed to other university laboratories. It also lacked the science-based electrical and chemical industries that promoted such research in the United States and Germany. The first advanced laboratory in Ireland was one at the Royal College of Science for Ireland, established by the British government in 1876.[47] Despite the work of Lloyd and Fitzgerald, Trinity lacked the network necessary to support the connection of science, industry, and the university.

Figure 7.3: The Physical Laboratory, Trinity College Dublin, Constructed in 1906 largely due to the efforts of George Francis Fitzgerald. ©Trinity College Dublin

[47]B.B. Kelham, "The Royal College of Science for Ireland (1867-1926)," *Studies* (1967) 56: 297-309.

7.3   The Maxwellians

While Trinity lagged in the development of laboratory facilities, this did not prevent Fitzgerald from working at the forefront of research in electromagnetism. He utilized his superior knowledge of mathematical theory and managed to secure some space in the Engineering School for simple experiments. Fitzgerald began by transforming Maxwell's theory of electromagnetic fields into a developed wave theory, able to explain optics as well as electricity and magnetism. Then Fitzgerald and others attempted to develop models of the ether that would explain Maxwell's abstract theory. At the same time, they searched for a method to produce and detect electromagnetic waves, viewing such waves as proof of the validity of Maxwell's theory. Finally, in the last decade of the century, electromagnetic theory led to questions about the inner workings of the atom and the relationship between ether and matter.[48]

James Clerk Maxwell's 1873 *Treatise on Electricity and Magnetism* did for electromagnetism what Newton's *Principia* had done for classical mechanics.[49] Maxwell transformed Faraday's qualitative field theory of electromagnetism into a sophisticated mathematical theory based on the hypothesis that electromagnetic effects are due to disturbances in an ether rather than forces acting at a distance. Perhaps most strikingly, Maxwell argued that light itself is an electromagnetic wave in the ether. Maxwell died in 1879, and his text was notoriously confusing. Nonetheless a small group of scientists known as the Maxwellians took up his ideas and forged them into a coherent theory.

George Francis Fitzgerald was a leading figure in this movement. In fact, historian Bruce Hunt describes him as "the soul of

---

[48] Jed Z. Buchwald, *From Maxwell to Microphysics: Aspects of Electromagnetic Theory in the Last Quarter of the Nineteenth Century* (Chicago: University of Chicago Press, 1985).

[49] Foreword to Hunt, *The Maxwellians*, p. ix.

the Maxwellian group."[50] From his first encounter with Maxwell's theory in the 1870s, Fitzgerald devoted himself to the mathematical and experimental development of the theory. Fitzgerald was very much a product of the Trinity College Dublin tradition. His father was Professor of Moral Philosophy at Trinity College and later Bishop of Cork. His mother was the sister of George Johnstone Stoney, a Trinity-educated scientist, and of Bindon Blood Stoney, a Trinity-educated engineer. As a child, Fitzgerald was tutored by the sister of George Boole, and he later married the daughter of John Hewitt Jellett, Trinity provost and mathematician. It was almost as if Fitzgerald was born to be a Trinity scientist. His brother Maurice taught engineering at Queen's College Belfast, and Fitzgerald, more than any other figure, combined Trinity's mathematical traditions with its new engineering interests. Fitzgerald graduated from Trinity in 1871 as first senior moderator in both mathematics and experimental science and immediately began reading for the fellowship examination. He won on his second attempt in 1877 and was appointed professor of natural and experimental philosophy and registrar of the engineering school in 1881.

---

[50]Hunt, *The Maxwellians*, p. 243.

Figure 7.4: George Francis Fitzgerald (1851–1901), Erasmus Smith
Professor of Natural and Experimental Philosophy, Registrar of the
Engineering School, and Pioneer of Electromagnetic Theory.
©Trinity College Dublin

Among the works that Fitzgerald studied for the fellowship examination were MacCullagh's papers on optics. Though ignored
outside of Trinity, they were still considered classics in the College.
Fitzgerald also began a study of Maxwell's *Treatise*, and by 1878
he realized that he could combine the two in an interesting way.
He found that by rewriting Maxwell's expressions for electrostatic
and electrokinetic energy in the same form as MacCullagh's expressions for the potential and kinetic energy of the ether, he could
apply Maxwell's theory to the reflection and refraction of electro-

magnetic waves.[51] Maxwell's theory, he demonstrated, could explain all ordinary optical phenomena at least as well as the old elastic solid theories of the ether and could also accommodate new electromagnetic phenomena like the Kerr and Hall effects.

Between 1879 and 1883, Fitzgerald began to ponder how to generate such electromagnetic waves together with Oliver Lodge, professor of physics at the University of Liverpool. Fitzgerald and Lodge shared their passion for Maxwell's theory with Oliver Heaviside, a former telegraph operator and self-taught mathematician who did most of his work while living at home with his parents.[52] At a time when even major figures like William Thomson discounted Maxwell's theory, these three forged it into a powerful theoretical and practical tool. Then in 1888 the German physicist Heinrich Hertz successfully produced and detected electromagnetic waves, demonstrating that they travel at the same velocity as light and undergo reflection, refraction, and diffraction, just as Maxwell's theory predicted. Hertz's experiments, Fitzgerald believed, had finally decided the great question of whether electricity and magnetism are forces that act at a distance or disturbances in an ether that pervades all of space. Fitzgerald himself never doubted Maxwell's theory or the existence of the ether, but Hertz's experiments appeared to provide the first unassailable proof.

Like the wave theorists of the nineteenth century, Maxwell based his theory on the existence of an ether with strange properties. Maxwell and his followers believed that the explanation of all optical and electromagnetic phenomena depended on understanding the complicated structure of the ether. Continuing in the tradition of Bryan Robinson, Fitzgerald believed that the ether held the solu-

---

[51] George Francis Fitzgerald, "On the Electromagnetic Theory of the Reflection and Refraction of Light," pp. 45-74 in *Scientific Writings*. [Originally published in *Philosophical Transactions of the Royal Society* (1880)]

[52] Ido Yavetz, *From Obscurity to Enigma: The Work of Oliver Heaviside, 1872-1891* (Boston: Birkhauser, 1995).

tion to all problems in physics, along with important metaphysical implications.

In order to better visualize the strange properties of the electromagnetic ether, Maxwell, Lodge, and Fitzgerald all constructed mechanical models. Fitzgerald, for example, actually built a model with rows of brass wheels on a mahogany board. Each wheel was connected to its four neighbors by rubber bands.[53] The spinning of the wheels represents the magnetic field, and the rubber bands represent the polarization in the field when some wheels turn at a different speed than others. Fitzgerald's model allowed him to better visualize what happens during, say, the charging of a condenser, by thinking in terms of the mechanical model. While mechanical models were useful as illustrations of the relationships between various elements of the theory, though, no one took them seriously as actual representations of the ether. Fitzgerald, however, also strove to understand the real structure of the ether.[54] Thomson and Helmholtz had both shown earlier in the century that atoms could be imagined as vortices in the ether. They had even demonstrated mathematically that such vortices would act like particles, retaining their identity even after colliding with other vortices. Fitzgerald was particularly attracted to this view because it reduced all natural phenomena to motions in the ether, and he offered his own 'vortex-sponge' model of the ether in 1885.

For Fitzgerald this view ultimately had a metaphysical significance. Fitzgerald ended an 1890 lecture at the Royal Institution with a description of the vortex theory. He concluded,

---

[53] Bruce Hunt, "'How My Model Was Right': G.F. Fitzgerald and the Reform of Maxwell's Theory," pp. 299-321 in R. Kargon and P. Achinstein (eds.), *Kelvin's Baltimore Lectures and Modern Theoretical Physics: Historical and Philosophical Perspectives* (Cambridge: MIT Press, 1987), Hunt, *The Maxwellians*, pp. 78-84.

[54] Howard Stein, "'Subtler Forms of Matter' in the Period Following Maxwell," pp. 309-40 in G. N. Cantor and M. J. S. Hodge (eds.), *Conceptions of Ether: Studies in the History of Ether Theories, 1740-1900* (Cambridge: Cambridge University Press, 1981), pp. 312-20.

This hypothesis explains the differences in Nature as differences of motion. If it be true, ether, matter, gold, air,
wood, brains are but different motions. Where alone can
we know what motion in itself is, that is, in our own
brains, we know nothing but thought. Can we resist the
conclusion that all motion is thought? Not that contradiction in terms, unconscious thought, but living thought;
that all Nature is the language of One in whom we live,
and move, and have our being.[55]

Matter, he argued, is merely a form of motion in the ether, and motion is a form of thought. Like William Rowan Hamilton, Fitzgerald shared a passion for the idealism of George Berkeley.[56] He had
apparently studied Berkeley in the course of preparing for the metaphysical section of the fellowship examination, and throughout his
life he reiterated his belief that all the phenomena of nature ultimately reduce to thought. Fitzgerald shared this philosophy with
his uncle G.J. Stoney, who presented it on a number of occasions.[57]
At the end of his Helmholtz Memorial Lecture, Fitzgerald similarly
moved from a long explanation of the vortex theory of atoms to his
peculiar brand of idealism. He speculated,

Is there not, then, reason in the suggestion that colour
and sound, nay, space, time, and substance are functions
of our consciousness, produced by it under the action
of what may be called an external stimulus, and that
the only part of the phenomenon which essentially corresponds to that stimulus is the always pervading motion?

[55] George Francis Fitzgerald, "Electromagnetic Radiation [1890]," pp. 266-76 in *Scientific Writings*, p. 276. [Originally a lecture to the Royal Institution in 1890]

[56] Hunt, *The Maxwellians*, pp. 98-100.

[57] G. Johnstone Stoney, "How Thought Presents Itself in Nature," *Proceedings of the Royal Institution* (1885) 11: 178-96, G.J. Stoney, "Studies in Ontology: The First Step," *Proceedings of the Aristotelian Society* (1889) 1: 1-9, G.J. Stoney, "Studies in Ontology, From the Standpoint of the Scientific Student of Nature," *Scientific Proceedings of the Royal Dublin Society* (1890) 6: 475-524.

And what is the inner aspect of motion? In the only place where we can hope to answer this question, in our brains, thought is the internal aspect of motion. Is it not reasonable to hold, with the great and good Bishop Berkeley, that thought underlies all motion?[58]

Reiterating Berkeley, he explained, "Nature is a language expressing thoughts, if we learn but to read them...."[59] Of course, Fitzgerald's idealism differed significantly from both Berkeley's and from Hamilton's. Unlike Berkeley, Fitzgerald believed that there are unobservable structures behind our sensations that can be discovered through scientific investigation. And unlike Hamilton, Fitzgerald devoted his life to the search for a physical theory of electromagnetism rather than abstract mathematical laws. But all three shared a religiously inspired belief that all the phenomena of nature are directly caused by God without any mediation by matter. All three reveled in a universe composed only of thought.

7.4   Aether and Matter

The apex of the Dublin mathematical tradition in the nineteenth century, however, occurred not in Dublin but in Cambridge with the work of Joseph Larmor, an Irishman who spent most of his career in England. Born in County Antrim in 1857, Larmor attended the Royal Belfast Academical Institution and then Queen's College Belfast. His mathematics teacher at the Belfast Academical Institution had been the senior wrangler at Cambridge and told the young Joseph Larmor that he had the makings of a senior wrangler himself. After winning first place on the Queen's University honors

---

[58] George Francis Fitzgerald, "Helmholtz Memorial Lecture [1896]," pp. 340-77 in *Scientific Writings*, p. 376. [Originally delivered before the Chemical Society of London in 1896.]

[59] Fitzgerald, "Helmholtz Memorial Lecture," p. 377.

examination in both mathematical science and experimental science in 1874, Larmor went on to Cambridge where he did indeed become senior wrangler in 1880 (beating out J.J. Thomson who would later became director of the Cavendish laboratory at Cambridge and winner of the Nobel prize in physics in 1906). Belfast had a proud tradition of producing senior wranglers at Cambridge. William Thomson (Lord Kelvin) had been senior wrangler in 1845. Andrew James Campbell Allen in 1879 and William McFadden Orr in 1888 would also go from Queen's College Belfast to Cambridge to win senior wrangler.

After taking his degree at Cambridge in 1880, Larmor left to take up the professorship of Natural Philosophy at Queen's College Galway. After just five years, however, he returned to Cambridge where he would spend the rest of his career, eventually succeeding his fellow Irishman George Gabriel Stokes as Lucasian Professor of Mathematics in 1903. (Larmor would hold the chair until 1932, giving Irishmen an unbroken 83 year lock on the most prestigious mathematics chair in the British Isles, perhaps the world.) While Larmor was trained at Cambridge and spent nearly his entire career there, certain aspects of his work were very self-consciously part of what he considered to be an Irish mathematical tradition. In his obituary of Larmor, A.S. Eddington wrote that Larmor "had a strong attachment to his native country, and generally spent part of his summer vacation in Ireland. It is no accident that *Aether and Matter* is so largely a development of the work of his countrymen MacCullagh, Hamilton, FitzGerald."[60] Larmor collaborated closely with Fitzgerald, adopted MacCullagh's rotational model of the ether, and made Hamilton's version of the principle of least action the foundation of his own work.

Larmor's research in the 1890s evolved out of the project begun by the Cambridge wave theorists in the 1830s and continued by

---

[60] A. S. Eddington, "Joseph Larmor, 1857-1942," *Obituary Notices of Fellows of the Royal Society* (1942) 4: 197-226, p. 206.

Fitzgerald and the Maxwellians in the 1870s and 1880s. Their goal was to unify all of the phenomena of light, electricity, and magnetism through a simple set of equations derived from the properties of the ether. Larmor showed little interest in ether theories until he was asked by the BAAS to review the research on the action of magnetism on light in 1893.[61] In studying the literature, Larmor came upon Fitzgerald's 1878 paper in which he created an electromagnetic version of MacCullagh's rotationally elastic ether. Larmor realized that MacCullagh's approach could be extended even farther than Fitzgerald had. In fact, it could accommodate permanent vortex rings of the kind Kelvin had proposed. Larmor submitted a paper on the "Dynamical Theory of the Electric and Luminiferous Medium" to the Royal Society in November of 1893, and Fitzgerald and Kelvin were asked to serve as referees. Ultimately, Fitzgerald ended up being more of a collaborator than a referee (he received permission from the Royal Society to break the practice of anonymity and to correspond directly with Larmor). In fact, upon seeing the final paper, Heaviside wrote to Fitzgerald, "[I am] inclined to think you are the virtual author of a good bit in L[armor]'s memoir."[62]

Larmor's critical breakthrough (spurred on by Fitzgerald) was to hypothesize that charges are points of rotational strain—the centers of the vortices that Kelvin and Fitzgerald had proposed. Positive charges, he hypothesized, swirl in one direction, while negative charges swirl in the opposite direction. Larmor initially called these point charges 'monads' until Fitzgerald suggested that they should be called 'electrons'.[63]

It was actually Fitzgerald's uncle, George Johnstone Stoney, who originally coined the term 'electron'. Stoney had graduated

---

[61]Bruce Hunt, *The Maxwellians* (Ithaca: Cornell University Press, 1991), pp. 210-39.

[62]Cited in Hunt, *The Maxwellians*, p. 216.

[63] Fitzgerald suggested it to Larmor in letter on July 19, 1894. G.F. Fitzgerald to J. Larmor (July 19, 1894), cited in Bruce Hunt, *The Maxwellians*, p. 220.

from Trinity College Dublin in 1848 and went on to work as an assistant to William Parsons, the Third Earl of Rosse, at Birr Castle, where the Earl had constructed the world's largest telescope. Stoney spent five years teaching physics at the Queen's College Galway before becoming Secretary to the Queen's University of Ireland and later superintendent of the civil service examinations in Ireland. Throughout his career Stoney remained active in experimental physics and played important roles in the Royal Dublin Society, the BAAS, and the Royal Society of London. Stoney first proposed the term 'electron' for the unit of electric charge in 1891 as part of his search for a system of natural physical units.[64] The electron was simply the smallest unit of electrical charge.

Larmor's "electron theory of matter" popularized Stoney's new word. It's important to note, however, that for Larmor electrons were not solid particles, they were singularities in the ether that could carry either a positive or a negative charge. Larmor described electrons as "freely mobile intrinsic strains". They did not even exist separately from the ether, rather, they moved through the ether like "a knot moving along a rope." Larmor, however, speculated that electrons were the sole constituents of matter. That is, what we perceive as solid matter is merely a collection of twists in the ether held together by electromagnetic forces. (In fact, Horace Lamb suggested that Larmor's book *Aether and Matter* could more accurately have been titled, *Aether and No Matter*.[65])

By turning matter into a property of the ether, Larmor's theory offered hope that all of physics could be converted into a problem of the electrodynamics of moving bodies. Kelvin had shown that a moving charge in a magnetic field accumulates extra mass due to self-induction (it generates a field that resists acceleration through the magnetic field just as mass resists acceleration). Lar-

---

[64] James G. O'Hara, "George Johnstone Stoney and the Concept of the Electron," *Notes and Records of the Royal Society of London* (1975) 29: 265-76.

[65] Horace Lamb, "Presidential Address to Section A of the BAAS" (1904).

mor suggested that perhaps the entire mass of the electron might be electromagnetic in origin. In other words, the ether might be responsible for mass as well as electromagnetic and optical phenomena.

Larmor's theory also offered an explanation for the puzzling lack of evidence for the earth's motion through the ether. While no one doubted the existence of the ether in the late nineteenth century (there was no other conceivable way to explain the propagation of electromagnetic waves), they struggled to understand how the earth moved through it. If the earth dragged the ether along with it, like a stick moving through molasses, observations of star light would be affected by the movement of the intervening ether, but no effect was observed. On the other hand, if the earth moves through the ether without dragging it along then light should appear to move more slowly in the direction of the earth's motion, since the earth is moving with respect to the ether. Experiments by the American physicists Albert Michelson and Edward Morley in 1887, however, failed to find any evidence that the speed of light is different in different directions.

Fitzgerald had suggested to Lodge in 1889 that perhaps the motion of bodies through the ether might cause them to contract.[66] In other words, light might move more slowly in the direction of the earth's motion through the ether, but the length of the apparatus measuring the speed of light might contract also, by just enough to cancel out the effect. While the suggestion sounded rather far-fetched at the time (even to Fitzgerald), it was not completely unfounded. Heaviside had calculated in 1888 that the electromagnetic field around a moving charge would shrink slightly along its line of motion. If the forces between the molecules of a measuring instrument are electromagnetic, they might contract as well in the direction of motion through the ether. Fitzgerald apparently did

---

[66]Bruce Hunt, "The Origins of the FitzGerald Contraction," *British Journal for the History of Science* (1988) 21: 67-76.

not take his own suggestion too seriously. He never published a paper on it, sending only a brief letter to *Science* magazine in 1889 (at the time a little known science journal in the United States). In fact, no one apparently read Fitzgerald's letter until a historian discovered it in 1967.[67] Lodge mentioned Fitzgerald's suggestion in an article in *Nature* in 1892 where it was read by a Dutch scientist named Hendrik Lorentz who had just had the same idea independently. For this reason, it became known as the Lorentz-Fitzgerald contraction. What Fitzgerald's friend R.T. Glazebrook described as "the brilliant baseless guess of an Irish genius"[68] ultimately became Fitzgerald's most famous contribution to physics. The Lorentz-Fitzgerald contraction plays a central role in the theory of special relativity, which is based on the assumption that light travels at the same speed in all frames of reference (implying that matter really does contract when its speed approaches that of the speed of light). In the 1890s, however, Fitzgerald's speculation was important because it saved the ether theory from destruction in the face of Michelson and Morley's null result. And Larmor's electron theory of matter provided a physical explanation for the contraction.[69]

Larmor's electron theory of matter capped nearly two centuries of ether theories. And yet unlike Maxwell, Kelvin, Stokes, Fitzgerald, and Lodge, Larmor was less interested in mechanical models of the ether and more interested in finding a dynamical model, that is, a mathematical model that does not depend on the underlying details of motion. Larmor sought an underlying mathematical rather than a physical unification, and he found it (as Hamilton had) through the principle of least action. According to A.S. Ed-

---

[67] Stephen G. Brush, "Note on the History of the FitzGerald-Lorentz Contraction," *Isis* (1967): 230-232.

[68] Hunt, "The Origins of the FitzGerald Contraction," p. 67.

[69] Andrew Warwick, "On the Role of the FitzGerald-Lorentz Contraction Hypothesis in the Development of Joseph Larmor's Electronic Theory of Matter," *Archive for History of Exact Sciences* (1991) 43: 29-91.

dington, Larmor had an "intense, almost mystical, devotion to the principle of least action" which he believed to be "the ultimate natural principle—the mainspring of the universe."[70]

In an essay on "Least Action as the Fundamental Formulation in Dynamics and Physics," Larmor explained, "The greatest *desideratum* for any science is its reduction to the smallest number of dominating principles. This has been effected for dynamical science by Sir William Rowan Hamilton, of Dublin...."[71] The goal of science, for Larmor, was to explain a wide range of observed phenomena with a handful of simple, underlying concepts. "[T]he most direct and compendious method of stating the mathematical conditions of a physical problem," he elaborated, "is to express it as a maximum or minimum relation, and... in dynamics this can always be accomplished by means of the Principle of Least Action."[72] Larmor sought to base physics on a mathematical principle rather than a mechanical principle, and he looked to Hamilton's work explicitly to achieve this goal. "The original ideas of William Rowan Hamilton...," he proclaimed, "have constituted an epoch in fundamental physics."[73]

While Larmor worshipped Hamilton and the elegance of the principle of least action, however, other physicists were revolted. Oliver Heaviside, trained as a practical telegraphist and impatient with anything that smacked of metaphysics, called least action "a golden or brazen idol" that the mathematical tutors at Cambridge force young men to fall down and worship.[74] Even Fitzgerald was skeptical, calling it an "analytical juggle" rather than a proper foun-

---

[70] Eddington, "Joseph Larmor," p. 204.

[71] Joseph Larmor, "Least Action as the Fundamental Formulation in Dynamics and Physics," *Proceedings of the London Mathematical Society* (1884) 15: 158-84, in Sir Joseph Larmor, *Mathematical and Physical Papers* (Cambridge: Cambridge University Press, 1929), pp. 31-70, p. 59.

[72] Larmor, "Least Action," p. 43.

[73] Larmor, "Historical Note on Hamiltonian Action (1927)," pp. 640-41 in Larmor, *Mathematical and Physical Papers*, p. 640.

[74] Hunt, *Maxwellians*, p. 227.

dation for physical theory, and worrying that it "makes the present depend on the future."[75] Larmor, however, continued to preach the gospel of least action. In 1900 he gave the presidential address to the Mathematical and Physical Section of the BAAS. He reiterated that the principle of least action "as a guide to physical exploration, remains fundamental. When the principles of the dynamics of material systems are refined down to their ultimate common basis, this principle of minimum is what remains."[76] In fact, for Larmor, Hamilton's dynamics was even more fundamental than the ether. "[W]e should not be tempted towards explaining the simple group of relations which have been found to define the activity of the aether, by treating them as mechanical consequences of concealed structures in that medium; we should rather rest satisfied with having attained to their exact dynamical correlation."[77] Eddington later commented that Larmor "was the first really to recognize the immaterial nature of the aether and to throw off the obsession that the ultimate explanation of things must fill the universe with whirring machinery 'like the nightmare of a mad engineer'."[78] Larmor even doubted that it would ever be possible to understand the micro-structure of matter. "[A]n exhaustive discovery of the intimate nature of the atom," he cautioned, "is beyond the scope of physics."[79]

Larmor's electron theory was in many ways the culmination of classical physics. It extended Maxwell's theory of electromagnetism into a full theory of matter, capping the long line of ether theories that went back to Bryan Robinson in the eighteenth century (see chapter 2). Larmor believed that he had unified all that was known about electricity and magnetism, showing that an enormous range

---

[75] Hunt, *Maxwellians*, p. 227.

[76] Joseph Larmor, "Address of the President of the Mathematical and Physical Section of the British Association for the Advancement of Science," *Science* (1900) 12: 417-36, p. 426.

[77] Larmor, "Address of the President," p. 423.

[78] Eddington, "Joseph Larmor," p. 205.

[79] Larmor, "Address of the President," p. 425.

of phenomena could be derived from the abstract principle of least action. Academic scientists were impressed by Larmor's achievement, but they were no longer the only ones whose opinion mattered. The rise of the telegraph and electrical industry had created a group of powerful 'electricians', and these professional engineers refused to acknowledge the dominance of mathematical theory.

7.5   Theory vs. Practice

While Maxwell's theory led in the last decades of the nineteenth century to the experimental exploration of the internal structure of matter in the laboratories of physicists (see chapter 8), at the same time it led to the development of technologies with enormous commercial value, most notably wireless telegraphy or radio. The precise relationship between theory and practice, however, remained disputed. While the Maxwellians asserted that wireless telegraphy depended on their theory and their laboratory techniques, their failure to commercialize the new invention appeared to some as evidence of the superior role of the practical engineer.

Maxwellian theory always remained close to the problems of telegraphy. In the late 1880s, however, as academic scientists began to make recommendations that contradicted the experience of practical telegraphists, a bitter debate ensued.[80] The struggle between theory and practice represented by Thomson and Whitehouse in the 1860s erupted once again in the 1880s. This time the theorists were Fitzgerald's colleagues Lodge and Heaviside, and the practical electrician was William Henry Preece, director of the Post Office's telegraph division. The dispute centered on long distance telephone lines and lightning protection. Heaviside and Lodge recommended heavy inductive loading in both cases based on Maxwellian theory, while Preece felt that their recommendations went against all ex-

---

[80]Hunt, "Practice vs. Theory."

perience. The struggle came to a head in the 1888 meeting of the British Association. While Fitzgerald chaired Section A (Mathematical and Physical Science) and announced Hertz's confirmation of Maxwell's theory, Preece chaired Section G (Engineering). In his address Preece asserted,

> The practical man, with his eye and his mind trained by the stern realities of daily experience, on a scale vast compared with that of the little world of the laboratory, revolts from such wild hypotheses, such unnecessary and inconceivable conceptions, such a travesty of the beautiful simplicity of nature.[81]

Practical men, he bellowed, will not be dictated to by theorists. Later in the meeting, Preece boasted that "he made mathematics his slave, and he did not allow mathematics to make him its slave."[82]

While the Maxwellians laid the theoretical and experimental groundwork for wireless telegraphy, Marconi gained international fame and fortune for it. For the rest of his life Lodge argued that he, not Marconi, deserved the credit.[83] Lodge did send telegraphic signals through the air in 1894, and Lodge also invented the coherers used to detect the waves. But Lodge was more interested in the consequences for Maxwell's theory and the connection between optics and electromagnetism than in the transmission of messages. Marconi transformed the laboratory apparatus into a commercial technology. Preece championed Marconi in part because he represented the practical as opposed to the theoretical.[84] In 1897 Marconi received a broad patent for wireless telegraphy, preventing Lodge from profiting from any of his inventions. Marconi's public success threatened the connection between university science

---

[81] Preece, quoted in Hunt, *The Maxwellians*, p. 169.
[82] Preece, quoted in Hunt, *The Maxwellians*, p. 171.
[83] Hong, "Marconi and the Maxwellians."
[84] Hong, "Marconi and the Maxwellians," p. 734.

and practical technology that the Maxwellians worked so hard to promote.

Fitzgerald responded to the arguments of men like Preece in his inaugural address as President of the Dublin Section of the Institution of Electrical Engineers in 1900. He argued for the essential role that theory had played in the history of electrical technology. Facing his critics head on, he explained,

> A recognized authority, who is fond of poking paradoxical fun at Professors, has recently stated that 'the progress of telegraphy and telephony owes nothing to the abstract scientific man.' I do not know exactly what he means by the abstract scientific man, but I do know that telegraphy owes a great deal to Euclid and other pure geometers, to the Greek and Arabian mathematicians who invented our scale of numeration and algebra, to Galileo and Newton who founded dynamics, to Newton and Leibniz who invented the calculus, to Volta who discovered the galvanic cell, to Oersted who discovered the magnetic action of currents, to Ampère who found out the laws of their action, to Ohm who discovered the law of the resistance of wires, to Wheatstone, to Faraday, to Lord Kelvin, to Clerk Maxwell, to Hertz. Without the discoveries, inventions, and theories of these abstract scientific men telegraphy, as it now is, would be impossible.[85]

Telegraphy, according to Fitzgerald, is the best example of the interdependency of theory and practice. The theory of electromagnetism had been forged in part from the problems of telegraphy, and through the second half of the nineteenth century, scientists had demonstrated their ability to contribute to technological progress.

---

[85] George Francis Fitzgerald, "The Applications of Science: A Lesson from the Nineteenth Century [1900]," pp. 487-99 in *Scientific Writings*, pp. 491-92. [Inaugural address to the Dublin Section of the Institution of Electrical Engineers (1900)]

The fact that Fitzgerald had to argue his position so forcefully, however, demonstrates that it was far from obvious.

The rise of electrical technology made science a part of everyday life. At the beginning of the nineteenth century, electricity was an experimenter's plaything, but by the end of the century it had spawned a number of important industries. This provided Fitzgerald with an even stronger justification for the sciences. Electrical engineering, he argued, "has revolutionized society and enabled high and low, rich and poor, to lead better lives, by making life less hard and grimy, and thus improved the well-being of man both materially and, what is far more important, morally as well."[86] For Fitzgerald, it was science that had revolutionized society. He argued that the work of underappreciated and poorly funded scientists like himself was in fact the real source of social progress. It was a bold claim at the end of the nineteenth century, but it was a position that quickly gained acceptance throughout the twentieth century as physicists and other scientists extended their power over technology, industry, and government.

7.6 Science, Industry, and Ireland

While Fitzgerald and the Maxwellians battled for the priority of theory over practice in the domain of engineering, Fitzgerald also took a broad view of the importance of scientific education, particularly in Ireland. Fitzgerald sat on a number of government commissions on education and played a role in the founding of the Kevin Street Technical College, one of Ireland's first technical colleges and the forerunner to the Dublin Institute of Technology. The fact that science had revolutionized society, according to Fitzgerald, meant that all members of society should learn science. Ireland's very future depended on it, he believed. Fitzgerald con-

---

[86] Fitzgerald, "The Applications of Science," p. 499.

stantly lamented the public's lack of understanding of science. In his address to the British Association in 1888 he complained, "The 'public' now are but the children of those who murdered Socrates, tolerated the persecution of Galileo, and deserted Columbus."[87] It was not simply that the public did not appreciate or support the activities of scientists; they lacked any understanding of the methods of science. Yet in a rapidly industrializing world, Fitzgerald believed, such a knowledge was essential. "All the classes of the country required this training," he explained in a speech to the Dublin University Experimental Science Association, "they would die without it."[88] Fitzgerald lambasted the proponents of classical education, who consistently excluded experimental science from education, describing them as, "those who would sacrifice the rising generation on an altar of so-called culture to starve and die, with their only comfort that they can describe their agony in well-expressed phrases."[89] Fitzgerald did not expect that all students would become scientists. Rather he argued that experimental science teaches reasoning, not just facts. Just as the mathematicians of the early nineteenth century argued that mathematics disciplines the mind, Fitzgerald and other proponents of experimental science offered laboratory work as a similar discipline.

Not only does the survival of industry depend on such scientific habits of mind, Fitzgerald warned, but the very existence of the Irish people does: "[F]or preventing extermination like that of the Maori and the Red Indian, a people must be provided with accurate information and with habits of accurate work, i.e. with scientific information and scientific methods."[90] The evolution of industry and civilization, he implied, threatened those who failed

[87]George Francis Fitzgerald, "Address to the Mathematical and Physical Section of the British Association [1888]," pp. 229-40 in *Scientific Writings*, p. 231.

[88]George Francis Fitzgerald, "Experimental Science in Schools and Universities [1886]," pp. 191-96 in *Scientific Writings*, p. 193. [Originally published in *Nature* (1886)]

[89] Fitzgerald, "Experimental Science," pp. 195-96.

[90] Fitzgerald, "Science and Industry," p. 395.

to adapt with extinction. The native Irish will face the same fate as the Native Americans if they do not adapt to a world of science and industry. Fitzgerald laid the blame for the problem with the Irish educational establishment. He charged that "the educational machinery of the country is controlled by a lot of very worthy old bookworms with more sympathy with the theory of equations and Greek verse than with the industrial welfare of the country."[91] Trinity's venerable reputation in mathematics and classics was in part responsible for Ireland's backwardness. Fitzgerald was particularly annoyed with George Salmon, Trinity's provost and one of the last great representatives of Trinity's mathematical tradition. Salmon was more interested in Protestant theology and pure mathematics than in experimental science and industrial progress. For Fitzgerald, traditional intellectuals like Salmon were responsible for Ireland's failure to industrialize. He also clashed with Catholic Bishops over their distaste for education in experimental science.

Those with an interest in practical education and industry in Ireland formed a small group. While in England technical education developed largely as a series of local initiatives designed to support existing industry, in Ireland the process was quite different. Technical education in Ireland developed primarily on the initiative of the British government and was based on British models. One of the earliest initiatives was the Museum of Economic Geology, established in Dublin in 1845 and directed by the chemist Robert Kane. In 1854 the Department of Science and Art was founded in South Kensington to encourage technical education throughout the United Kingdom. The Department set examinations and paid teachers by results. In 1867 the Royal College of Science for Ireland was founded. Few Catholics attended the new College, however, and most of its graduates emigrated since there were few jobs in Ireland for industrial chemists or electrical engineers.[92]

---

[91] Fitzgerald, "Science and Industry," p. 404.
[92] Kelham, "The Royal College of Science for Ireland," p. 307.

In England technical education developed in response to industrialization, while in Ireland it was intended to promote industrialization. Irish education, explains historian Richard Jarrell, "was designed for a fictitious population; it might have worked well in England or parts of Scotland, but the economic and social realities of Ireland did not match the system of education."[93] Technical education, however, served as an area where different political factions could find common ground. While Fitzgerald served on the Board of the Kevin Street Technical College and a Trinity graduate served as director, Charles Stewart Parnell and Michael Davitt were also involved. They all agreed on the importance of industry for Ireland's future, though they disagreed vehemently on what that future would look like. Fitzgerald himself was a firm Unionist.[94] In 1892 he wrote to Lodge, "with this Home Rule looming in the near future as possible I cannot afford to wast[e] more money than I can help as I shall almost certainly have to leave Ireland if it comes on."[95] In 1893, when it appeared that the British government might grant Home Rule, he wrote,

> Woe to Britain if for the sake of saving trouble over Irish
> squabbles that are only now beginning, if for the sake of
> puny local ends, for the sake of some supposed principle
> of so called popular government, she hands over the in-
> telligence, the industry, the whole people of Ireland to be
> a prey to these greedy vultures.[96]

Industrial education, then, played a significant role in Fitzgerald's Unionism. Nationalists and ultramontane Catholics, he believed, could not be trusted to rationally guide Ireland's future.

---

[93] Jarrell, "Some Aspects of the Evolution of Agricultural and Technical Education," pp. 114-15.

[94] Whyte, *Science, Colonialism and Ireland*, pp. 48-49.

[95] G.F. Fitzgerald to O. Lodge (July 11, 1892) quoted in Whyte, *Science, Colonialism and Ireland*, p. 48.

[96] G.F. Fitzgerald to O. Lodge (March 16, 1893) quoted in Warwick, "Sturdy Protestants of Science," p. 327.

Humphrey Lloyd and George Francis Fitzgerald struggled to promote a certain future for Ireland—scientific, industrial, and imperial. They articulated a particular vision of progress and emphasized the important role that Trinity might play in achieving that vision. But they faced difficulties on a number of fronts. Most engineers simply did not believe that university scientists had much of a role to play in industrial development. Lloyd and Fitzgerald failed to enlist industrialists and practical engineers in their project, and they also failed to convince the majority of the Irish population. The groups that came to power in Ireland in the early twentieth century did not share their vision of technology, industry, and empire.

Ireland was left behind as national governments in Britain, Germany, and the United States began to support large-scale scientific research as part of the war effort, and large corporations invested in advanced research and development laboratories. Science became increasingly competitive and increasingly expensive at a time when the Irish government and Irish industry were still struggling to survive. In other countries the amateur researchers who had played significant roles in science in the late nineteenth century gave way to professional researchers working in large well-funded institutions, but Ireland simply had no place for them to go. The majority of Fitzgerald's students made their careers outside of Ireland, as did subsequent generations of talented Irish men and women. In their work, many of Fitzgerald's dreams of scientific understanding, technical progress, and engineering education were realized. But his dreams for Ireland would have to wait until the end of the twentieth century.

Chapter 8

Two Revolutions

A Terrible Beauty Is Born

Easter Monday, 1916, "looked a day of peaceful thoughts if ever there was one,"[1] according to John Joly, Professor of Mineralogy and Geology at Trinity College Dublin. He was enjoying the cool bright weather on a holiday walk to visit a friend when he came upon Major Harris of Trinity's Officer Training Corps. Harris alerted Joly that rebels had seized the General Post Office (GPO) and St. Stephen's Green, shooting several police officers in the process. Cycling across Dublin, Joly found that all traffic had ceased, and the rebels were virtually in possession of the city. He inspected the situation at the GPO. The windows were shattered, and mailbags, chairs, and tables had been piled up as barricades. Above the building hung a huge green banner with the words "Irish Republic" in white letters. Armed rebels roamed the streets, and the police were nowhere to be seen.

Joly made his way to Trinity College, surprised that it had not already been captured by the rebels. Trinity's location was

---

[1] John Joly, "In Trinity College During the Sinn Fein Rebellion, by One of the Garrison," *Blackwood's Magazine* (1916) 200: 101-125. Joly mistakenly believed that Sinn Fein was responsible for the rebellion.

strategically important (it lies at the intersection of three main thoroughfares—Westmoreland Street, Dame Street, and Grafton Street). Moreover, the Officer Training Corps depot at Trinity had hundreds of rifles and thousands of rounds of ammunition. In fact, the rebels' strategic missteps convinced Joly that this was not a German plot (one of his first fears). Joly was quickly given a service rifle and joined the forty-three other men who would defend Trinity through the rebellion. He worried that it might be the last night of the ancient university. "So might perish Ireland's most priceless treasure," he feared, "the University of Berkeley, Goldsmith, Burke, Hamilton, and Lecky."[2]

The small garrison used Trinity's strategic location to divide the rebels and keep open some of the principal streets for the troops. When the British army finally did make it to Dublin a few days later, they used Trinity as their headquarters. Four thousand troops were stationed at the university (along with two eighteen pounder guns), and an emergency hospital was set up on campus. Joly proudly remembered, "There can be no doubt that the accurate fire maintained from the College was an important factor in the salvation of the City."[3] After centuries of defending Ireland from rebels through sermons, lectures, speeches, and pamphlets, Trinity was now defending its own vision of Ireland by force of arms. By Friday, the army had retaken the GPO, and on Sunday, Joly finally returned home, happy to put the rebellion behind him. The Rebellion, he remarked, "was in a fair way to be a thing of the past, and to take its place in Irish history as one of the many insane rebellions which constitute its principal episodes."[4]

To Joly, the Easter Rising was the attempt of a small group of young, misinformed rebels to take advantage of Britain's focus on the war in Europe. Ultimately, he believed, it was doomed

---

[2] Joly, "In Trinity College During the Sinn Fein Rebellion," p. 106.
[3] Joly, "In Trinity College During the Sinn Fein Rebellion," p. 111.
[4] Joly, "In Trinity College During the Sinn Fein Rebellion," p. 124.

Figure 8.1: The General Post Office After the Easter Rising (1916)
©National Library of Ireland

to failure like so many Irish rebellions in the past. In retrospect,
however, the Easter Rising marked a critical turning point in Irish
history. When most of its leaders were executed in the weeks that
followed, public opinion turned quickly in favor of the rebels. In
many ways, the fate of the Ascendancy and the Union had already been sealed. "Home Rule" (Irish political independence from
Britain) had already passed in 1914, and it was simply put on hold
at the beginning of World War I. The Anglican Church of Ireland
had been disestablished in 1871, and land reform had started the

process of transferring most of the land in Ireland from Protestant landlords to Catholic farmers. Between 1870 and 1921, the Anglican Ascendancy was almost completely dismantled. The Easter Rising simply accelerated a foregone conclusion.

At the very same time, another revolution was taking place— a revolution in physics. Young scientists in Copenhagen, Berlin, Zurich, and Göttingen were building new theories of the physical world that conflicted fundamentally with the basic principles on which physics had been based for centuries. In the span of fifteen years, Maxwell's theory of electromagnetism, Larmor's electron theory of matter, even Newton's theory of gravitation were fundamentally transformed. The ether that had served as the physical, mathematical, and philosophical basis for natural philosophy since the age of Newton was no more, leaving a generation of physicists in search of new foundations.

The end of the Ascendancy in Ireland coincided with the end of classical physics and the (temporary) end of the mathematical tradition at Trinity College Dublin. A new generation took over the reins of power in Ireland, and a new generation took over leadership in physics. Trinity was left abandoned, struggling to find its way in a world that was changed utterly. Trinity scientists who came of age in the last quarter of the nineteenth century refused to accept the new quantum mechanics just as they refused to accept the new political environment.

Joseph Larmor, the Irishman who served as Lucasian Professor of Mathematics at Cambridge from 1903 until 1932, represented the challenge his generation faced in adapting to the changed circumstances. By the 1930s, as historian Andrew Warwick relates, "Larmor was well aware that that the scientific 'values' he had imbibed in Ireland and Cambridge in the 1870s had been displaced by those of a new and international modern physics. In the final years of his life he felt increasingly isolated, the last survivor of a race of mathematical physicists that was now almost extinct."[5] Even as others abandoned the ether, Larmor clung to it, becoming increasingly out of touch with the rest of the scientific world. In 1941 he wrote to a friend that he had "outlived [his] generation" and "appeare[d] to be lingering in a world to which [he] did not belong."

Larmor's isolation was political as well as scientific. As a member of parliament for Cambridge University, he passionately defended the Union between Great Britain and Ireland, watching helplessly as that union dissolved before his eyes. In a sense, Larmor's fate mirrored that of the broader Anglo-Irish Ascendancy in the first decades of the twentieth century. They had outlived their generation and appeared to be lingering in a world to which they did not belong. As their numbers dwindled, and as the survivors struggled to adapt to a new world in politics and in science, Trinity and the Ascendancy tradition in science fell into a long decline. Trinity did not disappear, as Joly had feared during Easter week. In fact, it changed very little from 1916 to the 1960s. But as the Ascendancy withered, so did the scientific tradition at Trinity. The best scientists left, and those that stayed had no time for research in the new theories.

The man who did more than any other individual to shape the newly independent Ireland also played an important role in bringing the new physics to Ireland. While Joly was defending Trinity during the Easter Rising, a frustrated mathematics teacher by the name of Éamon de Valera was leading a group of rebels at Boland's Mill. De Valera had completed a degree in mathematics through the Royal University of Ireland while working as a teacher. He had even spent a year studying mathematics at Trinity College Dublin,

---

[5] Andrew Warwick, "'That Universal Aetherial Plenum': Joseph Larmor's Natural History of Physics," pp. 343-86 in Kevin C. Knox and Richard Noakes (eds.), *From Newton to Hawking: A History of Cambridge University's Lucasian Professors of Mathematics* (Cambridge: Cambridge University Press, 2003), p. 382.

entering in 1905 as one of thirty-four Catholics who were assigned
to the only Catholic fellow.[6] After twice failing to win a mathematics scholarship at Trinity, however, de Valera dropped out
and returned to teaching mathematics at various Dublin schools
while at the same time applying unsuccessfully for professorships
at the University Colleges at Cork and Galway. De Valera's focus
had gradually turned from mathematics to nationalist politics. He
helped to plan the Easter Rising and was one of the few participants to escape a death sentence (in part because he was a U.S.
citizen—he had been born in New York City—and in part because
of the mounting political pressure on the British government to
halt the executions). He would go on to become the most important political leader in modern Irish history.

Figure 8.2: Éamon de Valera (1882–1975), Mathematics Teacher and
Leader of the Struggle for Irish Independence. ©National Library of Ireland

---

[6]David Fitzpatrick, "Eamon de Valera at Trinity College," *Hermathena* (1982) 133: 7-14.

After his release from prison de Valera became president of the nationalist political party Sinn Féin (Irish for "We Ourselves"). The following year, the Sinn Féin delegation refused to take their seats in Westminster, forming an Irish parliament in Dublin (known as the Dáil Éireann). De Valera served as President of the Republic during the ensuing War of Independence (1919-21). Unhappy with the Treaty that ended the war (which left the six counties of Northern Ireland as part of Great Britain, kept Ireland part of the Commonwealth, and required Irish members of parliament to swear loyalty to the king), de Valera and a large majority of Sinn Féin left the Dáil, leading to the Irish Civil War (1922-23) in which Michael Collins' pro-Treaty Free State forces fought de Valera's anti-Treaty Irish Republican Army (IRA).

In 1926, after another stint in prison, de Valera founded a new political party, Fianna Fáil ("The Warriors of Destiny"), a party that (under de Valera's leadership) would dominate twentieth-century Irish politics. In 1932, de Valera became Prime Minister, and in 1937 he introduced a new constitution, creating Éire. The new constitution claimed the entire territory of Ireland (leading to continuing struggles over Northern Ireland); it made Irish the national language (though it recognized English as a second official language); and it declared the "special position" of the Roman Catholic Church, although it did not establish Roman Catholicism as the official state religion. De Valera went on to serve as Taoiseach (Prime Minister) on and off until 1959 and as President (a largely ceremonial position) from 1959 to 1973. At his retirement at the age of 90, he was the oldest serving head of state in the world.

Even as he devoted himself to politics, de Valera never lost his passion for mathematics, especially the work of Sir William Rowan Hamilton. According to his son, de Valera "read Hamilton with a disciple's zeal."[7] Sitting in Lewes Jail in February 1917 for his

---

[7]Vivion de Valera's description, cited in J.L. Synge, "Eamon de Valera," *Biographical Memoirs of Fellows of the Royal Society* (1976) 22: 634-653, p. 640.

role in the Easter Rising, de Valera spent his time working out the equations that govern the precession of the equinoxes, using a textbook on quaternions written by John Joly's cousin Charles Jasper Joly (who held Hamilton's position as Royal Astronomer of Ireland). During his stay in the Arbour Hill prison (1923-24) he tried to master relativity theory. Even in his old age, he continued to stay engaged with mathematics. An article in the *New York Journal* in 1965 (when de Valera was in his eighties and still serving as President) described his mathematical studies:

> [H]e has read to him from time to time articles on modern physics—atomic particles, quantum dynamics, etc., and books such as those by Professors Synge and Lanczos, Dr. McConnell, etc. etc. He works from time to time in Analysis—tensor analysis, biquaternions and octonions, Grassman's Algebra of Extension and Boolean and other algebras. The President uses dark green linoleum (found to be most effective from his eyesight point of view), covering the top of his large desk in his private study here in Árus an Uachtaráin [the President's house], as a blackboard on which with chalk he draws geometrical figures and pursues such algebraical expressions as he might find difficult to visualize otherwise.[8]

De Valera's personal interest in mathematical physics was the primary reason for the creation of the Dublin Institute for Advanced Studies (DIAS) in 1939, and de Valera himself recruited Austrian Nobel Laureate Erwin Schrödinger as the first director of its School of Theoretical Physics. DIAS quickly became an international center for mathematics and theoretical physics, and de Valera attended many of the scientific meetings there.

---

[8]Description of "President de Valera's mathematical reading and studies" given to the *New York Journal* in 1965, cited in Synge, "Eamon de Valera," pp. 638-39.

More than any other individual, Éamon de Valera defined the vision of twentieth-century Ireland as Catholic, Gaelic, politically and economically independent, and agrarian. It was a vision that sat uncomfortably with Trinity's tradition of Protestantism, unionism, imperialism, science, and technology. But Joly's fear that the Easter Rising might spell the end for his ancient university did not come to pass. Trinity survived the disestablishment of the Church of Ireland, the creation of the Irish Free State, and the birth of the Republic. But while Trinity survived, it did not flourish. The work of Hamilton, Lloyd, MacCullagh, and Fitzgerald would remain the height of Trinity's scientific accomplishment. Uncertain of its role in the new nation and lacking financial resources (as all Irish institutions did in the difficult decades of the early twentieth century), Trinity College entered a period of "low profile." At a time when universities in England, Germany, and the United States were investing in state-of-the-art research laboratories, post-graduate research positions, and partnerships with industry and government, Trinity's small staff struggled simply to meet their weekly lecture schedules.

8.1   The End of the Ascendancy and the Decline of Science

Joly's defense of Trinity during Easter week was in a sense the last stand of the Anglican Ascendancy. Joly may have won the battle, but the Ascendancy had already lost the war.[9] When W.E. Gladstone was elected Prime Minister in 1868, the disestablishment of the Church of Ireland, land reform, and Home Rule soon moved to the top of the legislative agenda. In 1871, the Anglican Church lost its status as the official religion of Ireland. The tithe was finally

---

[9]Michael McConville, *Ascendancy to Oblivion: The Story of the Anglo-Irish* (London: Quartet, 1986); Kurt Bowen, *Protestants in a Catholic State: Ireland's Privileged Minority* (Kingston: McGill-Queen's University Press, 1981).

ended, and most church lands were transferred back to the government. A series of Land Acts established mechanisms for arbitrating
disputed rent increases, made eviction more difficult, and encouraged landlords to sell.[10] Between 1870 and 1909, thirteen million
acres of land were transferred from large landlords to their tenants,
and by 1914, three fourths of former tenants had bought their hold-
ings, effectively destroying the economic and social foundation of
the Protestant Ascendancy. At the same time, the political foundations of the Ascendancy were under attack. While the first three
attempts to pass Home Rule bills failed, a Home Rule Bill finally
passed in 1914. It was quickly suspended at the outbreak of World
War I, but some form of independence for Ireland was a foregone
conclusion. Many Protestants left Ireland as their country houses
were burned down in the War for Independence and the Civil War.
The Protestant population in the southern twenty-six counties went
from 468,000 in 1861 to only 144,000 in 1961.

Cultural nationalism was an important complement to political
nationalism.[11] Organizations like the Gaelic Athletic Association
(organized in 1884) boycotted English games and promoted traditional Gaelic sports, while the Gaelic League (organized in 1893)
promoted the Gaelic language and Gaelic culture. Historian F.S.L.
Lyons explains, "The marriage between Catholicism and Gaelicism
was fatal to the hopes of Protestant Anglo-Irish protagonists of
cultural fusion. Catholicism and Gaelicism and the nationalism
they nourished were reacting primarily against England. It was
English manners and morals, English influences, English Protestantism, English rule, that they sought to eradicate."[12] There was
no longer a middle ground on which the Anglo-Irish could stand. It

---

[10]W.E. Vaughn, *Landlords and Tenants in Ireland, 1848-1903* (Dublin: Economic and
Social History of Ireland Press, 1984), pp. 11-12.

[11]John Hutchinson, *The Dynamics of Cultural Nationalism: The Gaelic Revival and the
Creation of the Irish Nation State* (London: Allen & Unwin, 1987).

[12]F.S.L. Lyons, *Culture and Anarchy in Ireland, 1890-1939* (Oxford: Clarendon, 1979),
p. 82.

became impossible to be Anglo-Irish, one had to be either English or Irish.[13]

And yet, even as they were being written out of the nationalist story of Irish history, Irish Protestants continued to play leadership roles in the Home Rule movement and the Celtic Revival. In the eighteenth century, Protestants such as Molyneux, Swift, and Berkeley had helped to define an early form of Irish nationalism, while the Ulster Presbyterians in the United Irishmen took those ideas to a revolutionary extreme. In the nineteenth century, Thomas Davis, William Smith O'Brien, and Isaac Butt all played critical roles in the nationalist movement, and Charles Stuart Parnell became by far the most powerful leader of the Irish Parliamentary party. In fact, in every phase of Irish history there had been Trinity men who had championed the cause of Ireland. J.A. Galbraith, a Trinity fellow, is even credited with coining the term 'Home Rule.'[14]

Irish Protestants also led the efforts to build an Irish cultural identity. The roots of the Celtic Revival lay in the antiquarian work of the Royal Irish Academy. Douglas Hyde, a Gaelic expert, served as Ireland's first president. The Irish Literary Theater, which became the Abbey Theater, was founded by Lady Augusta Gregory, W.B. Yeats, and J.M. Synge in an attempt to produce a truly Irish theater. "[T]he Anglo-Irish," historian George Boyce explains, "not only provided political precedents and precepts for Irish nationalists; they also, and especially after 1890, helped give Ireland a distinctive cultural identity, a sense of the individuality of the Irish nation, and of its particular linguistic, social and racial character-

---

[13]L.P. Curtis, "The Anglo-Irish Predicament," *Twentieth Century Studies* (1970) 4: 37-63, p. 41. See also Mark Bence-Jones, *Twilight of the Ascendancy* (London: Constable, 1987).

[14]Miguel DeArce, "The Parallel Lives of Joseph Allen Galbraith (1818-90) and Samuel Haughton (1821-97): Religion, Friendship, Scholarship and Politics in Victorian Ireland," *Proceedings of the Royal Irish Academy* (2011) 112C: 1-27.

istics."[15] But, while Protestants played crucial roles in the events leading to the creation of the Irish Free State in 1921, the end result was a conception of Irishness that explicitly excluded the Anglo-Irish.

The changing balance of power in Ireland played out in a series of debates around "the Irish university question." After centuries in which post-secondary education was dominated by the Anglican Church, the challenge was to design a set of educational institutions that could meet the needs of Ireland's Catholic, Anglican, and Presbyterian communities, and it would become one of the most controversial issues in British and Irish politics. The British government's first attempt to solve the problem was the establishment in 1845 of the Queen's Colleges in Belfast, Cork, and Galway. The Catholic Church, however, refused to accept these new secular institutions. Pope Pius IX himself criticized them, and in 1850 Cardinal Cullen forbade Catholic clergymen from holding positions in the new universities and required bishops to discourage the faithful from attending. As a result, the Queen's Colleges struggled. By 1851 they enrolled only 411 students across all three colleges, while Trinity enrolled 1,200 (half of whom lived outside Dublin).[16] A separate Royal College of Science for Ireland was created in 1867 to provide industrial education and teacher training, but it attracted only 20 to 30 students per year.[17] The Catholic Church founded its own university in Ireland in 1854, with John Henry Newman as rector, but without government recognition or funding, the Catholic University failed to thrive. In 1883 the Jesuit Order took over its facilities and founded University College Dublin (UCD).[18]

---

[15]D. George Boyce, *Nationalism in Ireland*, 2nd ed. (New York: Routledge, 1991), p. 228.

[16]T.W. Moody and J.C. Beckett, *Queen's, Belfast 1845-1949: The History of a University* (London: Faber and Faber, 1959).

[17]Brian B. Kelham, "The Royal College of Science for Ireland (1867-1926)," *Studies* (1967) 56: 297-309.

[18]Donal McCartney, *UCD: A National Idea-The History of University College, Dublin* (Dublin: Gill & McMillan, 1999).

The British public was not prepared to fund a Catholic university, and yet Irish Catholics would accept nothing less. In 1880, as a compromise, the Royal University of Ireland was established solely as an examining body (with duplicate Catholic and Protestant officers). It awarded degrees to students who studied at any of the Irish universities (including the Queen's Colleges at Belfast, Cork, and Galway, UCD, the Presbyterian Magee College in Derry, and the Catholic seminary at Maynooth).[19] In 1907, Chief Secretary Bryce proposed a solution that would bring all of the universities together under the umbrella of the University of Dublin. Trinity College Dublin had always been the only college within the University of Dublin (in fact the two names were used synonymously), but Bryce suggested bringing both the secular Queen's Colleges and the Catholic University College into equal positions with Trinity. The colleges could continue to provide denominational education, he suggested, while the university would teach the subjects with no religious implications. Bryce explained, "the University would provide teaching in advanced subjects which are non-controversial subjects, such as mathematics. You cannot have Protestant mathematics and Catholic mathematics."[20]

Bryce's proposal, however, had little appeal in Ireland. The Catholic Church was adamant that Catholic students must be taught by Catholic professors in a Catholic environment. The Reverend Professor D. Coghland of the Catholic seminary at Maynooth testified to a parliamentary committee in 1907, contradicting Bryce's claim that mathematics was above politics or religious controversy:

> There can be grave danger from non-Catholic teachers and fellow-students, even in purely secular or scientific

---

[19] Dermot Moran, "Nationalism, Religion and the Education Question," *Crane Bag* (1983) 7: 77-84.

[20] James Bryce, Speech of January 25, 1907, cited in Nicholas Whyte, *Science, Colonialism and Ireland* (Cork: Cork University Press, 1999), p. 44.

classes. A teacher could, for example, make a covert hostile allusion to the principle of authority in the church even when teaching mathematics, by remarking significantly that mathematical conclusions are not received on authority, that scientific work and authority are mutually incompatible.[21]

While Trinity professors scoffed at this notion that only Catholics could teach science to Catholics, for over two hundred years they had argued through lectures, textbooks, and sermons that science and mathematics are the best supports for Anglican natural theology. Trinity's defenders framed the debate about university education as the choice between scientific progress and blind obedience to authority, an argument that could have been taken almost verbatim from the seventeenth-century arguments of the Puritan planters who sought to replace scholasticism with Baconianism (see chapter 1).

The Trinity men no longer defended their special status with reference to Protestantism, however, but rather presented themselves as the defenders of intellectual liberty and scientific rationality. Joseph Larmor argued in parliament that imposing a Catholic college on Trinity College Dublin would "destroy the freedom of thought and learning which had been the cause of the marvelous success of the University of Dublin, which had given it a reputation and distinction which spreads far away from Ireland and over the whole world...."[22] The Trinity men also argued that they were non-sectarian. Trinity had accepted Catholic students since 1793 (long before Oxford and Cambridge), though they were not eligible for scholarships until the 1850s. In 1873, Trinity had abolished all religious tests, recognizing that this was the best strategy to protect itself from intervention by the rising political power of the

---

[21] E.P. Culverwell, *Mr. Bryce's Speech on the Proposed Reconstruction of the University of Dublin* (1907), p. 7.

[22] Joseph Larmor, House of Commons Debate, *Hansard* (October 21, 1912).

Irish Catholic Church. Trinity elected its first Catholic fellow in 1880, and in 1903 even asked the Catholic Church to nominate religious teachers for Trinity's Roman Catholic students. Trinity, they argued, was the home of religious tolerance.

Figure 8.3: John Joly (1857–1933), Professor of Geology and Mineralogy, Expert in Radioactivity. ©Trinity College Dublin

A decade before he took up arms to defend Trinity during the Easter Rising, John Joly had taken up his pen to defend her from Bryce's proposal. Trinity, he argued, represented "the liberal ideal:"

> In contradistinction to the Roman Catholic ideal, which recognizes the paramount influence of authority as based on the dogma of the Church, the education of Trinity College is liberal and undenominational. Students of every creed are on her books. . . . Similarly, curriculum, lectures, publications are marked with the breadth and freedom of

the liberal ideal; each subject is studied and considered
according to its own claims to credibility.[23]

Joly pointed to Trinity's strong record in science and her future
prospects (significantly improved by the recently completed Physical Laboratory), and he cautioned that topics such as the age of
the earth (his own area of expertise) could not be investigated if
the demands of the Catholic Church were allowed.

Ultimately, Bryce's scheme was defeated. In 1908, the Royal
University of Ireland was replaced by the National University of Ireland (with three constituent colleges in Dublin, Cork, and Galway,
and later Maynooth) and Queen's University Belfast was set up as
an independent university. While NUI was non-denominational on
paper, in practice its governance structure was designed to reflect
Catholic and nationalist perspectives. The board of Queen's University Belfast, on the other hand, was dominated by Presbyterians.
The putative objectivity and universality of scholarship could not
overcome the desire of each religious community to closely control
the education of their own flocks.

Trinity's victory, however, was a hollow one. Trinity avoided
the creation of a Catholic college within the University of Dublin,
and even maintained its independence (and its parliamentary representation) at the creation of the Irish Free State. In defending its
autonomy, however, Trinity ended up further isolating itself from
the main currents of Irish life. The Catholic hierarchy in Ireland
forbade Catholics from attending Trinity up through the 1970s,
preventing Trinity from attracting a truly representative sample of
Ireland's population. In 1920, a Royal Commission recommended
a £113,000 capital grant and annual subsidy of £49,000 for Trinity,
but when the Irish Free State was created in 1922, it declined to
continue the funding. Trinity struggled to get even small amounts

---

[23] John Joly, *An Epitome of the Irish University* (Dublin: Hodges, Figgis & Company,
1907), pp. 6-7.

of funding from the new government, which argued that UCD had a better claim on the government's limited resources due to its historic underfunding.[24] Trinity would not go back to the government for funding again until 1946.

Trinity's malaise was part of a broader decline of Irish science that occurred from the 1890s to the 1930s.[25] Historians have proposed a number of explanations. Some emphasize the growing power of the Catholic Church which feared the freethinking and materialistic tendencies of modern science. Others blame the Celtic Revival's anti-materialist and anti-modern form of romanticism that often portrayed science as antithetical to Irish values.[26] And some blame an Irish nationalism that saw science as elitist, Protestant, English, and irrelevant. Science, in fact, was removed from the primary school curriculum to make room for instruction in the Irish language and Irish history.[27]

While there is some truth to the view that the rising power of Catholicism and a romantic nationalism played a role in the decline

[24]Whyte, *Science, Colonialism and Ireland*, pp. 53-54, R.B. McDowell and D.A. Webb, *Trinity College Dublin, 1592-1952: An Academic History* (Dublin: Trinity College Dublin Press, 1982), p. 428.

[25]See Gordon L. Herries Davies, "Irish Thought in Science," pp. 294-310 in R. Kearney (ed.), *The Irish Mind: Exploring Intellectual Traditions* (Wolfhound Press: Dublin, 1985), pp. 306-10; Nicholas Whyte, "'Lords of Ether and of Light': The Irish Astronomical Tradition of the Nineteenth Century," *Irish Review* (1995) 17/18: 127-41, pp. 136-38; Nicholas Whyte, *Science, Colonialism and Ireland*, pp. 39-41, 184-6; Steven Yearly, "Colonial Science and Dependent Development: The Case of the Irish Experience," *Sociological Review* (1989) 37: 308-31, pp. 322-26.

[26]See Dorinda Outram, "Negating the Natural: Or Why Historians Deny Irish Science," *The Irish Review* (1986) 1: 45-49; Dorinda Outram, "Heavenly Bodies and Logical Minds: Banville, Science and Religion," *Graph: Irish Literary Review* (1988) 1: 9-11; John Wilson Foster, "Natural Science and Irish Culture," *Eire-Ireland* (1991) 26: 92-103; John Wilson Foster, "Against Nature? Science and Oscar Wilde," *University of Toronto Quarterly* (1993-94) 63: 328-46; John Wilson Foster, "Natural History in Modern Irish Culture," pp. 119-33 in Bowler and Whyte (eds.), *Science and Society in Ireland*, Sean Lysaght, "Science and the Cultural Revival, 1863-1916," pp. 153-65 in Bowler and Whyte (eds.), *Science and Society in Ireland*; Sean Lysaght, "Themes in the Irish History of Science," *Irish Review* (1996) 19: 87-97.

[27]D.H. Akenson, *A Mirror to Kathleen's Face: Education in Independent Ireland, 1922-1960* (Montreal: McGill-Queen's University Press, 1975), pp. 44, 48.

of science in Ireland, more fundamental challenges were at work.[28] The Catholic Church did oppose all forms of non-Catholic education and fought against the teaching of evolution and scientific materialism, but it gave an important role to science and medicine within the context of Catholic university education. (In fact, UCD would outpace Trinity in a number of areas of science in the first half of the twentieth century.) The Revivalists, for their part, did not criticize any particular science being practiced in Ireland. (And Hamilton's and Fitzgerald's idealism fit comfortably, in many respects, with Yeats' philosophical views.) But while neither the Catholic Church nor Irish nationalism actively opposed scientific research and education, neither made it central to their mission. In the context of the struggle for independence and the creation of a new state, science was a low priority.

A number of other factors actively discouraged the pursuit of science at the beginning of the twentieth century. Ireland was affected by war for much of the period, from World War I (1914-1918) to the War of Independence (1919-1921) to the Civil War (1922-1923). In terms of technology, Ireland at the beginning of the twentieth century was still an agrarian economy, and at the time of Independence the industrialized areas of Ulster remained within the United Kingdom. The Irish Free State protected indigenous industry with high tariff barriers, but the result was a lack of incentives to modernize and innovate.

Historian Nicholas Whyte has given a detailed account of how the major scientific institutions made the transition from the Union to Independence. The problem, as Whyte sees it, is that Ascendancy science always remained isolated from the main currents of Irish life. The Ascendancy had excluded Catholics from scientific education and participation in scientific institutions like the Royal Irish Academy. Moreover, they constantly stressed the connections

---

[28]See Whyte, *Science, Colonialism and Ireland*, pp. 151-81, and Greta Jones, "Catholicism, Nationalism and Science," *Irish Review* (1997) 20: 47-61.

between science and Protestant values. Science for them represented a component of the cultural and political union between Ireland and Great Britain. It is hardly surprising, therefore, that most nationalists saw little role for science and technology in their own concerns. Ultimately, however, science declined with the destruction of the propertied elite who had the time and resources to pursue science. Government and industry in Ireland failed to step in to fill that role. Whyte argues, "The common thread is not the downgrading of science but the cutting of government expenditure. ... The new Free State was not hostile to modernity. It was simply hostile to spending money."[29]

Underfunded and isolated, Trinity entered what historians of the college McDowell and Webb describe as a period of "low profile." After Fitzgerald's death in 1901, his best students left for other universities across Ireland, the United Kingdom, and the Commonwealth. Fitzgerald's successor, Billy Thrift, who held the Erasmus Smith Professorship in Natural and Experimental Philosophy from 1901 to 1929, did not distinguish himself as a researcher. R.W. Ditchburn was recruited from Cambridge to succeed Thrift but left at the outbreak of World War II to do war research in England. H.C. Plummer resigned as Royal Astronomer of Ireland in 1921 in order to take a better paying position in England, and Trinity decided not to replace him. Dunsink Observatory was closed. Even with the new Physical Laboratory completed in 1905, Trinity was hopelessly behind and would fall farther behind as other universities invested in state-of-the-art research laboratories.

8.2 Understanding the Atom

The death of George Francis Fitzgerald in 1901 not only took one of Trinity's most gifted theoretical physicists, it also marked an

---

[29]Whyte, *Science, Colonialism and Ireland*, p. 148.

abrupt end to the budding tradition of experimental physics at
Trinity. The career paths of Trinity's brightest physics graduates
at the beginning of the twentieth century indicate that the decline
of science in Ireland had little to do with direct interference from
the Catholic Church, the Celtic Revival, or Irish nationalism. Trinity continued to attract strong students (many from the North) and
to educate them without any outside influence, but after their degrees most students found better opportunities for research outside
of Trinity.[30] John Joly was one of the few to remain at Trinity. He
served as Fitzgerald's assistant before becoming professor of geology, doing pioneering research on radioactivity.[31] J.S.E. Townsend
went to the Cavendish Laboratory at Cambridge as one of its first
external research students in 1895 and later became professor of
experimental physics at Oxford. Frederick Trouton had assisted
Fitzgerald in recreating Hertz's experiments with electromagnetic
waves. He became professor of physics at University College, London.[32] E.E. Fournier d'Albe, another assistant, went on to transmit
the first wireless picture broadcast from London. John Burke went
to Cambridge as an advanced research student in 1900 and later
taught at Mason College in Birmingham and Owens College in
Manchester. Stephen Dixon became a professor of civil engineering in New Brunswick and later Imperial College London, rising to
Dean of the City and Guilds Engineering College. Thomas Ranken
Lyle became Professor of Natural Philosophy at Melbourne Uni-

---

[30] D. Weaire, "Experimental Physics at Trinity," pp. 239-54 in C. H. Holland (ed.) *Trinity
College Dublin & and the Idea of a University* (Dublin: Trinity College Dublin Press, 1991),
pp. 274-77.

[31] J.H.J. Poole, "John Joly," *Hermathena* (1958) 92: 66-73, John R. Nudds, "John Joly:
Brilliant Polymath," pp. 162-78 in Nudds *et al.* (eds.), *Science in Ireland*, Patrick Wyse
Jackson, "A Man of Invention: John Joly (1857-1933): Engineer, Physicist and Geologist,"
pp. 89-96 in D. Scott (ed.) *Treasures of the Mind* (London: Sotheby, 1992).

[32] Andrew Warwick, "The Sturdy Protestants of Science: Larmor, Trouton, and the
Earth's Motion Through the Ether," pp. 300-343 in J. Z. Buchwald (ed.) *Scientific Practice:
Theories and Stories of Doing Physics* (Chicago: University of Chicago Press, 1995); J.W.
Fox, "From Lardner to Massey: A History of Physics, Space Science and Astronomy at
University College London, 1826-1975," www.phys.ucl.ac.uk/department/history/BFox1.
html#Fox140

versity and erected the first physical laboratory south of the equator.[33] Thomas Preston became professor of natural philosophy at the Royal University of Ireland.[34] Unable to find positions or resources at Trinity, her best scientists went to other institutions.

Trinity's decline, ironically, coincided with a period of rapid new developments in all of the areas of science that Fitzgerald had studied including radio waves, vacuum tubes, spectroscopy, and cathode rays, and his students went on to make important contributions in all of these areas (just not at Trinity). In the last decade of the nineteenth century and the first three decades of the twentieth century, scientists made a series of breakthroughs that began to reveal for the first time the mysteries of atomic structure. They discovered subatomic particles (electrons, protons, and neutrons) and used them to probe the atomic nucleus itself. Their experiments not only led to the radical new theory of quantum mechanics, but they also lay the foundation for many of the most important technologies of the twentieth century: the cathode ray television screen, the x-ray, the transistor, microwaves, radar, and even nuclear energy.

All of these experiments required increasingly expensive equipment—high voltage generators, electromagnets, vacuum pumps, diffraction gratings, precision measuring instruments, as well as supplies of glass, metal, and chemical compounds. They also demanded dedicated laboratory space and full-time research staff. Even with the new Physical Laboratory, Trinity's facilities could not compare with those of the Cavendish Laboratory at Cambridge or even the laboratories of the Royal University of Ireland. (Oxford, long a laggard in experimental science, soon surpassed Trinity thanks in large part to the work of a Trinity graduate, J.S.E.

[33] Denis Weaire, "Thomas Ranken Lyle" in Charles Mollan, Brendan Finucane and William Davis (eds.), *Irish Innovators in Science and Technology* (Dublin: Royal Irish Academy and Enterprise Ireland, 2002).

[34] D. Weaire and S. O'Connor, "Unfulfilled Renown: Thomas Preston (1860–1900) and the Anomalous Zeeman Effect," *Annals of Science* (1987) 44: 617-44.

Townsend.) While other universities were creating new research positions and graduate fellowships, Trinity remained short-staffed and under-resourced well into the 1950s.

The new lines of research grew out of the long history of the study of light. Molyneux had found in the eighteenth century that geometrical optics was critical to understanding perception and validating the evidence of the newly invented telescope and microscope. Hamilton and the wave theorists of the early nineteenth century developed sophisticated mathematical techniques for explaining a wide range of optical phenomena, while in the late nineteenth century, Fitzgerald and the Maxwellians succeeded in unifying the forces of electricity and magnetism, showing that light is actually an electromagnetic wave. They devoted their work to modeling the ether, trying to explain the underlying mechanisms that support all of the phenomena of electricity, magnetism, and optics. At the end of the century, Larmor integrated ether and matter, hypothesizing that electrons are actually twists in the ether. All physical phenomena, he believed, could be explained by the characteristics of the ether.

Fitzgerald's students would use the study of light to probe the composition of matter. In the late nineteenth century, the Germans Kirchhoff and Bunsen had discovered that each element produces a unique color when burned or sparked. When passed through a prism that color generates a distinct set of lines at different wavelengths of light. Fitzgerald's uncle G.J. Stoney, professor of natural philosophy at Queen's College Galway, had suggested in 1868 that light is produced by electrons moving around within an atom—and the frequency of light depends on the frequency of electron motion.[35] The patterns of light, therefore, provide a map to the interior of the atom. Sir Walter Hartley, an Englishman who served as professor of chemistry at the Royal College of Science in Dublin

---

[35] John Joly, "George Johnstone Stoney," *Proceedings of the Royal Society* (1912) 86A: xx-xxxv.

from 1879 to 1912, pioneered the systematic study of these spectral lines. He was the first to establish a relationship between the spectral lines emitted by an element and its position in the periodic table, confirming that the specific wavelengths of light emitted by an atom indicate its unique atomic structure.[36]

In 1896 Dutch physicist Pieter Zeeman had found that the spectral lines of certain elements are split when they are placed in a strong magnetic field. The magnetic field, he believed, must somehow change the motion of the electrons within an atom, causing them to emit light at different frequencies. Zeeman's experiment was seen as an important confirmation of the electron theory of matter. Thomas Preston, one of Fitzgerald's first students and laboratory assistants, set out to verify Zeeman's results, hoping the experiment would provide "some clearer insight into the structure of matter itself."[37] In 1891 Preston had become Professor of Natural Philosophy at University College Dublin and also a fellow of the Royal University of Ireland. While UCD lacked the facilities Preston required to pursue his research, his affiliation with the RUI gave him access to the laboratories used for examinations. By 1899 Preston had actually found a more complicated set of spectral lines. While Zeeman's observations could be explained by Larmor's theory (the magnetic field changes the orientation of the orbits of the electrons), Preston's so-called Anomalous Zeeman effect could not be explained by any existing theory. In fact, its explanation would require quantum theory.

While Preston focused on the electromagnetic waves emitted by different elements, others were looking at the radiation produced when electric currents are passed through glass tubes filled with gas at low pressure. As they pumped air out of these vacuum tubes,

---

[36]P.K. Carroll and S. O'Connor, "Walter N. Hartley: Pioneering Spectroscopist," *Irish Chemical News* (1998) 12: 41-46.

[37]Thomas Preston, "Radiating Phenomena in a Strong Magnetic Field, Part II: Magnetic Perturbations of the Spectral Lines," *Scientific Transactions of the Royal Dublin Society* (1899) 7: 7-22, p. 20.

they found that at a certain pressure the gas in the tube will glow (the principle behind neon signs). In the 1870s, the English scientist William Crookes found that by reducing the pressure even further, the gas in the tube went dark, but the glass at the end of the tube (opposite the negative electrical terminal or cathode) began to glow. The phenomenon, he explained, was caused by "cathode rays" which travel in a straight line from the cathode and hit the glass at the other end of the tube, causing it to fluoresce (a phenomenon that became the basis for cathode ray televisions). It was while experimenting on a modified version of Crookes' vacuum tubes in 1895 that the German physicist Wilhelm Röntgen discovered a new form of radiation that could create an image of the bones in his hand on a photographic plate. He called them x-rays. In various forms, therefore, these vacuum tubes became the fundamental technology underlying fluorescent light, television screens, x-ray machines, the first transistors, and even, as we shall see, particle accelerators. In the 1890s, however, the nature of these rays was a mystery that obsessed experimenters around the world. Were they electromagnetic waves or were they particles? Were they different frequencies of the same underlying phenomena or were they distinct?

Röntgen published his discovery of x-rays in December of 1895, and by February of 1896 John Joly had made his own x-ray photographs (the first in Ireland). He exhibited x-rays of his spectacles inside their case, a coin inside a wooden box, and the bones of his hand to the Dublin University Experimental Science Association. That same year, French scientist Henri Becquerel discovered that uranium salts emit rays that resemble x-rays, but while x-rays were produced by passing large electrical currents through vacuum tubes, these rays were emitted spontaneously without any external source of energy. Marie Curie soon found that another element, radium, emitted radiation that was significantly more powerful than uranium. By 1903, scientists had distinguished three types of radiation: alpha, beta, and gamma rays. Alpha particles had a positive charge and could be stopped by a sheet of paper. Beta particles had a negative charge and were stopped by an aluminum plate. Gamma rays had no charge and could only be stopped by lead.

Marie and Pierre Curie eventually recognized that the radiation from radium damages human cells. (In fact, Marie Curie died from her exposure to radiation.) Interestingly, however, they also found that radiation destroys tumor cells faster than healthy cells. It could therefore be used to treat cancer, and Joly became one of the leading exponents for the therapeutic potential of radiation. He convinced the Royal Dublin Society to found a Radium Institute in 1914 and in collaboration with Dr. Walter Stevenson invented a method for treating cancer by inserting radium in hollow needles into tumors (known as 'the Dublin method').[38]

When Joly became Professor of Geology and Mineralogy at Trinity in 1897, he realized that radioactivity could also help in the study of the age of the earth. He recognized that spherical colorations in certain minerals (known as pleochroic haloes) were the result of alpha particles emitted from small amounts of radioactive uranium and thorium in the samples. The diameter of the colorations depends on the amount of time that has elapsed since the rock was formed. He was able to calculate that the rocks must have been formed at least 400 million years ago—implying that the earth was dramatically older than physicists like Lord Kelvin had argued (in their attempt to discredit Darwin's theory of evolution).[39]

While Joly continued to contribute to science at Trinity, by the 1890s, the center for experimental physics in Britain, if not the world, was the Cavendish Laboratory at Cambridge University. In

---

[38] D. Murnaghan, "History of Radium Therapy in Ireland: The 'Dublin Method' and the Irish Radium Institute," *Journal of the Irish Colleges of Physicians and Surgeons* (1988) 17: 174-6.

[39] Joe D. Burchfield, *Lord Kelvin and the Age of the Earth* (Chicago: University of Chicago Press, 1990).

1897, its director J.J. Thomson finally discovered the nature of cathode rays. He concluded that the rays were actually composed of very small, light particles with a negative charge. Thomson called them "corpuscles," but Larmor quickly identified them as the electrons of his theory.[40] In what came to be called the "plum pudding" model, Thomson envisioned atoms as a positively charged mass with small electrons orbiting within (the plums in the pudding).

Thomson assigned one of his postgraduate research students the task of measuring the charge of the newly-discovered electron. That student was a Trinity College Dublin graduate named John Sealy Edward Townsend. Townsend, the son of a professor of civil engineering at Queen's College Galway, had completed his degree at Trinity in 1890. He had unsuccessfully competed on the fellowship examination at Trinity four times between 1892 and 1895.[41] Luckily, in 1895, the Cavendish offered fellowships for external students for the first time. They accepted two students in that first year, Townsend and a young New Zealander named Ernest Rutherford. Townsend found that bubbling a charged gas through water produces a fine mist of water droplets as the water vapor condenses on the charged particles. By measuring the total charge of the mist and dividing by the (laboriously counted) number of droplets, he ended up with a measure of the unit of charge of 1.7×10^{-19} coulombs (close to modern value of 1.6). His work inspired the famous oil-drop experiment that earned the American Robert Millikan the Nobel Prize. It also presaged C.T.R. Wilson's develop-

---

[40]Jed Z. Buchwald and Andrew Warwick (eds.), *Histories of the Electron: The Birth of Microphysics* (Cambridge: MIT Press, 2001); Isabel Falconer, "Corpuscles, Electrons and Cathode Rays: J.J. Thomson and the 'Discovery of the Electron'," *British Journal for the History of Science* (1987) 20: 205-242.

[41]David Attis, "John Sealy Edward Townsend," pp. 202-3 in Charles Mollan, Brendan Finucane, and William Davis (eds.), *Irish Innovators in Science & Technology* (Dublin: Royal Irish Academy and Enterprise Ireland, 2002).

ment of the cloud chamber, an early particle detector that used clouds of vapor condensing around particles to show their paths.

Townsend continued his research at the Cavendish until 1900 when he was appointed the first holder of the new Wykeham Chair of Experimental Physics at Oxford where he oversaw the construction of a new physical laboratory.[42] At Oxford he continued his study of electricity in gases, and his theory of ionization by collision became the basis for the 1908 particle detector developed by Rutherford and Hans Geiger (later the Geiger-Müller Counter for detecting radioactivity). During World War I, he worked for the Royal Naval Air Service on wireless telegraphy and the cavity magnetron (a high frequency transmitter essential to development of radar). Despite his critical role in the experimental developments of the 1910s, however, after WWI, Townsend became increasingly isolated scientifically. Townsend was in his fifties when the new physics first appeared, and he never accepted either quantum mechanics or relativity.

The year after Townsend arrived at the Cavendish, another Irishman joined him. John Alexander McClelland was from Coleraine and had attended Queen's College Galway before coming to the Cavendish on a research fellowship. (He may have chosen Galway over Trinity College Dublin because the Professor of Natural Philosophy in Galway, Alexander Anderson, was a Coleraine man who had succeeded Joseph Larmor.) At the Cavendish, McClelland worked on x-rays, cathode rays, and the electrical conductivity of gases. He returned to Ireland in 1900 to take over the professorship of physics at University College Dublin after the death of Thomas Preston. In many ways it was an odd choice given that McClelland was an Ulster Presbyterian, and UCD was still under the direct control of the Jesuits (until 1908 when it became part of the National University of Ireland, though it remained de facto under Catholic

---

[42]B. Bleaney, "Two Oxford Science Professors: F. Soddy and J.S.E. Townsend," *Notes and Records of the Royal Society of London* (2002) 56: 83-88.

control). Asked by a Royal Commission about why he accepted the professorship, McClelland explained (in terms that will sound familiar to many academics) that these professorships "are not of great value but still Professorships are not so easily obtained that they can be treated with disrespect."[43] Besides, Preston had been able to accomplish important work on the anomalous Zeeman effect in the position.

UCD's laboratories were terribly equipped, but in 1901 McClelland (following Preston's example) won a fellowship at the Royal University of Ireland. He had to examine students, but he also got access to the relatively well-equipped laboratories used for practical examinations. McClelland extended his research on the electrical conductivity of gases to the study of the ionization of atmospheric air. In fact, McClelland's research at UCD is credited with the creation of long and proud tradition of Irish research on atmospheric electricity and aerosols (mixtures of particles or liquids in a gas such as clouds). Three of his students C.J. Power, P.J. Nolan, and John McHenry won NUI travelling studentships and spent time at the Cavendish. From 1934 to 1973 the chairs of experimental physics in the NUI constituent colleges at Cork, Dublin, Galway, and Maynooth were all held by people who had done their initial postgraduate research in atmospheric electricity and aerosols.[44] While Trinity languished, UCD under McClelland became "the largest and most active physics department in the country in the early part of the twentieth century."[45]

Ironically, Trinity's (and Ireland's) only Nobel Prize winner in science did his work during this low period in Trinity's scientific history. While he spent nearly his entire life at Trinity, Ernest

---

[43] Thomas C. O'Connor, "John A. McClelland, 1870-1920," pp. 176-85 in Mark McCartney and Andrew Whitaker (eds.) *Physicists of Ireland: Passion and Precision* (Bristol: IOP Publishing, 2003), p. 180.

[44] Thomas O'Connor, "The Scientific Work of John A. McClelland: A Recently Discovered Manuscript," *Physics in Perspective* (2010) 12: 266-306.

[45] O'Connor, "The Scientific Work of John A. McClelland," p. 288.

Walton did his ground-breaking research during a brief stint as a post-graduate student at the Cavendish. Walton was the son of a Methodist minister and spent most of his childhood in Northern Ireland.[46] He came to Trinity in 1922, finished an MSc in 1927, and then decided to go to Cambridge, still the best place to study physics. By this time, the Cavendish had already produced four Nobel Prizes.

Ernest Rutherford, Townsend's fellow student back in the 1890s, was now the director of the Cavendish and one of the most important members of the global physics community, having won a Nobel Prize in 1908. Rutherford had discovered that alpha particles are positively charged nuclei of helium atoms (later to be identified as two protons and two neutrons). While these particles could penetrate a thin gold foil, Rutherford found surprisingly that a very small percentage bounced back. He explained, "It was almost as if you fired a 15-inch shell at a piece of tissue paper and it came back and hit you."[47] He quickly came to understand that the alpha particles had hit something smaller than an atom but much more dense—the nucleus. Contrary to Thomson's plum pudding model, the positive charge in an atom wasn't spread out, it was highly concentrated in the center. The alpha particles that struck the nucleus ricocheted, while the majority that missed the tiny nucleus passed right through. In fact, Rutherford later found that some of the alpha particles that hit the nucleus stuck, adding their own mass and charge to the nucleus. By bombarding nitrogen (atomic weight 14) with alpha particles, he was able to transform some of the atoms into oxygen (atomic weight 17) plus a proton that was ejected. Rutherford had fulfilled the ancient alchemists' dream of transmuting one element into another.

---

[46]Vincent J. McBrierty, *Ernest Thomas Sinton Walton, 1903-1995: The Irish Scientist* (Dublin: Trinity College Dublin Press, 2003).

[47]Cited in McBrierty, *Walton*, p. 22.

The era of nuclear physics had begun, but Rutherford was limited in his ability to break up nuclei because the naturally occurring alpha particles emitted by radium that he was using simply did not have enough energy. In his Presidential Address to the Royal Society in 1927 he described how an artificially produced stream of very fast particles would open up a whole new set of possible nuclear experiments. Ten days later, when the young Walton had his first meeting with Rutherford, he proposed just such an experimental approach, having no idea that this was already Rutherford's top research priority. Walton, working with the Englishman John Cockroft, was tasked with building the world's first particle accelerator. Walton and Cockroft, however, faced a seemingly insurmountable challenge.[48] Their calculations showed that it would require millions of volts of electricity to accelerate hydrogen nuclei (protons) to the speeds required to overcome the electrical repulsion of the nucleus. (Since the nucleus is positively charged and protons are positively charged, they repel each other.) Rutherford believed that it would require eight million volts, significantly higher than any sustained voltage produced in a laboratory before. They began building their apparatus, but they were skeptical that they would be able to reach the voltages necessary. A breakthrough came when a visiting Russian physicist, George Gamow, calculated that a weird quantum effect might allow some of the protons to "tunnel" through the electrical barrier surrounding the nucleus. While the probability for each proton was very small, if they could produce a large stream of particles artificially, they could be confident that some would get through. It turned out that they would only need to produce 300,000 volts of electricity rather than millions. Suddenly, success seemed within their grasp.

Building the apparatus was still a challenge. Working with enormous voltages safely in the laboratory was their greatest obstacle.

---

[48]Brian Cathcart, *The Fly in the Cathedral: How a Group of Cambridge Scientists Won the International Race to Split the Atom* (New York: Farrar, Straus and Giroux, 2004).

The need for x-ray machines for medical purposes had led to the development of electrical transformers that could operate at up to 500,000 volts. At the same time, the commercial electric companies that were building power networks across England had realized that transmitting electricity over long distances is more efficient at very high voltages. Companies like Metropolitan Vickers (Metro-Vick) in Manchester had their own research laboratories to develop high voltage technologies. Walton's partner John Cockroft was from Manchester. He had a degree in electrical engineering from Manchester College of Technology, and he had worked for Metro-Vick. In fact, Metro-Vick had funded Cockroft's first degree at Cambridge. Cockroft's continued connections to the company ended up being critical for the high voltage experiments at the Cavendish.

The apparatus that Cockroft and Walton built was a twelve foot high tower with a hydrogen discharge tube at the top to supply the protons. A large Metro-Vick transformer (paid for with a £500 grant from the university) generated up to 800,000 volts of alternating current electricity, while a custom-built rectifier (a series of evacuated glass cylinders with electrodes and hot filaments) transformed the alternating current to a direct current that could be used to accelerate the protons. The protons were aimed at a lithium target and scintillation screens were arranged to detect any particles that might be ejected.

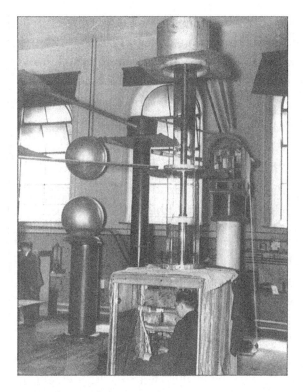

Figure 8.4: E.T.S. Walton and His Particle Accelerator (1932). ©Walton Family

On April 14, 1932, Walton made his first successful observation.[49] When he observed alpha particles coming from the lithium, he knew what had happened, and he immediately called Rutherford in to confirm the discovery. The lithium nucleus had an atomic mass of seven (three protons and four neutrons). When a proton penetrated the nucleus, it increased the atomic mass to eight, but the nucleus immediately split into two alpha particles (two protons and two neutrons each), each with a mass of four. Cockroft and Walton had become the first to split the nucleus. Their discovery was important for a number of reasons. First, the adjustable pro-

---

[49]J.D. Cockroft and E.T.S. Walton, "Disintegration of Lithium by Swift Protons," *Nature* (1932) 129: 242.

ton beam that they had created enabled physicists to much more directly control the number and energy of particles that they could use to bombard various targets. American physicists quickly used this demonstration to raise millions of dollars in funding from large foundations to build larger and larger particle accelerators. Second, they had verified Gamow's theory that protons could tunnel into the nucleus. And third, they had demonstrated that it is possible to destroy matter and create energy, just as Einstein had predicted. The two alpha particles produced weighed slightly less than the original proton and lithium nucleus (the difference was just 0.02 units of atomic weight). In other words, the disintegration process had actually destroyed some of the mass of the original particles. The missing mass, however, was precisely equal to the increased energy of the alpha particles that were released (about sixteen million electron volts). They had provided the first artificially-generated experimental proof of Einstein's equation, $E = mc^2$.

Even the popular press recognized the importance of the experiment. *Reynold's Illustrated News* called it "the greatest scientific discovery of the age."[50] The *New York Times* covered the story on its front page. The *Irish Independent* reported "Atom 'Split' by Irishman—A Discovery of Great Scientific Importance." Cockroft and Walton instantly became celebrities. Rutherford had to tamp down press reports that the discovery would lead to a new source of energy. In an article in the *Times*, he cautioned, "It was a very poor and inefficient way of producing energy, and anyone who looked for a source of power in the transformation of the atoms was talking moonshine."[51] Since the apparatus used more energy than it created, it could not be used as a power source.

Ironically, that very article in which Rutherford dismissed the most radical implications of Cockroft and Walton's experiment, spurred another scientist to recognize the horrific potential of nu-

---

[50]Cathcart, *The Fly in the Cathedral*, p. 246.
[51]"The British Association–Breaking Down the Atom," *The Times* (September 12, 1933).

clear disintegration. Leó Szilárd, a Hungarian physicist in London, read Rutherford's account on September 12, 1933. Later that day, a realization dawned on him. While protons require significant energy to penetrate the nucleus (because of their positive charge), neutrons (discovered in 1932 by Cockroft and Walton's Cavendish colleague James Chadwick) can penetrate the nucleus much more easily. "[It] suddenly occurred to me," Szilárd later recalled, "that if we could find an element which is split by neutrons and which would emit *two* neutrons when it absorbs *one* neutron, such an element, if assembled in sufficiently large mass, could sustain a nuclear reaction."[52] In other words, the nuclear reaction once started would continue on its own, releasing enormous amounts of energy from a very small input. Splitting much heavier elements (like uranium) would generate more energy than lithium as well as lots of neutrons. Szilárd's insight set off its own chain reaction that resulted thirteen years later in the detonation of the first two atomic bombs over Hiroshima and Nagasaki.

Walton very consciously played no role in these subsequent developments. While he could have had his pick of positions at any number of top laboratories around the world, he decided to return to Trinity College Dublin. He had failed to pass the fellowship examination in 1928, but now he was elected a fellow under a new ordinance that enabled candidates with "distinguished merit as shown primarily by published work" to bypass the examination altogether. Walton dedicated the rest of his career to teaching at Trinity. He was so dedicated that when Chadwick invited him to join a group of people going to the United States to do top secret war research, Walton politely declined. As he later explained, "I wasn't very keen on going and I consulted the Provost about it. Provost Alton wouldn't hear of me going. ... If I had gone it would have left two people to run the department. ... Well of course later

---

[52]Richard Rhodes, *The Making of the Atomic Bomb* (New York: Simon & Schuster, 1986), p. 28.

on I realized what this job meant; it was the organization working on the atomic bomb."[53]

Just before he returned to Trinity, Walton was invited to attend the Solvay Conference in Brussels in October of 1933. The invitation-only conference brought together the top physicists in world: Niels Bohr, Werner Heisenberg, Marie Curie, Erwin Schrödinger, Enrico Fermi, Paul Dirac, Wolfgang Pauli, E.O. Lawrence, and others. Half of the men and women in attendance had already won or would go on to win a Nobel Prize. These were the architects of the quantum revolution.[54] In less than thirty years they had completely changed the way we understand the laws of nature. Walton must have felt uncomfortable in such august company. Not only was he just a postgraduate student, but he was an experimentalist rather than a theorist. The Cavendish emphasized the experimental investigation of natural phenomena. Rutherford once quipped, "Theorists play games with their symbols, but we, in the Cavendish, turn out the real solid facts of Nature."[55] Walton was a tool-maker. He later recalled, "Tools have always had a fascination for me.... They could be used to put new ideas into concrete form and they could produce machines and instruments not available on the market."[56] His work on the experimental apparatus made it difficult for him to keep pace with the barrage of theories.[57]

The new theories were not only mathematically complex, they also represented a universe radically different from our everyday experience. Energy is not continuous; it is released in small packets

---

[53] McBrierty, *Walton*, p. 46.

[54] Suman Seth, "Quantum Physics," pp. 814-59 in Jed Z. Buchwald and Robert Fox (eds.), *The Oxford Handbook of the History of Physics* (Oxford: Oxford University Press, 2013).

[55] A. S. Eve, *Rutherford: Being the Life and Letters of the Rt. Hon. Lord Rutherford, O. M.* (New York: Macmillan, 1939), p. 304.

[56] Charles Mollan, "Ernest Thomas Sinton Walton (1903-1995)," pp. 1575-95 in Charles Mollan, *It's Part of What We Are: Some Irish Contributors to the Development of the Chemical and Physical Sciences, Vol. II* (Dublin: Royal Dublin Society, 2007), p. 1580.

[57] Cathcart, *The Fly in the Cathedral*, p. 189.

or quanta. Light is both a wave and a particle. Electrons do not emit light continuously as they orbit within the atom, but only when they jump from one energy state to another. Just as light can behave like a particle, electrons can behave like waves. In fact, electrons don't orbit at all in the traditional sense, rather they exist in a probability cloud. It is impossible to say precisely where an electron (or any particle) is.

While quantum mechanics revealed the bizarre nature of the world within the atom, Einstein's theory of relativity exposed the unexpected structure of space-time. Starting with the assumption that the speed of light in a vacuum is constant for all observers, he found that objects that move at speeds close to the speed of light must contract in length, increase in mass, and even slow down in time. The Fitzgerald-Lorentz transformation did not describe the effect of ether as Fitzgerald had originally proposed, but it did predict precisely how observers moving at different speeds will measure time and distance differently. Einstein's general theory of relativity began with the assumption that accelerated motion and being at rest in a gravitational field are equivalent. From that one simple hypothesis, Einstein found that space is curved and light is bent by gravitational fields—another example of how the study of light held the key to both the composition of matter and the structure of the universe.

By the Solvay Conference in 1933, it was clear that a revolution had occurred in physics. The ether was abandoned, as was the concept of Euclidean space, the fundamental difference between waves and particles, even the very idea of deterministic laws of physics. Everything that scientists knew before about 1910 would come to be called 'classical physics' while everything since then— specifically quantum mechanics and relativity—would be known as 'modern physics'.[58] Walton had a ring side seat for this revolution,

---

[58]Graeme Gooday and Daniel John Mitchell, "Rethinking 'Classical Physics'," pp. 721-64 in Buchwald and Fox (eds.), *The Oxford Handbook of the History of Physics*.

and yet, after the Solvay Conference, he returned to Trinity and never again published any important work in physics.

8.3 The Hamiltonian Revival

Trinity's great tradition in mathematical and theoretical physics had evaporated—its leading exponents dead, their students scattered around the British Empire, their theories overturned by quantum mechanics and relativity, and their proud tradition now marginalized in a new independent state. And yet, by the 1930s, after both the Irish and the quantum revolutions had been secured, it was possible to look back and see continuity, to reclaim the traditions that had been lost. In 1940, Edmund Whittaker, an English mathematician at the University of Edinburgh, pointed to a 'Hamiltonian Revival' that had taken place.[59] Whittaker himself had briefly held Hamilton's own position as Royal Astronomer of Ireland and professor of astronomy at Trinity College Dublin from 1906 to 1911. (In fact, de Valera attended his lectures at the time, and the two remained friends for many years.) Since the beginning of the twentieth century, he explained, "one after another, the significance of his great innovations has been appreciated: and today he is placed among the greatest, in the class inferior only to Newton."[60] William Rowan Hamilton's work resurfaced not only as the mathematical key to the new sciences but also as the standard bearer for a reemerging Irish mathematical tradition. Whittaker, in fact, linked the Hamiltonian revival to Ireland's national rebirth. He described the publication of Hamilton's collected mathematical papers as "one of the most agreeable fruits of the renaissance

---

[59] E.T. Whittaker, "The Hamiltonian Revival," *The Mathematical Gazette* (1940) 24: 153-58.

[60] Whittaker, "The Hamiltonian Revival," p. 154.

which has followed the achievement of national independence in Ireland."[61]

Hamilton's mathematics turned out to be surprisingly useful in both quantum mechanics and relativity theory. In the hands of Erwin Schrödinger, the Hamiltonian equation became the basic formulation for quantum mechanics. Schrödinger explained, "The central conception of all modern theory in physics is 'the Hamiltonian'. If you wish to apply modern theory to any particular problem, you must start with putting the problem 'in Hamiltonian form.'" Schrödinger went on, "I daresay not a day passes—and seldom an hour—without somebody, somewhere on this globe, pronouncing or reading or writing or printing Hamilton's name.... [62] And while Hamilton's quaternions failed to become the universal language of physics as Hamilton had hoped, they did keep cropping up in unexpected places. Whittaker pointed out that Pauli's spin matrices "are simply Hamilton's three quaternion units i, j, k,"[63] and quaternions could also be used to express Dirac's equation for the spinning electron.[64] Quaternions turned out to be useful in relativity theory as well because they can be applied to rotations in four-dimensional space. They led to a particularly elegant expression for the Fitzgerald-Lorentz transformation. Whittaker suggested that the quaternion algebra of 1843 "may even yet prove to be the most natural expression of the new physics."[65]

Hamilton had once again become a mathematical hero, but he also became an Irish hero (even appearing on an Irish postage

---

[61] Whittaker, "The Hamiltonian Revival," p. 154.

[62] "The Hamilton Postage Stamp: An Announcement by the Irish Minister of Posts and Telegraphs," in D.E. Smith (ed.), *A Collection of Papers in Memory of Sir William Rowan Hamilton* (New York: Scripta Mathematica, 1945), p. 82.

[63] Whittaker, "The Hamiltonian Revival," p. 158.

[64] J. Riversdale Colthurst, "The Influence of Irish Mathematicians on Modern Theoretical Physics," *The Mathematical Gazette* (1943) 27: 166-70.

[65] Whittaker, "The Hamiltonian Revival," p. 158.

Figure 8.5: William Rowan
Hamilton Postage Stamp
(1943)

stamp in 1943). Hamilton became a rallying point for a diverse
group of mathematicians and physicists working in Ireland in the
1940s. The greatest symbol and instrument of the Hamiltonian revival was the Dublin Institute for Advanced Study (DIAS). DIAS
was the brainchild of Éamon de Valera, who had idolized Hamilton
since his days as a mathematics teacher. In July 1939, fresh from
his success in framing the Irish constitution, de Valera introduced
a bill into the Dáil to create an Institute for Advanced Study in
Dublin. The Institute was modeled after the Institute for Advanced
Study in Princeton, New Jersey (founded in 1930), where Einstein
and a number of other European scientists had settled after fleeing
the Nazi's. The Irish Institute, however, would bring together two
fields which at first sight might seem wholly incompatible—Celtic
Studies and Theoretical Physics. De Valera saw the School of Theoretical Physics as a way to regain Ireland's reputation for scientific
research. He told the Dáil,

> This is the country of Hamilton, a country of great math-
> ematicians. We have the opportunity of establishing now

a school of theoretical physics...which I think will again
enable us to achieve a reputation in that direction comparable to the reputation which Dublin and Ireland had
in the middle of the last century.[66]

For de Valera, DIAS would be a means to rebuild Ireland's intellectual reputation. Just as Robert Burrowes had argued in the
inaugural volume of the *Transactions of the Royal Irish Academy* in
the late eighteenth century (see chapter 3), de Valera believed that
science and mathematics just like the Irish language were critical
components to the cultural development of the Irish nation. The
Irish Times welcomed the proposal as evidence "that this country is leaving behind at last its parochialism, its suspicions and
its petty jealousies; that instead of prattling childishly any longer
about imagined slights and ill-treatments, it intends to take its
entitled place as a free adult among neighbour nations!"[67]

Few others, however, shared de Valera's enthusiasm. John Dillon, one of the founders of the opposition Fine Gael party, attacked
DIAS as "a Machiavellian scheme...being done for the purpose of
obtaining cheap and fraudulent publicity for a discredited administration.... In a year when bread was rationed and milk was a
rarity, the establishment of an Institute of Advanced Studies recalled Gulliver's travels to Laputa."[68] Even de Valera's own party
was only lukewarm in its support. A scientist at DIAS later recalled, "I have never heard that any of his political colleagues had
personal enthusiasm for the proposed Institute, and I believe that
they let him have his way as a tribute to the high respect they had
for him as a great leader."[69]

---

[66] *Dail Debates* (July 6, 1939) 76: 1969-70.

[67] "Irish Culture," *The Irish Times* (1939).

[68] Walter Moore, *Schrödinger: Life and Thought* (Cambridge: Cambridge University Press, 1989), p. 434.

[69] Synge, "Eamon de Valera," p. 646.

For de Valera, the Institute was more than just a vanity project. He offered a number of justifications for the focus on theoretical physics. First, he acknowledged, Ireland simply did not have the resources to support world-class experimental physics. Laboratory research, he explained, "would require equipment altogether beyond our means." But to excel in theoretical physics, he argued, "all you want is an adequate library, the brains and the men, and just paper."[70] Secondly, de Valera believed that with a relatively small investment Ireland could attract a handful of great scientists, re-establishing Ireland's reputation as a world-leading center for theoretical physics. Third, by establishing a group of postgraduate students he hoped to provide an opportunity for top Irish students to continue their advanced work in Ireland and hopefully stem the wave of emigration of the most talented. And finally, de Valera wanted to create a neutral place where scientists from all of Ireland's universities (North and South) could leave behind historical animosities and collaborate.

Among de Valera's key advisors on the creation of DIAS were three prominent mathematicians from his student days. E.T. Whittaker, now a professor at Edinburgh, had taught de Valera when he was a student at Trinity. Paddy Browne, professor of mathematics at Maynooth and later President of UC Galway, had been one of de Valera's students, and Arthur Conway, professor of mathematics at UCD and later its president, had taught de Valera at Maynooth. Whittaker described Conway as "the most distinguished Irish Catholic man of science of his generation."[71] Born in Wexford, Conway had attended UCD and then went on to Oxford where he won the highest honors in mathematics. In 1901, he returned to UCD as professor of mathematical physics and senior fellow at RUI. He also lectured on mathematics at Maynooth from

---

[70] *Dail Debates* (July 6, 1939) 76: 1969-70.
[71] E.T. Whittaker, "Arthur William Conway, 1875-1950," *Obituary Notices of the Fellows of the Royal Society* (1951) 7: 329-40, p. 329.

1903 to 1910. Conway's research ranged from spectroscopy, atomic models, and quantum mechanics to relativity, and like de Valera, he shared a special fondness for Hamilton and quaternions.[72] Conway played a fundamental role in shaping UCD, serving as registrar from 1909 and 1940 and as President from 1940 to 1947. Conway was appointed the first Chairman of the Board of the School of Theoretical Physics.

Planning for DIAS also brought de Valera into contact with A.J. McConnell, the provost at Trinity College Dublin and a well-known mathematician in his own right. In fact, after Conway's death McConnell became de Valera's informal mathematical advisor, taking late night calls whenever de Valera had a question about the finer points of some equation. De Valera's friendship with McConnell was an important step towards re-integrating Trinity back into Irish society. DIAS would be a place outside the bitter debates about Irish university education and ecclesiastical control, where physicists from many different backgrounds could come together. According to McConnell, de Valera "was very conscious of the divisions that existed in the field of higher education in Ireland and was concerned in particular at the isolation, due to historical and religious reasons, in which Trinity College found itself after the foundation of the new state...."[73] In 1947, not long after the creation of DIAS, Trinity received its first annual grant from the Irish state.

---

[72] A.W. Conway, "The Influence of the Work of Sir William Rowan Hamilton on Modern Mathematical Thought," *Scientific Proceedings of the Royal Dublin Society* (1931) 20: 125-128.

[73] Synge, "Eamon de Valera," p. 648.

Figure 8.6: First Meeting of the Governing Board of the School of Theoretical Physics at the Dublin Institute for Advanced Studies (November 21, 1940). From left: Erwin Schrodinger, A.J. McConnell, Arthur Conway, D. McGrianna (Registrar), Eamon de Valera, William McCrea, Monsignor Patrick Browne, Francis Hackett. ©Dublin Institute for Advanced Study

In July of 1942, the School of Theoretical Physics held its first colloquium.[74] Erwin Schrödinger was there as the first director. Visiting physicist P.A.M. Dirac from Cambridge gave five lectures on quantum electrodynamics, and A.S. Eddington lectured on the unification of relativity theory and quantum theory. Locals including A.W. Conway, T.E. Nevin, Cormac O'Ceallaigh, and E.T.S. Walton also gave lectures. De Valera attended the proceedings, and afterwards all of the participants were invited to meet President Douglas Hyde. Subsequent years brought additional distinguished visitors: Max Born, Paul Ewald, Kathleen Lonsdale, Wolfgang Pauli, Leon Rosenfeld, James Chadwick, P.M. Blackett, Neville Mott, Rudolf Peierls, and others. De Valera had secured

---

[74]DIAS, "History of the School of Theoretical Physics," (www.dias.ie)

the support of Ireland's top mathematicians (who also happened to be the leaders of UCD, Trinity, and UC Galway) as well as the global community of physicists, but the nucleus of the new Institute was a complete outsider to Ireland—the Austrian physicist Erwin Schrödinger.

It was Whittaker who had originally suggested to de Valera that Schrödinger, under increasing pressure from the Nazi's, might be interested in coming to Ireland. De Valera had met secretly with Schrödinger in September of 1938 in Geneva, while de Valera was serving as President of the League of Nations. Schrödinger arrived in Dublin in October of 1939, and he ended up staying until 1956, the longest period that Schrödinger spent in any one place in his career. In fact, Schrödinger and his wife became citizens of Ireland in 1948. (According to his biographer, Schrödinger even made "a serious, but ultimately unsuccessful effort to learn the Irish language."[75])

Schrödinger was important not just because of his international fame and his network of collaborators but also because it was Schrödinger more than any other physicist who had rehabilitated Hamilton's reputation. The Hamiltonian equation could describe both particles and waves (see chapter 6), and Schrödinger, inspired by Hamilton's optical-mechanical analogy, made it the foundation of his wave mechanics. He wrote a Hamiltonian equation that described how the wave function for a particle or system evolves over time.

Schrödinger's similarities with Hamilton extended well beyond his use of the optical-mechanical analogy and the Hamiltonian function. Schrödinger also wrote poetry throughout his life, enjoyed socializing with artists, and shared a fondness for German metaphysics. Schrödinger, like Hamilton, believed that mathematical simplicity, intellectual beauty, and scientific truth are intimately

---

[75]Moore, *Schrödinger*, p. 373.

connected. The English mathematician Paul Dirac said of him, "Schrödinger and I both had a very strong appreciation of mathematical beauty and this dominated all our work. It was a sort of act of faith with us that any equations which describe the fundamental laws of Nature must have great mathematical beauty in them. It was like a religion with us."[76]

While Schrödinger remained scientifically active in Dublin, he devoted the years 1943 to 1951 to the unsuccessful search for a unified theory of gravitation and electromagnetism. Einstein had been working on the problem for 25 years without apparent success, and the two great scientists ended their careers frustrated by their inability to reconcile the two great theories of modern physics. Interestingly, Schrödinger's greatest contribution to science during his time in Dublin was to biology rather than physics.

One of the few obligations of the professors at the Institute for Advanced Study was to give a public lecture every year. These were typically well-attended affairs, bringing out Dublin's elite to hear about the latest scientific theories. In 1943, it was Schrödinger's turn to deliver a lecture at Trinity College Dublin. He decided to discuss the application of physics and chemistry to human biology, more specifically to genetics. The lectures, entitled "What Is Life?," became a sensation in Dublin at the time. The first lecture drew more than 400 people. *Time* magazine in the United States reported, "Only in the precarious peace of Eire could Europe today provide such a spectacle. At Dublin's Trinity College last month crowds were turned away from a jampacked scientific lecture. Cabinet ministers, diplomats, scholars and socialites loudly applauded a slight, Vienna-born professor of physics."[77] De Valera himself attended, and turnout was so large that Schrödinger had to repeat each of his Friday lectures on the following Mondays.

---

[76]Moore, *Schrödinger*, p. 384.
[77]"Schrödinger," *Time* (April 5, 1943).

Schrödinger began his lectures with a fundamental question: "How can the events in space and time which take place within the spatial boundary of a living organism be accounted for by physics and chemistry?"[78] In other words, how can science explain life? According to the second law of thermodynamics, all natural systems tend to disorder, and yet life is able to preserve its form, even over multiple generations. How can we explain the stability of life based on the laws of physics? The answer, Schrödinger argued, lies in the chromosome molecule. The chromosome must be a molecule that contains "in some kind of code-script the entire pattern of the individual's future development and of its functioning in the mature state."[79] Life is not some mystical force, he argued, it is a mathematical code.[80]

Schrödinger used the example of Morse code. Even a small number of molecular "letters" could create a code capable of specifying the enormous diversity of biological forms and functions. "The two different signs of dot and dash in well-ordered groups of not more than four," he explained, "allow of thirty different specifications. Now, if you allowed yourself the use of a third sign, in addition to dot and dash, and used groups of not more than ten, you could form 88,752 different 'letters'...."[81] The chromosome structures, however, are more than just a code, they must also provide the instructions for how to read the code: "They are law-code and executive power... architect's plan and builder's craft... in one."[82] The genetic code is the key to understanding life.

Schrödinger's lectures were primarily a popular exposition of work done by other biologists, particularly Max Delbrück (who

---

[78] Erwin Schrödinger, *What Is Life? The Physical Aspect of the Living Cell* (Cambridge: Cambridge University Press, 2013), p. 3.

[79] Schrödinger, *What Is Life?*, p. 21.

[80] See Evelyn Fox Keller, *Refiguring Life: Metaphors of Twentieth-Century Biology* (New York: Columbia University Press, 1995), esp. pp. 66-78.

[81] Schrödinger, *What Is Life?*, p. 61.

[82] Schrödinger, *What Is Life?*, p. 22.

had actually been student of Niels Bohr before focusing on biology). Schrödinger's emphasis on the gene as an information carrier, however, was a revelation, and inspired many scientists. Two of those scientists were James Watson and Francis Crick. Watson first read Schrödinger's book in 1946 while an undergraduate student at the University of Chicago, and from that moment, he recalled, "I became polarized towards finding out the secret of the gene."[83] Crick completed a degree in physics at University College London, and Schrödinger's book was a major factor in his leaving physics and developing an interest in biology.[84] He explained, "This book very elegantly propounded the belief that genes were the key components of living cells and that, to understand what life is, we must know how genes act."[85] Only a decade later, on April 25, 1953, Watson and Crick (working at the Cavendish Laboratory) discovered the DNA helix and its four letter code that contains the answer to Schrödinger's question.

While Schrödinger spent most of his time in Dublin mired in the search for a unified theory, DIAS attracted others who could broaden the scope of its research. Walter Heitler, who had worked with Schrödinger as well as Planck, Einstein, and Sommerfeld, arrived in 1941. He specialized in the study of cosmic rays and played a critical role in establishing the School of Cosmic Physics at DIAS (housed in a reopened Dunsink Observatory). Heitler succeeded Schrödinger as director in 1946. Cornelius Lanczos, a Hungarian Jew and expert in relativity theory who had worked with Einstein in Berlin, came to DIAS in 1952 fleeing McCarthyism in the United States.[86] He started as a visiting professor and would later become

---

[83]James D. Watson, Lecture at Indiana University (October 4, 1984) cited in Moore, *Schrödinger*, p. 403.

[84]James D. Watson, *The Double Helix: A Personal Account of the Discovery of the Structure of DNA* (New York: W.W. Norton, 1980), p. 12.

[85] Watson, *The Double Helix*, p. 12.

[86]Barbara Gellai, *The Intrinsic Nature of Things: The Life and Science of Cornelius Lanczos* (American Mathematical Society, 2010).

director. DIAS was not simply a haven for European refugees, however. A number of Irish-trained physicists joined the staff as well. Cormac Ó Ceallaigh, a UCD graduate who had studied in Paris and Cambridge, was appointed a senior professor in the School of Cosmic Physics in 1953, and Lochlain Ó Raifertaigh, another UCD graduate, returned to DIAS in the 1960s and built up an internationally respected group of theoretical physics.

DIAS was also critical in luring back to Ireland another native son who had abandoned a fellowship at Trinity College Dublin to pursue opportunities in Canada and the United States. John Lighton Synge who, according to his Royal Society obituary, was "arguably the greatest Irish mathematician and theoretical physicist since Sir William Rowan Hamilton,"[87] joined DIAS as Senior Professor in 1948 and stayed until he retired in 1972. Under Synge, DIAS became one of the top centers in the world for the study of relativity theory.

The event that precipitated Synge's return was a perceived slight to Ireland's scientific tradition. In February of 1947, Schrödinger had announced a breakthrough in his search for a unified field theory (prematurely, it turned out). Schrödinger's status, however, was such that the announcement was covered in *Time* magazine. The article began, "Last week, from nonscientific Dublin, of all places, came news of a man who not only understands Einstein, but has bounded like a bandersnatch far ahead (he says) into the hazy, electromagnetic infinite."[88] Synge, living in Pittsburgh at the time, furiously penned off a letter to the editor. "Your reporter...needs a swift kick in the pants," he blasted. He pointed out that Schrödinger's formula (printed in the article) is based on Hamilton's Principle. "Who was this Hamilton?" Synge asked rhetorically, "Born, lived, and worked (1805-1865) in 'nonscientific

---

[87] Petros Florides, "John Lighton Synge, 1897-1995," *Biographical Memoirs of Fellows of the Royal Society* (2008) 54: 402-24, p. 403.

[88] "Science: Einstein Stopped Here," *Time* (February 10, 1947).

Dublin, of all places'!"[89] The *Time* magazine editor, however, couldn't resist having the last word. He added below Synge's letter, "Let Dublin-born Higher Mathematician Synge call to mind his city's great ghosts (among them, his uncle's—author of The Playboy of the Western World), and admit that Dublin is a writer's town." The Board at DIAS was delighted by the letter. They wrote to Synge to see if he would consider returning to Dublin, and the following year he did (though his daughter claimed that one of the deciding factors was his hay fever, which was worse in America).[90]

As the *Time* editor had noted, John Lighton Synge was indeed the member of a distinguished Anglo-Irish family. While the playwright John Millington Synge was perhaps the most famous member, the Synge family had deep roots in Ireland. Edward Synge had settled in Ireland in 1620s, becoming Bishop of Cork, Cloyne, and Ross, and many of his descendants had attained high office in the Anglican Church of Ireland. John Lighton Synge was perhaps most proud of relative Hugh Hamilton (1729-1805), Erasmus Smith Professor of Natural Philosophy at Trinity College Dublin and an accomplished mathematician. (Synge's daughter Cathleen Synge Morawetz continued the family tradition in mathematics, becoming the first woman to direct the Courant Institute of Mathematical Sciences at New York University and the second woman to be elected President of the American Mathematical Society.) Synge's father was a land agent for Lord Gormanston's estate. In fact, during the land war of the 1880's, Edward Synge acquired a certain amount of notoriety for the harsh methods by which he evicted tenants from the estates he was managing. John Lighton Synge, however, distanced himself both from his family's tradition of Church service and from his family's Ascendancy politics. The Easter Rising happened just days before Synge's scholarship exam-

---

[89]"Letters to the Editor," *Time* (March 3, 1947).

[90] Florides, "Synge," p. 418.

inations at Trinity. Synge joined the Gaelic League and spent two
months at an Irish College learning Gaelic and "breaking out of my
Ascendancy cocoon," as he described it.[91] While he never joined
Sinn Féin, he was sympathetic to the cause. He even stole a rifle
from the College and handed it over to the rebels.

Synge graduated in 1919 with a double senior moderatorship
in mathematics and experimental physics and became a lecturer
in mathematics in 1920. That year, Trinity announced that it
would choose a new fellow on the basis of a thesis rather than the
traditional fellowship examination, and Synge lost out to his friend
C.H. Rowe. Synge instead took a position at the University of
Toronto (he says that he was hired largely because the department
chair was a fan of his uncle's plays). When a fellowship position
opened at Trinity in 1925, Synge easily won it, becoming Erasmus
Smith Professor of Natural Philosophy.

Synge had intended to spend the rest of his life in Dublin until
he was invited to set up and head a new Department of Applied
Mathematics at the University of Toronto. He resigned his fellowship in 1930 and returned to Toronto, where he successfully
launched the department and made significant contributions to
radar waveguides, antenna theory, and ballistics during WWII.
Perhaps one of his greatest legacies from his time at Toronto, however, was his role in creating the Fields Medal, the most prestigious
award in mathematics. In 1932 Toronto professor J.C. Fields fell
seriously ill and told Synge he wanted to endow an international
medal for mathematics (since there is no Nobel Prize for mathematics). Synge made all of the arrangements to launch the Fields
Medal.[92] Synge left Toronto in 1943 and spent most of the next ten
years heading the mathematics departments at Ohio State Univer-

---

[91] Florides, "Synge," p. 409, citing Synge's unpublished autobiography.

[92] Elaine McKinnon Riehm and Frances Hoffman, *Turbulent Times in Mathematics: The Life of J.C. Fields and the History of the Fields Medal* (American Mathematical Society, 2011).

sity and the Carnegie Institute of Technology (now Carnegie Mellon University). It was in Pittsburgh that he saw the fateful *Time* magazine reference to "unscientific Dublin" that would bring him back to where he had started.

Over the course of his career, Synge published over 200 papers and 11 books, covering fields as diverse as classical mechanics, geometrical optics, hydrodynamics, electrical networks, differential geometry, and the theory of relativity.[93] His most influential work involved the application of geometrical methods to general relativity. Synge also trained a number of students who went on to become important mathematicians. In his early days at Trinity, A.J. McConnell and E.T.S Walton were among his students. In Toronto, he supervised three Chinese students, Guo Yonghuai, Chien Wei-zang and Chia-Chiao Lin, who later became leading applied mathematicians in China and the United States. And two of his students in Pittsburgh went on to become some of the most renowned mathematicians of the twentieth century: Raoul Bott and John Nash, winner of the 1994 Nobel Prize in economics and the subject of the book (and movie) "A Beautiful Mind."[94]

While Synge helped to make DIAS a world-class center for research in relativity, he also played a very personal role in the 'Hamiltonian revival' that Whittaker had described. Synge was one of the editors of the first volume of Hamilton's collected mathematical papers. The idea for a collected volume actually originated with Synge's older brother, Edward Hutchinson Synge (known as 'Hutchie').[95] Hutchie had inherited half of his uncle J.M. Synge's considerable estate and abandoned his promising college career (he had won a mathematics prize in his first year at Trinity). Hutchie went on to publish a handful of important scientific papers, but

---

[93]"Bibliography of J.L. Synge," in L. O'Raifeartaigh (ed.), *General Relativity: Papers in Honour of J.L. Synge* (Oxford: Clarendon Press, 1972).

[94]Sylvia Nasar, *A Beautiful Mind* (New York: Simon and Schuster, 1998).

[95]J.F. Donegan, D. Weaire, P. Florides (eds.), *Hutchie: The Life and Works of Edward Hutchinson Synge, 1890–1957* (Atascadero: Living Edition, 2012).

he remained socially and intellectually isolated. Ultimately, he would spend the final two decades of his life confined to a mental institution. But like many of his mathematical contemporaries, Hutchie had developed "a somewhat mystical reverence" for Hamilton.[96] Lacking academic standing, Hutchie wrote directly to Albert Einstein in 1922, asking for his support for publishing Hamilton's collected papers. Einstein replied, "I am very sympathetic to the publication of Hamilton's works.... Many, perhaps the greatest ideas of Hamilton have been transferred into the consciousness of all physicists and mathematicians, and the working out of their implications continues with unremitting force. But the genesis of these ideas is hard to grasp: they appear to have fallen from heaven." He added, "the Hamiltonian method was found to be valuable in the formal framework of Relativity theory, and therefore Hamilton's way of thinking is particularly close to me."[97]

Writing in the *The Irish Statesman* in 1924, Hutchie argued that "Hamilton was not only the greatest of Irish mathematicians but was also one of the greatest—if he was not indeed the greatest— mathematician in Europe during the past century." And yet, he complained, "Hamilton has been treated by his country... with a neglect which no contemporary mathematician of any other European nation, who at all approaches his rank, has suffered."[98] Hutchie asked his younger brother to take on the enormous task of collecting and editing Hamilton's papers. Synge later recalled that he "found it impossible to refuse, partly because it was intrinsically right that Hamilton's papers should be published and partly because Hutchie dominated over me...."[99] In 1927 Synge and A.W. Conway were appointed by the Royal Irish Academy to edit the first volume, covering geometrical optics. It was published in 1931, just

[96] Donegan, *Hutchie*, p. 16.

[97] Albert Einstein to Edward Hutchinson Synge (May 4, 1922), Donegan, *Hutchie*, p. 16.

[98] Donegan, *Hutchie*, p. 18.

[99] Donegan, *Hutchie*, p. 19.

after Synge had left Dublin to return to Toronto. Synge's work on Hamilton's papers significantly shaped his own research. He published ten papers on the application of Hamilton's method as well as two books: *Geometrical Optics: An Introduction to Hamilton's Method* (1937) and *Geometrical Mechanics and de Broglie Waves* (1954). Conway also co-edited the second volume of Hamilton's papers (on dynamics) with A.J. McConnell. Volume II was completed in 1942, while volume III (on algebra) was released in 1967, edited by H. Halberstam (a Czech refugee from the Nazi's who served briefly as professor of mathematics at Trinity College Dublin) and R.E. Ingram (a Jesuit mathematics professor at UCD). The fourth and final volume was not completed until 2001, edited by B.K.P. Scaife (an engineering professor at Trinity College Dublin).

8.4   Science and the Economy

The Dublin Institute for Advanced Study was the embodiment of de Valera's view of mathematics. Like his intellectual idol Hamilton, de Valera believed that mathematical science should be pure, independent, and above politics (even when it was being used for political purposes). They both saw mathematics as a form of high culture, unsullied by practical or commercial concerns. The elaborate mathematical theories of fundamental particle physics discussed at the DIAS seminars may have improved Ireland's intellectual reputation, but they did nothing to help the masses of the poor and unemployed. As John Dillon had charged when de Valera first proposed the Institute, DIAS did bear a striking resemblance to Swift's island of Laputa—isolated from daily life and focused on pure thought while the population lived in abject poverty. The 1950s in Ireland are almost universally described by historians with terms such as stagnation, drift, and malaise. While other major economies were experiencing a technology-fueled post-war economic boom, Ireland was stuck with flat household incomes, high unemployment, massive emigration, and cultural stagnation.

In 1957 Ernest Walton wrote to de Valera with a proposal for a very different approach to science.[100] Walton was well aware of the way in which advances in science had led to new technologies and new industries such as aerospace, pharmaceuticals, electronics, and plastics. The massive investments in scientific research and development made during WWII and continued into the Cold War, coupled with demographic growth and the rebuilding of Europe and Japan, made the 1950s a period of dramatic growth and prosperity in the U.S., the UK, Japan, and other industrialized economies. Ireland had instead put up steep tariff barriers to protect domestic industries, and ended up keeping out foreign multinationals that might have brought new investment and new technologies. Ireland was falling farther and farther behind.

Echoing the arguments of George Francis Fitzgerald and other proponents of technology-based economic development, Walton implored de Valera to consider another approach: "There seems to be no doubt that the only way to raise substantially the standard of living in this or any other country is to make adequate use of science in industry, agriculture, and medicine.... Better still, we should be in a position to initiate and develop new lines ourselves."[101] And yet Ireland was failing in this pursuit. Walton pointed out that no institution in Ireland had a mass spectrograph, an electron microscope, nuclear magnetic resonance equipment, or microwave spectroscopy equipment, all rapidly becoming critical in a range of fields. He noted that the total sum for post-graduate students available from government funds for the entire country was only £1,200 annually. Not surprisingly, the best graduates were leaving Ireland.

---

[100] E.T.S. Walton, "The Need for More Attention to Science in Éire with some Suggestions for its Encouragement (1957)," Appendix V in McBrierty, *Walton*, pp. 83-87.

[101] Walton, "The Need for More Attention to Science in Éire," p. 83.

While de Valera had argued that experimental science was too expensive for Ireland, Walton countered that in fact there was a much higher cost for not engaging in experimental science. Investing in science, Walton claimed, could put Ireland on the leading edge in a handful of scientific areas with important economic consequences. He pointed to the recent development of transistors in the radio industry (the semiconductor-based replacements for vacuum tubes). Walton explained, "The research work leading up to them did not require abnormally expensive equipment. Their manufacture could have been undertaken here, for the quantities of raw materials required are trivial. There is no basic reason why they should not have been first manufactured here. However, there was no possibility of this happening for no research was being done in this country on semi-conductors."[102] Walton's equation—that basic research leads to new technologies which lead to new industries which raise the standard of living—would become an article of faith in the second half of the twentieth century. By the end of the 1950s, Ireland would begin to embark on a new strategy, based largely on the approach that Walton suggested. Protectionist tariffs came down, foreign companies were invited—even courted—and high-technology manufacturing became the focus of industrial policy. By the 1990s, Ireland had transformed itself into the "Celtic Tiger" economy—another revolution that would not only reshape Ireland but also Trinity College Dublin. And mathematics would for the first time come to be seen as the foundation of national prosperity.

---

[102] Walton, "The Need for More Attention to Science in Éire," p. 87.

Chapter 9

The Knowledge Economy

Virtual Worlds

It is a beautiful, clear day in October 2009. A young man is giving a bicycle tour of Dublin. He takes off from St. Stephen's Green, still in bloom despite the time of year. Passing the famous Shelbourne Hotel, he turns up Grafton Street, past fancy stores and Bewley's coffee shop. He goes by the front gates of Trinity College, normally swarming with students and tourists but now strangely quiet, and then stops briefly at the Blarney Stone Pub, already full of people dancing and socializing. He traverses the Ha'Penny Bridge and then passes a park where U2 recently performed a live performance. Crossing back over the Liffey, he goes by Christ Church Cathedral and then finally ends his tour at the Guinness Brewery. These landmarks will be familiar to anyone who has visited Dublin, and yet while the scenery looks familiar, something is slightly off. The streets are nearly empty. Along the entire tour there is but a single pedestrian, and the Blarney Stone seems to be the only establishment with any customers. What's more, the streets are foreshortened. Major landmarks loom large, but there is little in between.

This Dublin exists in Second Life, an online virtual world where users interact with each other through avatars and can create (and buy and sell) virtual buildings or items. Our cyclist is called Xeon Scribe. In real life ("RL" as they say in Second Life), his name is Eneas McNulty, an M.Sc. student in E-Learning at the University of Edinburgh.[1] Virtual Dublin is the brainchild of Ham Rambler, aka John Mahon, an Irishman living in London. Mahon first built the Blarney Stone to recreate the feel of an Irish pub in Second Life, and it soon became the most popular pub in the online community. He then purchased some virtual real estate and began to develop Second Dublin as the proper setting for the pub. Mahon has since developed a business partnering with companies to promote their brands in Second Life, working with clients such as Irish Tourism and Guinness.[2] IDA Ireland (the government agency responsible for industrial development in Ireland) opened a Technology and Innovation Showcase in Second Dublin, and U2 (through their avatars) played a live concert there in 2008.[3]

Second Life depends on thousands of connected computer servers with a capacity of more than 100 terabytes to simulate complicated environments in real-time, allowing 30 million "residents" from all over the world to communicate, socialize, collaborate, create, and trade. Underlying it all is a "physics engine," a computer program that ensures that residents and objects in Second Life follow the laws of physics—bicycles stop when they hit a wall, balls bounce

---

[1] Eneas McNulty, "Micro-Ethnography Study of the Virtual Dublin Community in Second Life," Eneas' E-learning and Digital Cultures Blog, `digitalculture-ed.net/eneasm/micro-ethnography-study-of-the-virtual-dublin-community-in-second-life/` (Accessed December 11, 2013)

[2] Ellie Doyle, "Irishman Creates A Virtual Dublin On The Website Second Life," Prosperity Blog, `/blog.prosperity.ie/2007/11/12/irishman-creates-a-virtual-dublin-on-the-website-second-life/` (Accessed December 11, 2013)

[3] "IDA Ireland launches Technology & Innovation Showcase in Second Life," (01/04/09) `www.idaireland.com/news-media/press-releases/ida-ireland-launches-tech/` (Accessed December 11, 2013), "U2 in SL Live Concert in Second Life (2008.3.29)," `www.youtube.com/watch?v=K4LE6VBAPd8`

Figure 9.1: Trinity College Dublin in Second Life

when they hit the ground, trees sway in the wind. The physics engine for Second Life was created by a company called Havok, founded by Steven Collins and Hugh Reynolds in the Computer Science department at Trinity College Dublin. They use a range of mathematical tools (including Hamilton's quaternions) to rapidly calculate the motions of millions of virtual objects. In addition to Second Life, their software has been used in over four hundred video games as well as in movies such as *The Matrix.* In 2007 Havok was sold to Intel for €76M.[4] In the knowledge economy, even the laws of physics can be bought and sold.

From the 1990s to the 2010s, the world became digital—not only video games, movies, and music, but also healthcare, business, engineering, physics, and astronomy were all transformed as computer processing power improved exponentially, the cost of data storage plummeted, low cost sensors became pervasive, and global networks emerged to link all of these devices together. Simulation and com-

---

[4]"Havok Plays Master Chief at Innovation Awards," *Silicon Republic* (December 20, 2007), Interview with Steve Collins (September 3, 2013).

putation became critical tools as computers were used to model everything from weather to brain function to financial markets to supply chains. The resulting data deluge placed a new priority on the mathematical sciences.

The Irish were among the most aggressive (and successful) at capitalizing on the global mobility of information, capital, and goods beginning in the 1980s and 1990s. The so-called Celtic Tiger economy was built on investments from high tech multinational companies, evolving rapidly from electronics and pharmaceutical production to business services and financial engineering as banks, hedge funds, and internet companies made Dublin their home. Ireland became a prime global location for "intellectual property"— the patents and trademarks that had become the most important sources of wealth in the knowledge economy.

The Celtic Tiger economy finally came crashing down in 2008. It turned out that Ham Rambler was not the only Irishman making money selling virtual real estate. Ironically, the crash had nothing to do with sophisticated financial engineering, global competition, or fickle multinational corporations. It was an old fashioned real estate bubble that would have been familiar even to Sir William Petty back in the seventeenth century. Land, long the most concrete and valuable of assets in Ireland—the source of centuries of unrest and war, the heart of the Irish imagination—became the object of speculation and "imaginary" wealth. But the loss of those virtual assets caused very real harm to millions of people.

The crash, however, reinforced the importance of science and innovation to the future of the Irish economy. Constructing virtual realities was ultimately a more solid economic foundation than constructing homes and office buildings, or so the government hoped. Mathematics (in the form of "big data" and business analytics) and physics (in the form of nanoscience) became the new hope for Ireland's future. As its economic importance grew, the scientific enterprise came to look more and more like the multinational corporate enterprises that supported the Celtic Tiger economy. The large global teams, expensive facilities, multimillion euro budgets, and corporate partnerships changed the role of the scientist. The traditional roles of the mathematician as the keeper of truth, the creator of beauty, the defender of the faith, and the shaper of minds receded in the face of the new image of the mathematician as corporate consultant.

Perhaps no single individual better represents how the Ascendancy tradition adapted to new circumstances than Michael Purser, a lecturer in mathematics at Trinity College Dublin from the 1980s to the 2000s. Purser's family tree is firmly rooted in Trinity's history.[5] Gabriel Stokes, the eighteenth-century surveyor, instrument maker, and Trinity professor of mathematics is an ancestor on his mother's side, as is the Cambridge mathematician Sir George Gabriel Stokes as well as at least six other Stokes' with distinguished careers in science. Purser's mother was the granddaughter of the Trinity mathematician J.H. Jellett and a niece of George Francis Fitzgerald. On his father's side of the family, four Pursers gained fame in science in Ireland in the nineteenth and early twentieth century.

Purser's own career demonstrates the new opportunities open to mathematicians in the second half of the twentieth century. Born in Dublin, Purser took a degree in mathematics at Cambridge University and pursued research briefly at the Dublin Institute for Advanced Studies where he worked with J.L. Synge on hydrodynamics. From that point on, however, Purser's career focused primarily on industrial problems. In 1962 he went back to the UK to work for a computer business on process control programs. In

---

[5] Michael Purser, *Jellett, O'Brien, Purser and Stokes: Seven Generations, Four Families* (Dublin: Prejmer Verlag, 2004).

the late 1960s, he went to Peru to work in the mining industry.[6] Then he advised the Chilean Electricity Company on the computer control of hydroelectric stations, dealing with issues such as optimizing generators, minimizing use of water, and spreading load. In the 1970s he returned to the UK where he programmed document readers for a British software company, and finally in 1976, he returned to Dublin because he wanted his children to go to school there.

Purser became a lecturer at Trinity, first in the computer science department and then in the mathematics department. He started a company called Baltimore Technologies, primarily as a means to offer his students at Trinity work experience.[7] In the early days, Baltimore worked on the design and optimization of the computer networks that were rapidly spreading across Europe. "We started having an interest in the telecommunication area," he explains, "and an interest in real time systems branched out into an interest in networks, and not only the use of networks but the design of networks, the optimization of network traffic, the measurement of network traffic...."[8] Eventually he focused on network security and cryptography, protecting digital information with hard-to-solve mathematical problems such as factoring very large numbers.[9] Baltimore Technologies was acquired in 1996 by a team financed by Dermot Desmond and led by Fran Rooney. It listed on the NASDAQ exchange in 1999 and reached a market

---

[6] Purser has written (and illustrated) three books on his experiences in Latin America: *Beyond Buenos Aires: Stories of Peru and Chile (1)* (Dublin: Prejmer Verlag, 2011), *La Rosa con El Clavel: Stories of Chile and Peru (2)* (Dublin: Prejmer Verlag, 2011), *From Lvov to Lima: Stories of Argentina, Chile and Peru (3)* (Dublin: Prejmer Verlag, 2012).

[7] Cormac Sheridan, "Baltimore Technologies," *Technology Ireland* (January 1999), pp. 16-17.

[8] Interview with Michael Purser, Founder of Baltimore Technologies, Lecturer in Mathematics at Trinity College Dublin (June 3, 1999).

[9] See Sarah Flannery with David Flannery, *In Code: A Mathematical Journey* (London: Profile Books, 2000), an elementary introduction to cryptography and the autobiography of the sixteen-year-old girl who won the Irish Young Scientist of the Year Award in 1999 for mathematical work done in association with Michael Purser.

capitalization of more than $13 billion before the internet bubble burst in March 2000.

At each step, Purser found the application of mathematics to be the key to the solution of a range of technical and industrial problems. For Purser, many of the most challenging technical problems of today are essentially mathematical problems and require people with mathematical training to solve them. "I started hiring mathematicians as opposed to software people," he explains. "Because if you have mathematical products—cryptographic products—it is essential that the people writing them understand. My view was that mathematicians can learn software but software writers cannot necessarily learn mathematics."[10] As the key to manipulating digital information, mathematics assumed a new importance in the 'information economy'.

9.1   The Rise and Fall of the Celtic Tiger

Ireland during the three decades between 1980 and 2010 witnessed one of the most dramatic economic transformations in world history. De Valera's Ireland was focused on political and economic independence and the traditional values of Catholic, agrarian Ireland, and under de Valera, Ireland remained the poor man of Europe. Historian Joe Lee notes that across the twentieth century, "No other European country, east or west, north or south, for which remotely reliable evidence exists, had recorded so slow a rate of growth of national income."[11] As recently as the 1980s, one million Irish people—a third of the population—lived below the poverty line. By 2006, however, Ireland had become one of the richest countries in the world with a poverty rate below 6%.[12] Ireland be-

---

[10] Purser interview.

[11] J.J. Lee, *Ireland, 1912-1985: Politics and Society* (Cambridge: Cambridge University Press, 1989).

[12] Michael Lewis, "When Irish Eyes Are Crying," *Vanity Fair* (March 2011).

came perhaps the world's best example of rapid technology-enabled economic development.

The foundations for Ireland's economic rise were laid in the late 1950s.[13] The government began to encourage foreign direct investment by phasing out restrictions on foreign ownership and instituting a zero tax on profits from increased manufacturing exports. In 1959 Seán Lemass succeeded de Valera as Prime Minister in a symbolic passing of the torch. "The historic task of this generation," Lemass proclaimed, "is to ensure the economic foundation of independence."[14] From this point on economic concerns would gradually eclipse the traditional emphasis of Irish nationalism on issues like partition and the revival of the Irish language. The change in strategy, however, did not lead immediately to a smooth course of economic development. In 1987 the *Economist* ran a story portraying the Irish economy as a disaster.

Over the following decade, however, Ireland's economy doubled in size, making it the fastest growing economy in the world. In 1994 a Morgan Stanley analyst termed it the 'Celtic Tiger' in analogy to the fast developing Asian Tigers—South Korea, Taiwan, Hong Kong, and Singapore. And for a period after 1995, the Irish economy actually outperformed the Asian Tigers. Ireland's phenomenal success hinged on its ability to attract high tech multinational corporations in areas such as pharmaceuticals, electronics, software, and financial services. Ireland's entry into the European Economic Community in 1973 was a critical turning point. Membership in what would become the European Union turned Ireland from a

---

[13]See John Kurt Jacobsen, *Chasing Progress in the Irish Republic: Ideology, Democracy and Dependent Development* (Cambridge: Cambridge University Press, 1994); Denis O'Hearn, *Inside the Celtic Tiger: The Irish Economy and the Asian Model* (London: Pluto Press, 1998); Paul Sweeney, *The Celtic Tiger: Ireland's Economic Miracle Explained* (Dublin: Oak Tree Press, 1998); Ray MacSharry and Padraic White, *The Making of the Celtic Tiger: The Inside Story of Ireland's Boom Economy* (Cork: Mercier Press, 2000). For historical background, see Mary E. Daly, *Industrial Development and Irish National Identity, 1922-1939* (Syracuse: Syracuse University Press, 1992).

[14]Quoted in MacSharry and White, *The Making of the Celtic Tiger*, p. 13.

small peripheral country to an entry point to the world's largest consumer market. Moreover, the EU provided investment funds to build the foundations for a new economy. Over the next twenty-five years, Ireland received more European subsidies per capita than any other member state.

Directing Ireland's phenomenal growth was the Industrial Development Authority or IDA. Essentially, the IDA's job was to jumpstart Ireland's economy by selling Ireland as a prime location for multinational corporations. The IDA's strategy was to focus on those multinationals that it believed would be the fastest growing and most profitable. Beginning in the 1960s and 1970s the IDA went after the pharmaceutical companies, and by 2000 sixteen of the top twenty global pharmaceutical companies had manufacturing operations in Ireland.[15] In the late 1970s they began to target the new electronics companies, and by 2000 twenty of the top twenty-five global electronics companies had set up in Ireland. A third of all personal computers sold in Europe were being manufactured in Ireland.

During the 1980s, however, the Irish began to realize that their economy was dangerously dependent on foreign-owned manufacturers attracted to Ireland by its relatively low wages and generous tax incentives.[16] The IDA and the Irish government introduced a new strategy—to bring higher skilled jobs, particularly in service industries, and to increase Irish science and technology capabilities in order to develop the innovative abilities of indigenous industries to more strongly anchor foreign multinationals. As a result, Ireland in the late 1980s began to target software development and financial services companies. By 2000 Ireland was the largest exporter of software in the world. In 2010 almost all of the world's leading software companies, including IBM, Oracle, Microsoft, Google, and Facebook, had a significant presence in Ireland.

---

[15] For current statistics see the IDA's website, www.idaireland.com.
[16] See Lee, *Ireland, 1912-1985*, pp. 531-36, Sweeney, *The Celtic Tiger*, pp. 139-40.

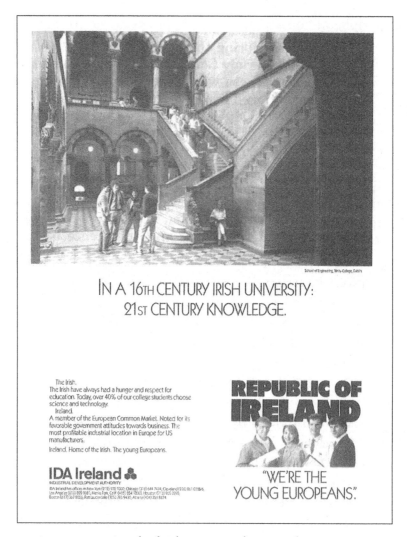

Figure 9.2: IDA Ireland Advertisement featuring the Engineering
Building at Trinity College Dublin.

The deregulation and computerization of the world's financial
markets in the 1980s allowed bankers and traders to buy and sell
securities regardless of their physical location. The IDA built an
International Financial Services Centre (IFSC) in Dublin in 1987

in an attempt to make Dublin a hub for global finance. The strategy was stunningly successful. By 2005, the IFSC accounted for 75% of all foreign investment in Ireland.[17] It had attracted half of the world's top fifty banks and top twenty insurance operations, making Ireland the main global location for money market mutual funds. In fact, Dublin's investment funds industry surpassed London's.[18]

Ireland's primary attraction for financial services was a low tax rate (10% before 2005 and then 12.5%) and an almost complete lack of regulation.[19] The flow of international funds dramatically outpaced the actual creation of goods and services within Ireland. Apple Computer, for example, had established its first European base in Cork in 1980, but by the late 1990s it had moved all production to cheaper locations outside of Ireland. But Apple continued to realize enormous profits in Ireland. Thanks to a negotiated income tax rate of less than two percent, one Irish subsidiary of Apple (with no physical presence or employees) realized $36 billion in income. Apple's Irish companies held reserves of cash and investments worth about $100 billion—more than the entire budget of the Irish government.[20] Journalist Fintan O'Toole observed, "These were virtual entities, existing in a financial version of Second Life."[21]

This intangible economy ultimately came to dwarf the physical economy. Ireland evolved from agriculture to high tech manufacturing to a knowledge economy in a single generation. But while the risk of dependency on highly mobile global corporations and a "race to the bottom" regulatory environment were clear, ultimately

---

[17]Fintan O'Toole, *Ship of Fools: How Stupidity and Corruption Sank the Celtic Tiger* (New York: Public Affairs, 2010).

[18]Irish Funds Industry Association (www.irishfunds.ie.).

[19] O'Toole, *Ship of Fools*, p. 129.

[20]Simon Bowers, "Apple's Multi-Billion Dollar, Low-Tax Profit Hub: Knocknaheeny, Ireland," *The Guardian* (May 29, 2013).

[21] O'Toole, *Ship of Fools*, p. 133.

the roots of the economic crash that ended the Celtic Tiger economy in 2008 were almost entirely domestic. When exports started to fall after 2002, a rise in real estate and construction kept the economy going. Real estate eventually grew to 25% of GDP and generated a third of government revenue. More than a fifth of the Irish workforce was employed building houses. From 1994 to 2006, the average price for a Dublin home rose more than 500 percent.[22] While few would acknowledge it at the time, such growth was clearly unsustainable.

It was the global financial crisis that began in late 2007 that finally popped the bubble, leading in September 2008 to the collapse of the stocks of the three main Irish banks, Anglo Irish, A.I.B., and Bank of Ireland. The Irish government controversially decided to guarantee all of the bank losses, pushing it almost to bankruptcy and requiring a rescue by the EU and IMF in 2010. The Irish budget went from surplus to a deficit equal to 32 percent of GDP, the highest by far in the history of the Eurozone. The average Irish family lost almost half of its financial assets, and household debt rose to the highest levels in the EU. As the economy collapsed, hundreds of thousands of people lost their jobs, and government austerity led to significant cutbacks.

While the collapse of the real estate bubble ended the phenomenal growth of the Celtic Tiger, the underlying economic strengths built up over two decades remained. Foreign direct investment recovered rapidly. The science-based multinational corporations remained and more kept coming. In fact, some argued that the slowdown in the financial services sector had freed up talent and led to a vibrant culture of startup companies. The Celtic Tiger may have been over, but Ireland was not about to return to an agrarian economy. In fact, the crash put even more emphasis on science and innovation-based development. The Irish government's Action

---

[22]Lewis, "When Irish Eyes Are Crying."

Plan for Jobs 2013 described the importance of transitioning "from a failed model built around debt, construction and housing to one based on innovation, enterprise and exports." The government's mandate was to "deliberately reconfigure and retool our economy so that long term sustainable jobs are created for our people and so that the disastrous mistakes of the past will not be repeated."[23] The number one priority in the report: "Make Ireland the leading country in Europe for Big Data." Mathematics in the form of business analytics had become the savior for Ireland's economy.

9.2   Science and Economic Development

Ireland's economic success resulted from a conscious attempt to attract industries at the forefront of technological innovation, and from the beginning Irish science policy was subordinated to this industrial policy. The origins of Irish science policy lie in a 1966 OECD-sponsored report, *Science and Irish Economic Development*. Its main proposal was "that a national science policy should be formulated and integrated with economic policy so that science contributes to, and makes possible, desirable technological change."[24] Throughout the 1980s and the 1990s, report after report reasserted the link between science, technology, and economic development. A 1995 report was appropriately entitled "Making Knowledge Work for Us."[25] "Innovation, based on the application of science and technology," claimed the report, "is now the mainspring of international economic competitiveness."[26]

---

[23]Department of Jobs, Enterprise and Innovation, *2013 Action Plan for Jobs* (February 2013).

[24]*Science and Irish Economic Development* (Dublin: Stationery Office, 1966), p. xxvi.

[25]Science, Technology and Innovation Advisory Council, *Making Knowledge Work for Us: A Strategic View of Science, Technology and Innovation in Ireland* (Dublin: Stationery Office, 1995), p. 15.

[26]*Making Knowledge Work for Us*, p. 22.

The following year the government issued its first ever White Paper on science and technology policy, arguing that science policy should be located "firmly within the framework of wider industrial, economic and national development policy."[27] In contrast to what the report described as earlier 'elitist' approaches to science (no doubt a swipe at the Ascendancy tradition), it noted, science and technology "will be evaluated by its ability to contribute to wider national goals, as a means of achieving them rather than as an end in itself."[28] The White Paper described a revolution that was taking place. As information technology transforms the global economy, it explained, so too will it transform Ireland. With echoes of Yeats' "Easter 1916," the White Paper intoned, "Everything has changed utterly and nothing remains the same."[29] The Information Revolution had replaced the Easter Rising as a new turning point in Irish history.

While the rhetoric of science and technology as central to economic development dominated official reports, however, the actual amount of Irish funding for scientific research remained far below the European average.[30] The vast majority of Irish science was funded by European Union programs—more than 80% at one point.[31] The success of foreign, high technology multinationals also masked the fact that indigenous Irish industry remained stagnant and technologically backward. Neither foreign multinationals nor indigenous firms did any significant research and development in Ireland. In other words, contrary to the claims of the government reports, the economy had taken off without any strong base

---

[27]Department of Enterprise and Employment, *Science, Technology and Innovation: The White Paper* (Dublin: Stationery Office, 1996), p. iii.

[28]*The White Paper*, p. 2.

[29]*The White Paper*, p. 40.

[30]Steven Yearly, "The Social Construction of National Scientific Profiles: A Case Study of the Irish Republic," *Sociology of Science* (1987) 26: 191-210.

[31]T.P. Hardiman and E.P. O'Neil, "The Role of Science and Innovation," pp. 253-84 in Fíonán Ó Muircheartaigh (ed.), *Ireland in the Coming Times: Essays to Celebrate T.K. Whitaker's 80 Years* (Dublin: Institute for Public Administration, 1997), p. 267.

of scientific research. Yet, still, the government came to believe that indigenous research was critical to economic growth. Scientific research rather than an end in itself was simply another way to market Ireland as a location for multinational investment.

By the late 1990s, it appeared as though the government was finally beginning to take the support of research seriously. Indicative of the shift was the Technology Foresight Exercise initiated in 1998. The Irish Council for Science, Technology, and Innovation was asked "to identify areas of strategic research and the emerging technologies likely to yield the greatest economic and social benefit."[32] The panel recommended in particular making Ireland a center of excellence in information and communication technologies and in biotechnology. In response to the report, the government announced the creation of a €646 million Technology Foresight Fund, which it described as "the largest single investment in research in the history of the State."[33] The fund was to serve as "a vital element of Government's essential strategy to move Ireland higher up the value chain."[34] Science Foundation Ireland (SFI) was established in 2000 to administer the Technology Foresight Fund. Its stated mandate was to "invest in academic researchers and research teams who are most likely to generate new knowledge, leading edge technologies and competitive enterprises in the fields of science and engineering."[35] It began with the focus areas identified in the Technology Foresight report—biotechnology and information and communications technology—later expanding its remit to include sustainable energy technologies. While SFI dramatically increased the amount of Irish funding for scientific research, Irish researchers still depended heavily on the EU. In 2011, for example,

---

[32] Forfás, *Technology Foresight Ireland: An ICSTI Overview* (April 1999), p. 5.

[33] Department of Enterprise, Trade and Employment, *Technology Foresight Fund: Investing in Future Competitiveness* (March 2000).

[34] *Technology Foresight Fund.*

[35] Science Foundation Ireland, "What We Do," www.sfi.ie/about/what-we-do/ (Accessed December 11, 2013)

SFI funded €154 million in grants, but SFI researchers also secured € 156 million of funding from non-SFI sources (the largest being €56 million from the EU).

The financial crisis that began in 2008 took a serious toll on all areas of government funding, but even the de facto bankruptcy of the Irish state could not fully dampen the belief in the critical importance of funding scientific research. It did, however, force an even tighter link between scientific research and economic benefit. A Research Prioritisation Steering Group reported in November 2011 on the need to "set a new over-riding national objective to accelerate the delivery of specific economic outcomes from our investment in research" and to "focus investment in those areas that are most likely to give demonstrable returns in the medium term" (defined as five years).[36] Research investments would need to be evaluated much like investments in equities or new companies— based on their ability to generate measurable financial returns over the course of a few years.

Perhaps the best example of the new mindset was a series of reports by the accounting firm Deloitte entitled, "The Importance of Physics to the Irish Economy."[37] In a context where all investments in scientific research were evaluated in economic terms, finding ways to measure the economic impact of science became critical. In 2012, Deloitte calculated that "physics-based sectors" contributed more than €7 billion to the Irish economy. These sectors included a broad range of companies involved with materials physics, nanotechnology, optics and photonics, biophysics, high energy physics, semiconductor physics, and thermodynamics. Together they generated 5.9% of total Irish economic output, supported more than 86,000 jobs, and exported more than € 23 billion worth of goods and services annually. In fact, direct employment in physics-based

---

[36] Forfás, "Report of the Research Prioritisation Steering Group" (November 2011).

[37] Institute of Physics, "The Importance of Physics to the Irish Economy" (November 2012).

sectors, the report argued, was comparable to that of all of the finance, banking, and insurance sectors combined. Perhaps most importantly, these sectors were relatively unaffected by the downturn. Physics had evolved from a fundamental branch of human knowledge—one of the most inspiring examples of man's ability to understand the universe—to an economic sector to be evaluated by employment, gross value added, and exports.

9.3   Education for the New Economy

The education sector played a critical role in the rise of the Celtic Tiger. Multinational corporations were attracted to Ireland not just because of low tax rates and lax regulation but also because of a young, well-educated, English-speaking (and relatively low wage) workforce. But multinational corporations had very specific needs, and the Irish education system was hardly prepared to meet them. Between 1976 and 1979, for example, the IDA had negotiated 18,000 jobs in electronics, but Ireland produced only 100 electrical engineers and 200 electrical technicians each year.[38] As a result, the government created a vocational education system. Between 1970 and 1980, the Irish government established nine Regional Technical Colleges as well as National Institutes of Higher Education in Limerick (later University of Limerick) and Dublin (later Dublin City University) as well as the Dublin Institute of Technology. The Dublin Institute of Technology was by 1997 the largest third level institution in Ireland with over 20,000 students. Education came to be seen in terms of the production of 'human capital', defined as "the stock of people with scientific or technological skills available to the State for economic and social development."[39] To better manage the educational system, the

---

[38]MacSharry and White, *The Making of the Celtic Tiger*, p. 284.

[39]Eolas, *National Framework of Research Needs: Human Capital and Mobility* (November 1993).

government also created the Higher Education Authority in 1968, a national coordinating and planning agency to monitor and fund university budgets.

As experts began to realize that Ireland would not be able to compete on the basis of low wages forever, and industrial policy began to shift in the 1990s, higher education was asked not simply to turn out cost effective technicians but also state-of-the-art researchers. Higher education, many argued, should be understood as part of a 'national system of innovation' working together with industry and government to create marketable forms of knowledge, particularly in the areas of science and technology.[40] By the mid-1990s the state became increasingly involved in the university sector. As crucial resources for national development, even the relatively autonomous universities had to be brought in line with national needs. Universities were required to submit detailed strategic plans to the Higher Education Authority for approval and funding. The abolition of undergraduate fees in 1995 served to make Ireland's universities even more dependent on government funding.

Vincent McBrierty, a physics professor at Trinity and one of the earliest proponents of university-industry links in Ireland, together with Raymond Kinsella, a professor of finance, defined what they termed "the new techno-academic paradigm."[41] They described knowledge as "the new form of equity" and focused on issues of intellectual property and technology transfer, particularly the role of universities in the generation, transfer, and commercialization of knowledge. "Knowledge, information and associated skills," they argued, "have displaced labour as the primary source of productivity and competitiveness."[42] McBrierty and Kinsella saw the uni-

---

[40]See James Wickham, "An Intelligent Island?," pp. 81-88 in Michel Peillon and Eamonn Slater (eds.), *Encounters with Modern Ireland: A Sociological Chronicle, 1995-1996* (Dublin: Institute of Public Administration, 1998).

[41]Vincent J. McBrierty and Raymond P. Kinsella, *Ireland and the Knowledge Economy: The New Techno-Academic Paradigm* (Dublin: Oak Tree Press, 1998).

[42] McBrierty and Kinsella, *Ireland and the Knowledge Economy*, p. 8.

versity almost entirely as an economic institution and knowledge itself as a form of property.

Trinity College Dublin is no longer the only university in Ireland, or the largest in terms of student numbers, or even the university with the most research funding, and it has arguably changed more over the past twenty years than in the entire four hundred years that came before. It is no longer an ivory tower or a bastion of Protestant privilege. It has completely shed its religious character and even elected its first Catholic provost in 1991. Rather than relying on rents from its former landholdings, Trinity must now raise funds from the Irish government, the European Union, industry, and private donors. The new Irish focus on economic development through technological innovation has offered a means by which Trinity College can finally overcome its outsider status and play a full role in the Irish national project.

As the Irish economy globalized and grew, so did Trinity College. By 2013, Trinity had 17,000 students and 785 academic staff supported by 655 research staff and 1,500 library, technical, and administrative staff.[43] Its annual budget was €270 million, including € 65 million in research expenditures. A student body that had remained overwhelmingly male, Anglo-Irish, and Protestant up through the 1970s became primarily female (58%), Catholic (70%), and increasingly international (21% of students and 40% of staff).

Like all universities in the knowledge economy, Trinity's impact has been quantified and tabulated. Trinity remains the preeminent university in Ireland, ranking 18th in Europe in 2013 and 61st in the world in the QS World University Rankings.[44] Trinity is ranked among the top 50 universities in the world in four subjects (En-

---

[43]Trinity College Dublin, "Student Numbers 2011/12," www.tcd.ie/Communications/Facts/student-numbers.php. (Accessed December 11, 2013.)

[44]QS World University Rankings, www.topuniversities.com/node/2319/ranking-details/world-university-rankings/2013. (Accessed April 30, 2014.)

glish, Language and Literature; History; Geography; and Politics and International Studies) and among the top 100 in a further 14 areas (among them Biological Sciences; Chemistry; Computer Sciences and Information; Economics & Econometrics; Mathematics; Physics and Astronomy).[45] Trinity is also actively engaged in supporting industry with over 400 collaborations with companies, 76 campus companies formed since 1985, and over 300 invention disclosures in the last ten years.[46]

Trinity's adaptation to the new economy is clear from a tour of the campus in 2013.[47] While Trinity was founded in the sixteenth century on a hill near Dublin, it is now very much in the center of the city. The famous Georgian West Front of the College looks out over College Green, one of the city's busiest intersections. The statue of William of Orange that once adorned the Green has since been replaced by one of Thomas Davis, a Protestant nationalist and graduate of Trinity. Statues of Edmund Burke and Oliver Goldsmith flank the narrow main gate, which is invariably mobbed with tourists. The Book of Kells, a medieval illuminated manuscript held in the Trinity library, draws over 400,000 visitors a year, making it one of Ireland's most popular tourist destinations.

If the West End of campus is given over to tradition and guided tours, however, the East End represents Trinity's new relationship to technology and innovation. Buildings named for great Trinity scientists stand proudly alongside buildings named for Celtic Tiger industrialists. Eighteenth century rooms connect to 21st century research facilities. The School of Physics now includes over 28 academic staff, 50 postdoctoral fellows, and 100 graduate students spread primarily across three buildings. The Fitzgerald Building

---

[45]Trinity College Dublin, "Annual Report, 2011-12," www.tcd.ie/about/content/pdf/tcd-annual-report-1112.pdf. (Accessed December 11, 2013).

[46] Trinity College Dublin, "Towards a Future Higher Education Landscape" (July 2012).

[47]For a 14 episode podcast of a scientifically themed walking tour of Trinity's campus see, "Science Safari—The Trinity Trail," www.tcd.ie/visitors/sciencesafari/trail/. (Accessed December 11, 2013.)

was completed in 1905 and named for George Francis Fitzgerald, who played such a crucial role in its construction. The Sami Nasr Institute of Advanced Materials, named for a geology graduate who became managing director of a large oil company, opened in 2001. The newest is the Naughton Institute, named for Martin Naughton, Chairman of the Glen Dimplex Group, the world's largest manufacturer of electric heating appliances. It is a €100 million facility, which houses the Centre for Research on Adaptive Nanostructures and Nanodevices (CRANN) and the Science Gallery.

The School of Physics continues Trinity's long history of research into the properties of light and matter.[48] Materials science—the physics of polymers, semiconductors, superconductors, magnets, and other advanced materials—is not only an area of great scientific interest but also a field singled out in the Technology Foresight Report as strategic for Ireland's economic development. Materials science and solid state physics are essential for the development of microelectronics, optical networking, and biotechnology, the central technologies of the information economy. Michael Coey, one of the leaders in the field at Trinity, explains, "It is in materials science that the throbbing pulse from physics to technology is most clearly felt."[49] In 2010 Ireland ranked eighth globally for materials science research, and 70% of the research cited was by CRANN researchers. It is Trinity's largest research institute, including over 300 researchers from a range of disciplines. CRANN's state-of-the-art facilities are used by academics as well as more than one hundred corporate partners.

Nearby, a sculpture titled "Throwing Shapes" displays the Weaire-Phelan structure, the most efficient arrangement ever found for packing shapes together (beating Lord Kelvin's proposal by

[48] D. Weaire, "Experimental Physics at Trinity," pp. 239-54 in C. H. Holland (ed.), *Trinity College Dublin & and the Idea of a University* (Trinity College Dublin Press: Dublin, 1991), pp. 277-9.

[49] J.M.D. Coey, "Physics and New Materials," *Hermathena* (1982) 133: 47-57, p. 51.

0.3%). Not only was its discovery at Trinity in 1993 by Erasmus
Smith Professor of Natural Philosophy Denis Weaire a major sci-
entific advance, it was also adopted by the architects of the Water
Cube at the 2008 Beijing Olympics.[50]

The School of Engineering that Humphrey Lloyd, James Mac-
Cullagh, and Thomas Luby launched in 1842 now has internationally leading research groups in bioengineering; digital media;
energy, transport, and the environment; and telecommunications.
Mechanical and Manufacturing Engineering are located in the Parsons Building (named for Charles Parsons, the famous nineteenth-
century engineer). And across Westland Row in the historic Dunlop
Oriel Building (the former site of Dunlop Rubber where the original
pneumatic tire patent was drafted), sits the Centre for Telecommunications Value-Chain Research (CTVR), a national research center based at Trinity that investigates wireless networking, optical
networking, agile radios, antenna design, tuneable optical components, and photonics.[51]

Biomedical research is centered on the new Trinity Biomedical Sciences Institute, a €130 million facility opened in 2011 along
Pearse Street. It will ultimately bring together over 700 researchers
in the areas of immunology, cancer, and medical devices. On cam-
pus sit the Moyne Institute of Preventive Medicine (named for
Walter Edward Guinness, the first Baron Moyne), the Smurfitt In-
stitute of Genetics (funded by the Irish paper magnate Michael
Smurfitt), and the Panoz Institute for Pharmacy (donated by Don
Panoz, the Irish American founder of the biotechnology company
Elan), as well as a number of industrial biotechnology laboratories.

---

[50]Henry Fountain, "Bubble Question Frames Design for Olympic Aquatics Center," *New
York Times* (August 5, 2008); Denis Weaire, "Sand and Foam," pp. 227-39 in Eoin P.
O'Neill (ed.), *What Did You Do Today, Professor? Fifteen Illuminating Responses from
Trinity College Dublin* (Living Edition: Pöllauberg, Austria, 2008); Tomaso Aste and Denis
Weaire, *The Pursuit of Perfect Packing*, 2nd Ed (New York: Taylor & Francis, 2008), pp.
107-10.

[51]CTVR-The Telecommunications Research Center, "About Us," ctvr.ie/about-us/.
(Accessed December 11, 2013.)

The School of Mathematics is located in an older building on Westland Row. By 2013 it had 20 academic staff and 13 graduate students. In pure mathematics, research areas include algebra and number theory, complex analysis and geometry, functional analysis, and partial differential equations. In applied mathematics and theoretical physics focus areas include string theory, lattice quantum chromodynamics, quantum field theory, and general relativity. The William Rowan Hamilton building, named for Trinity's (and Ireland's) most famous mathematician, was completed in 1992 with help from the European Union and houses the science library as well as lecture halls named after Trinity's great scientists. On the north side of the Hamilton building stands the O'Reilly Institute for Communications and Technology, built in 1989 with funds donated by Irish media mogul Tony O'Reilly. The Lloyd Institute (named for nineteenth-century physicist and provost Humphrey Lloyd), includes Computer Science and Statistics, Management Science and Information, and the Institute of Neuroscience. It is also the location for the Trinity Centre for High Performance Computing with researchers and technical staff with expertise in the areas of numerical modelling, risk analysis, molecular dynamics, and visualization.

The mathematics department has played an important role in computing and networking since their origins. Trinity entered the computer age in 1962 when the Engineering School purchased its first computer and started a postgraduate course in computer science. A computer science department was created in 1969, and by 1999 it was by far the largest at Trinity with more than forty-five faculty members. The mathematics department at Trinity, however, continued to play an important role in computing. In 1980 it began running the first Unix system in Ireland, and in 1981 it set up the first Irish node on what later became the internet. From the beginning students ran the network in the mathematics department. In fact, it was an undergraduate student, James Casey, who set up the `www.maths.ie` address (one of the first in Ireland) after he returned from summer internship at CERN (where Tim Berners-Lee invented HTML and the world wide web in 1990). Tim Murphy, who played a key role in the development of the mathematics department computer system, saw a strong connection between mathematics and computing. "Computing without mathematics," he explains, "reduces to a sort of cookery course."[52] Many of the mathematics students from the 1990s went to work in the rapidly emerging software industry.

The vast scale of scientific research at Trinity requires not only new facilities but also new administrative roles. In 1991 a Dean of Research was appointed for the first time to plan, coordinate, and manage the College's research efforts. Essentially, research had become too important to be left to the individual. Just as the university's research efforts had to be coordinated with national industrial strategy, so individual researchers now had to coordinate their work with the university's strategy. John Hegarty, Professor of Optoelectronics, Dean of Research, and Provost from 2001-2011, saw the creation of the office as part of a general trend facing all universities. As governments became more demanding in terms of the relevance and effectiveness of the research they funded, they began to hold researchers more accountable. It then became necessary to have structures in place both to navigate the byzantine systems of governmental funding structures and to account for funds received. As Hegarty put it, "Individuals cannot raise enough money to support themselves. If you want to convince the government to support graduate students or biomedical research or classics or new courses, you have to have a pretty clear strategy. Otherwise no one will give you anything."[53] In order to develop such a plan, in 1997 Trinity College created a Working Party on Research to take stock of all

---

[52]Interview with Tim Murphy, Department of Mathematics, Trinity College Dublin (June 2, 1999).
[53]Interview with John Hegarty, Dean of Research (April 21, 1999).

Trinity research and to propose a strategy for the next decade. The committee's recommendations asserted the necessity of socially and commercially relevant research.[54] And it introduced a system to measure the research performance of staff. It also stressed the importance to the College of "using its knowledge and expertise in the interests of society by developing intellectual property, protecting it by copyright or patent, and realising maximum financial benefit through licensing, formation of campus companies, and joint ventures."[55]

Professors are not typically known for their business acumen, and Trinity realized that, if they expected faculty to patent discoveries and launch companies, they would need to support them. They established an Innovation Centre in 1986. Eoin O'Neill, its first director, explains,

> It was a conscious decision by some people to change the perception in society about the utility of research and also to make research relevant. ... The feeling being that in the early 1980s there was a general disillusionment with the prospect of having expensive universities which trained graduates to PhD level in biotechnology or whatever and then they promptly went to California and lived there the rest of their lives. It was felt to be a drain on resources. ... [56]

The Innovation Centre and the strategy of university-industry links, therefore, was a means to improve the image (and the funding) of the university as well as a means to help the Irish economy. High technology multinationals, according to O'Neill, "provided an opportunity by which we could demonstrate that we were in fact

---

[54]"Report of Working Party on Research: Executive Summary" (May 1998), www.tcd.ie/Physics/People/John.Hegarty/local/executiv.htm (Accessed April 19, 1999.)

[55]"Report of Working Party on Research."

[56]Interview with Eoin O'Neill, Director of Innovation Services, Trinity College Dublin (June 1, 1999).

fulfilling a useful role...."[57] In addition to helping researchers at Trinity, the Innovation Centre worked closely with the IDA to sell Ireland to outside investors and to establish confidence in Ireland's infrastructure for high technology industry.

Trinity's most famous start-up—and one of Ireland's most successful—is Iona Technologies, a software firm started in 1991 by Chris Horn, Sean Baker, and Annrai O'Toole, three lecturers in Trinity's Computer Science Department. Iona was launched in Trinity's Innovation Centre and went on to raise $60 million when it first listed on the NASDAQ exchange in New York in 1997, making it the fifth largest IPO at the time. It became one of the top 10 software-only companies in the world and spun out more than 20 other companies over fifteen years.

9.4 Simulating Physics

The economic relevance of nanotechnology, telecommunications, and biomedical research are clear, but in the Celtic Tiger even particle physics became part of the knowledge economy. In the 21st century, physics still strives to describe the entirety of the universe with a handful of mathematical symbols, but that endeavor now requires the largest, most complex, and most expensive machinery that humanity has ever constructed. The closer we get to simple equations, the more complicated the enterprise of science becomes. The lone thinker ("Newton with his prism and silent face" in Wordsworth's iconic phrase) has been replaced by a global network of thousands of researchers, postdocs, graduate students, and technicians.

The Large Hadron Collider (LHC), opened in 2008, is the most impressive example yet.[58] Under the auspices of CERN (the Eu-

---

[57] O'Neill Interview.

[58] CERN, "The Large Hadron Collider," `home.web.cern.ch/about/accelerators/large-hadron-collider`. (Accessed December 11, 2013.)

ropean Organization for Nuclear Research), the collaboration involves over 10,000 scientists and engineers from more than 100 countries. It accelerates protons at 0.999998 the speed of light through a tunnel 27 km in circumference beneath the Franco-Swiss border near Geneva, Switzerland. Monitoring the collisions of the protons are 150 million sensors (some the size of a five-story office building) delivering data 40 million times a second and generating about 15 petabytes of data per year—equal to the bandwidth allotted to half a million simultaneous telephone conversations. Processing that data requires the Worldwide LHC Computing Grid, which was by 2012 the world's largest computing grid, comprising over 170 computing facilities across 36 countries. The total cost to construct the facility was €4.6 billion. E.T.S. Walton was able to win the Nobel Prize for physics working with a single collaborator in a small room in the Cavendish laboratory. A century later experimental physics was truly "big science".

It is not just the experimentalists who work in large-scale multinational collaborations on millions of euros worth of equipment. While string theorists maintain the long tradition of pencil and paper calculations, another group of theoretical physicists has adopted the tools of large-scale computational analysis to explore a field called lattice quantum chromodynamics (QCD), the theory that explains how protons and neutrons interact.

By the 1930s, it was clear that atoms (and therefore all matter) are composed of three particles —electrons, protons, and neutrons. The theory of electrons and light (photons), first glimpsed in Maxwell's electromagnetic theory and transformed by the quantum theory, developed by the 1960s into quantum electrodynamics (QED), the most precisely verified theory that humans have ever produced.[59] Understanding the nucleus of the atom turned out to be more of a challenge. Protons and neutrons had been dis-

---

[59]Richard Feynman, *QED: The Strange Theory of Light and Matter* (Princeton: Princeton University Press, 1985).

covered by bombarding atoms with alpha particles and watching the result. Soon physicists were building more sophisticated particle accelerators and watching the scattered particles trace paths in cloud chambers. Rather than ejecting the fundamental building blocks of protons, however, these accelerators just produced a long list of new particles—positrons, muons, pions, kaons, neutrinos. As Frank Wilczek, an American physicist who shared the Nobel Prize in physics in 2004 for his work on QCD, observed, "It's as if you smashed together two Granny Smith apples, and got three Granny Smith apples, a Red Delicious, a cantaloupe, a dozen cherries, and a pair of zucchini!"[60]

In the early 1960s, American physicists Murray Gell-Mann and George Zweig realized that it was possible to explain the entirety of this "particle zoo" by imaging each particle as composed of a combination of three constituents. Gell-Mann proposed to call these new entities "quarks", taking a line from Joyce's *Finnegan's Wake*, "Three quarks for Muster Mark."[61] The quarks came in three "flavors," which he termed up, down, and strange. A proton, for example, could be described as composed of two up quarks and one down quark. Later, three more flavors of quarks were added— top, bottom, and charm. Gell-Mann did not believe that quarks were actual particles but rather a useful mathematical abstraction (which is perhaps why he named them so flippantly). When later scattering experiments indicated that protons were in fact composed of smaller constituents, quarks began to seem more like real

---

[60]Frank Wilczek, *The Lightness of Being: Mass, Ether, and the Unification of Forces* (NY: Basic Books, 2008), p. 31.

[61] Gell-Mann later wrote a letter to the editor of the Oxford English Dictionary explaining his original use of the word. The allusion to three quarks seemed perfect since originally only three quarks had been discovered. Gell-Mann, however, wanted to pronounce the word to rhyme with 'cork' rather than with 'Mark', as Joyce seemed to indicate. Gell-Mann got around that "by supposing that one ingredient of the line 'Three quarks for Muster Mark' was a cry of 'Three quarts for Mister...,' heard in H.C. Earwicker's pub. Murray Gell-Mann, *The Quark and the Jaguar: Adventures in the Simple and the Complex* (New York: Heny Holt, 1994).

particles, but quarks have never been observed as individual free particles. They are always bound up inside other particles.

The force that holds quarks together inside a proton or neutron (and holds protons and neutrons together in the nucleus of an atom) is called the strong force, and it turns out to be much more complicated to understand than the electromagnetic force. While photons carry the electromagnetic force, the strong force is carried by particles called gluons. But while there is only one charge in electromagnetism (the electric charge), with the strong force there are three (known as colors—red, green, or blue). These "colors" (which have nothing to do with visible colors) are why the theory is known as quantum *chromo*dynamics.

According to QCD most of the mass of protons and neutrons comes from the interactions between quarks and gluons. The three quarks inside a proton account for only about 1% of the proton's mass (and the gluons have no mass)—all the rest of the mass comes from the energy that binds the quarks together. Far from being solid particles, the building blocks of matter look like patterns of motion. Berkeley, Hamilton, and Fitzgerald would all be delighted to learn that in the latest theories, materialism has been replaced by energy flows.

Moreover, in quantum theory, empty space itself is a dynamic medium. Wilczek even describes nature at the smallest scale as an ether. He explains, "the most important lesson we learn from QCD is that what we perceive as empty space is in reality a powerful medium whose activity molds the world."[62] Pairs of particles and antiparticles spontaneously emerge only to collapse into each other again. But these short-lived fluctuations can be measured and have an impact on the interactions between particles. Perfectly empty space is unstable, alive with spontaneous and unpredictable quantum activity. This activity has a very small (but

---

[62] Wilczek, *The Lightness of Being*, p. 73.

precisely measurable effect) in QED, but it has a critical effect in QCD. The equations that describe how quarks and gluons interact can be expressed in just a few lines of mathematics. Solving those equations, however, has proven to be very difficult for three major reasons. First, while electromagnetism follows an inverse square law (the force between two charged particles falls off rapidly as the distance increases), the force between quarks and gluons remains constant (and large) at long distances. In fact, the energy needed to separate two quarks grows without limit as you pull them apart. This explains why no one has ever observed free quarks. As you try to pull two quarks apart, the force between them grows, eventually generating enough energy to create another quark to pair up with the quark you just tried to free. Second, while photons are electrically neutral (they interact with charged particles but not with each other), gluons not only interact with quarks but also with each other. Third, in quantum mechanics, the state of a physical system is expressed as a probability distribution. Calculating the interaction between two particles requires taking into account every possible interaction between the two, assigning a probability to each, and then adding up the probabilities. In QED, the simplest interactions are the most probable, and most of the possible interactions make very small contributions to the calculation. In QCD, all of the possible exchanges (even those virtual quarks that arise spontaneously from empty space) make large contributions to the overall interaction.

Calculating the motions of electrons using QED can be done with pencil and paper. For QCD, even a rough approximation requires a supercomputer.[63] The only way to solve QCD is through simulation. Because a potentially infinite number of paths interact

---

[63] Brian Hayes, "Getting Your Quarks in a Row," *American Scientist* (November-December 2008) 96: 450-454; Sinéad Ryan, "Quarks from Top to Bottom: Particle Physics at the Energy Frontier," pp. 73-88 in Eoin P. O'Neill (ed.), *What Did You Do Today, Professor?*

to produce the motions of quarks within a proton, scientists must first simplify the problem. In the model, they imagine a grid—a lattice of points (hence the name lattice QCD) that makes the calculation more manageable. The quarks can exist only at the nodes of the lattice, and the gluons can interact only along the links between the nodes. This reduces an infinite number of possible states to a finite (but still very large) number. It also eliminates all interactions at distances shorter than the lattice. The computer then generates random paths and calculates their probability based on the equations of QCD. To give some sense of the magnitude of the calculation: A teraflop computer (one that carries out one million calculations every millionth of a second) operating continuously for a month can calculate what a single proton does in 10^{-24} seconds.[64] Ultimately the process is similar to the way the Havok physics engine works in Second Life, simulating the universe frame by frame. Solving the equations of QCD is critical to verifying the theory. The equations enable physicists to calculate the mass of the proton and other particles from first principles and then compare the predicted value to experimentally measured values. Wilczek describes QCD's calculation of the mass of the proton as "one of the greatest scientific achievements of all time."[65]

The study of QCD is a global endeavor, involving hundreds of scientists at dozens of universities and national laboratories around the world, and Trinity College Dublin has become an important node in that network due to a combination of determination, opportunism, and historical accident. Jim Sexton is widely acknowledged as the father of lattice QCD and high performance computing in Ireland. Sexton was an undergraduate at Trinity College Dublin and won the gold medal in mathematics and theoretical physics in 1978. He left to do his PhD at Columbia University, followed by a postdoc at Fermilab and two years at the Institute

---

[64] Wilczek, *The Lightness of Being*, p. 122-127.
[65] Wilczek, *The Lightness of Being*, p. 127.

for Advanced Study in Princeton. He then went to work with Don Weingarten at IBM in New York.[66] Weingarten was interested in building parallel computers that could run many different calculations simultaneously. The first problem for the new computer—the one it was constructed to solve—was calculating the mass of a particle from QCD. The calculation was hailed as "the largest single numerical calculation in the history of computing."[67]

In 1992, Sexton returned to Trinity (though he continued to work at IBM over summers and holidays). Frankly, Ireland could not offer the same scientific opportunities as the United States, but Sexton's wife (a violinist) had been offered a job with the symphony in Dublin. Sexton's work in Dublin really began to take off in 1997 when an anonymous American donor pledged IR £750,000 (€950,000) to support a new high performance computing initiative. The money was used to purchase Ireland's first supercomputer (an IBM SP-2 with a peak performance of 30 billion calculations per second) to be run jointly with Queen's University Belfast and to establish the Trinity Centre for High Performance Computing. It was at the time the largest investment that Trinity had ever made in a single research project.[68] From the start, the computer was intended to support business as well as science. The anonymous donor mandated that 20% of its processing time be used for industrial problems.

In 1999, Sexton brought in Sinéad Ryan, a graduate of University College Cork who had completed her PhD in lattice QCD in Edinburgh and then worked at Fermilab. Ryan had not planned to return to Ireland, but when Sexton offered her the opportunity to set up a master's degree program in high performance computing

[66]Interview with Jim Sexton, Director of High Performance Computing and Lecturer in Mathematics (June 1, 1999).

[67]Malcolm Browne, "448 Computers Identify Particle Called Glueball," *New York Times* (December 19, 1995).

[68]Karlin Lillington, "Super-Duper Computers," *Technology Ireland* (October 1998), pp. 18-21.

and to continue her work on lattice QCD, she jumped at the opportunity.[69] Ryan was soon followed by Mike Peardon, an Englishman who had also completed his PhD in lattice QCD in Edinburgh and who became the course director for the master's program in high performance computing. Sexton ultimately returned to IBM's Watson Research Center in New York in 2005, but the lattice QCD team at Trinity is still going strong. In 2005, Ryan and Peardon organized an International Lattice Conference, which brought more than 400 scientists from around the world to Trinity.[70]

How and why did Ireland become a leader in an obscure field like lattice QCD with little direct relevance to Irish economic competitiveness? The answer lies in the tight connection between QCD and high performance computing (HPC). Jim Sexton's work with IBM illustrates how QCD played a critical role in the development of HPC. Lattice QCD was the perfect proving ground for new computers, and many of the early high performance parallel machines were designed and built to solve QCD. In the early 2000s, working closely with IBM, research teams from the University of Edinburgh, Columbia University, RIKEN in Japan, and Brookhaven National Laboratory in the U.S. created a new design for a massively parallel computer.[71] The resulting "QCD on a chip" (QCDOC) computers then became the foundation for IBM's Blue Gene supercomputer, a faster, cheaper, and more energy efficient machine introduced in 2004.[72]

---

[69]Interview with Sinéad Ryan, Head of the School of Mathematics (July 18, 2013).

[70]"School of Mathematics Organises International Lattice Conference at TCD," *Trinity Research News* (November 2005), www.tcd.ie/research_innovation/assets/pdf/NewsletterNov05.pdf.

[71]"QCDOC Architecture," Department of Physics, Columbia University, phys.columbia.edu/~cqft/qcdoc/qcdoc.htm. (Accessed December 11, 2013.)

[72] IBM, "Blue Gene Overview", www-03.ibm.com/ibm/history/ibm100/us/en/icons/bluegene/. (Accessed December 11, 2013.), U.S. Lattice Quantum Chromodynamics, "Computing Hardware and Software: The Blue Gene/Q," www.usqcd.org/BGQ.html. (Accessed December 11, 2013.)

Blue Gene was originally built to help biologists simulate protein folding and gene development (hence the name). It was a significant improvement over IBM's previous supercomputer Deep Blue (which made headlines by beating chess world champion Garry Kasparov in 1997). While Deep Blue could calculate about 200 million potential chess moves per second, Blue Gene could handle 280 trillion operations per second. Although it was based on a design created to solve QCD, Blue Gene found applications across a broad range of areas, from mapping the human genome, to simulating the human brain, to predicting climate trends and identifying oil and gas deposits.

Mastering QCD required mastering supercomputing, and the skills developed in supercomputing were highly transferable to a broad range of problems across the sciences and industry. In seeking to understand quark behavior, the physicists had become experts in big data and simulation. And Sexton in particular realized how valuable that could be. He explained, "I want to solve QCD. That's my fundamental driving goal. And along the way there are a lot of interesting spin-offs."[73] Those interesting spin-offs, he recognized, were the key to finding support for basic physics research. When Trinity upgraded its HPC facilities in 2006, the Minister for Education and Science announced, "These facilities now make a significant contribution to TCD's ability to compete on the international stage in the area of materials and biomaterials modelling and visualization." These types of initiatives, she explained, help to ensure that Ireland "remains the location of choice for manufacturing and international services in areas such as electronics, pharmaceuticals and financial services."[74]

But what makes mathematical physicists qualified to solve industrial problems? As industry became more technologically so-

---

[73] Sexton interview.

[74] Trinity College Dublin Communications Office, "Minister for Education Launches IITAC's Visualisation and Supercomputing Facilities at TCD," (September 14, 2006).

phisticated, more globally integrated, and more competitive, its problems began to look more and more like the problems that mathematicians knew how to solve. Sexton explains, "You wouldn't necessarily think that a company that transports packages from Singapore to Dublin as its mainstay actually has an interesting computational problem, but they do."[75] Eamonn Cahill, former technical manager for high performance computing at Trinity, notes, "In many respects a network of telecommunications links and the traffic over them is very analogous to a physical system of molecules or charged particles. So we can carry over that expertise into a real world problem very easily."[76] The mathematical description of a system of charged particles is quite similar to the mathematical description of traffic in a telecommunications network. Mathematicians who are trained to solve the problem for particles can transfer this expertise to the telecommunications problem.

Another important example at the end of the twentieth century was the field of finance. A key turning point was the creation of the Black-Scholes equation for pricing options in 1973. Its success not only transformed the field of finance into a mathematical science (and won a Nobel Prize for Myron Scholes and Robert Merton), it also changed dramatically how people bought and sold stocks and other securities.[77] Mathematical finance enabled the global flows of capital required by multinational corporations looking to hedge risks across dozens of countries, and the IFSC in Dublin became a global center for this kind of work.

It turns out that the Black-Scholes equation is really just a form of the Navier-Stokes equation, developed by Claude-Louis Navier and George Gabriel Stokes (an Irish mathematician at Cambridge) in the nineteenth century to describe fluid dynamics. So solving a

---

[75] Sexton Interview.

[76] Eamonn Cahill, from Interview with Audrey Crosbie, ICeTACT, and Eamonn Cahill, technical manager for high performance computing at Trinity (April 19, 2000).

[77] Donald Mackenzie, *An Engine, Not a Camera: How Financial Models Shape Markets* (Cambridge: MIT Press, 2008).

problem in fluid dynamics can also solve problems in finance. As Cahill explains,

> The Black-Scholes is quite difficult to solve. It's singu-
> larly perturbed. Most numerical schemes behave quite
> badly on it. So if we could find a very effective new way
> of solving it, that would have a major impact through-
> out the financial community because anyone who has a
> better way of solving the Black-Scholes is going to make
> a killing.[78]

As the masters of mathematical modeling, mathematicians and physicists have therefore come to be important players in the world of finance. Since the 1970s, investment banks and hedge funds have been hiring such 'quants' or 'rocket scientists' to develop sophisti-cated computer programs to improve profits.[79]

The new centrality of mathematics, therefore, lies in the abil-ity of mathematicians to interpret a dizzying range of industrial, financial, and biological systems in terms of the familiar models of statistical physics or fluid dynamics. Mathematics, on this view, is neither the study of the Platonic forms, nor the intuition of pure time, nor the creation of purely formal systems. In the context of the Celtic Tiger, those mathematicians who have championed a view of mathematics as the science of modeling have been most suc-cessful at garnering support from government and industry. With the help of computers, extremely complicated mathematical mod-els and simulations can be created, modified, and applied in a way that would otherwise be impossible.

Not only could the physicists help to solve many problems in in-dustry, they could also train others to do so. And that was the idea behind the master's of science program in high performance com-

---

[78] Cahill Interview.

[79] See Nicholas Dunbar, *Inventing Money: The Story of Long-Term Capital Management and the Legends Behind It* (New York: John Wiley & Sons, 2000).

puting. The program is run by Mike Peardon, a lattice QCD physicist, but it takes in students with a wide variety of backgrounds—not just mathematics, physics, and chemistry, but also economics, finance, and computer science. The course website lists a broad range of applications including electronic chip design, circuit board layout, drug design, financial market forecasting, database mining, flow analysis, and resource management.[80]

During the last six months of the course students must complete a project, and many of the projects come from companies in the financial sector such as hedge funds or derivative companies who jointly supervise the work.[81] The program teaches them how to solve large-scale numerical problems on parallel computers and then has them apply those skills to real problems in industry. Not surprisingly, there is zero unemployment among graduates of the program. While many go into academic research positions across a range of disciplines, the skills they built in modeling and numerical simulation are very much in demand by companies such as Accenture, IBM, Google, PayPal, and the banks.

Sexton's strategy has been crucial to the growth of the mathematics department. The funding for the MSc made a full-scale graduate program in mathematics and theoretical physics feasible for the first time. When Sexton returned to Trinity in 1992, there were six graduate students in mathematics. When he left in 2005, there were 21 PhD students and 18 MSc students. In a very real sense, study of the fundamental forces of nature at Trinity therefore depends on the Celtic Tiger economy—the anonymous donor who funded the first supercomputer but mandated that 20% of its time be used for industry, the grants from the Irish government based on the importance of HPC to materials science, finance, and other strategic industries, and the demand by multinational employers for students trained in big data analysis.

---

[80]From www.maths.tcd.ie/hpcmsc/describecourse.htm (Accessed July 31, 2000).

[81]Interview with Mike Peardon, Associate Professor in Mathematics (August 27, 2013).

Mathematics and physics have long been considered 'fundamental' sciences. From the seventeenth to the early nineteenth century they were presented as the paradigms of certain knowledge. Now their claim to fundamental status lies with their ability to manage big data, deal with complexity, or find approximate solutions to equations that cannot be solved through traditional means. These techniques turn out to have relevance in a broad range of business and industrial settings. Physicists have become creators of the basic infrastructure for science and industry—not just basic concepts and mathematical tools but also software and hardware. In a sense they are discovering the fundamental laws of data as well as the fundamental laws of nature.

The connection between basic research and business application no longer takes 150 years. In the case of high energy physics, the computers designed by the lattice QCD researchers led almost immediately to IBM's Blue Gene computer. The computational needs of the physicists at CERN led to the development of the World Wide Web and grid computing. The optimization techniques for lattice QCD can be applied across network analysis and financial engineering. At a certain basic level, the scientific problems and the business problems are the same. That was the philosophy behind Jim Sexton's work. The ultimate goals may be very different, but the tools to reach those goals are the same.

It's not just that mathematicians have decided to study business problems, it's also the case that the way business works has changed fundamentally. Big business now runs on big data. Weather, traffic flows, financial transactions, even consumer sentiment are now all captured in real-time. During a single cross-country flight, a Boeing 737 generates 240 terabytes of data.[82] Wal-Mart handles more than one million customer transactions every hour, generat-

---

[82]McKinsey Global Institute, *Big Data: The Next Frontier for Innovation, Competition, and Productivity* (June 2011).

ing a database estimated at 2.5 petabytes.[83] Increasingly every interaction with a retailer, a bank, a government agency, even a friend online generates a trail of data. The rise of smart phones put a digitization device in the pocket of a significant percentage of people in the developed world. By 2013 over half of the Irish population owned a smart phone. As GPS capability became standard on most mobile phones, billions of phone users were transformed into a global sensor network.

A 2011 report from the European Science Foundation made the link between mathematics and the knowledge economy explicit,

> Knowledge has become the main wealth of nations, companies and people. Hence, investing in research, innovation and education is now the key leverage for competitiveness and prosperity in Europe. At the heart and foundation of this challenge, mathematics plays a crucial role as it provides a logically coherent framework to industry and a universal language for the analysis, simulation, optimisation, and control of industrial processes.[84]

Mathematics was seen as the key to managing large multinational corporations and global supply chains and therefore as the key to national competitiveness and rising standards of living.

9.5   The Search for Symmetry

While one group within the School of Mathematics at Trinity is working on calculating solutions to QCD, another group is part of the search for an even more fundamental theory of physics. The standard model of particle physics has now held up to more than forty years of experimental testing. Its description of the strong,

---

[83]"Data, Data Everywhere: A Special Report on Managing Information," *The Economist* (February 27, 2010).

[84]European Science Foundation, *Mathematics and Industry* (2011), p. 4.

weak, and electromagnetic forces and how they act on 12 different fundamental particles (six quarks and six leptons) has been verified with incredible precision, but physicists do not believe that the standard model can be the final answer. First of all, it cannot explain gravity; it fails to unify all four of the fundamental forces. But the standard model also faces important aesthetic objections— it doesn't look the way that physicists believe a fundamental theory should. American physicist Leonard Susskind explains,

> "I often feel a discomfort, a kind of embarrassment, when I explain elementary-particle physics to laypeople. It all seems so arbitrary—the ridiculous collection of funda- mental particles, the lack of pattern to their masses, and especially the four forces, so different from each other, with no apparent rhyme or reason. Is the universe 'ele- gant'...? Not as far as I can tell, not the usual laws of particle physics anyway."[85]

Physicists strongly believe that a fundamental 'theory of every- thing' should be simple, elegant, and unique. It should look nec- essary, as though the universe could not have been framed in any other way. The standard model may be precise and predictive, but it lacks the fundamental necessity that physicists seek. Just as Maxwell unified all of the laws of electricity and magnetism, and Einstein derived his general theory of relativity from the sim- ple assumption that the laws of physics are the same in all frames of reference, physicists are not satisfied until they can explain a wide range of complicated phenomena from a handful of simple equations.

The most important attempt to explain the known laws of physics at a more basic and simpler level over the past thirty years is string theory. String theory imagines that the fundamental build-

---

[85]Leonard Susskind, *The Cosmic Landscape: String Theory and the Illusion of Intelligent Design* (New York: Back Bay Books, 2006), p. 202.

ing blocks of matter (the quarks and the leptons) are not point particles but one-dimensional strings. All of the different particles are just different modes of vibration of a single fundamental object—a string. The different modes of vibration give the strings different properties, such as mass, charge, or spin. Vibrations of open strings generate electromagnetic and nuclear forces (the so-called gauge fields) while the vibrations of closed loops generate gravitons. This is perhaps string theory's greatest strength, the fact that it very naturally unifies gravity with the other three forces. In fact the theory won't work without all of the forces. Moreover, the motion of the strings follows a very simple law—the area created by a string moving through space-time is minimized. Like Hamilton and Larmor, physicists still see minimization as both a powerful mathematical tool and a compelling aesthetic criterion.

Even after thirty years of development, however, string theory still faces significant challenges. In fact, a number of physicists believe that their field faces an unprecedented crisis. For the mathematics to work the strings must vibrate in ten dimensions, but we only experience three dimensions of space and one dimension of time, physicists speculate, because the other six dimensions are curled up or 'compacted' so that we don't notice them. Lee Smolin argues, for example, that "the theory makes no predictions that are testable by current—or even currently conceivable—experiments."[86] How can we call it a scientific theory, he asks, if it cannot be tested? Others point out that string theory is not even a single theory. In fact, string theory offers a vast landscape of theories (10^{500}), each of which seems equally plausible. No version of string theory achieves the uniqueness or inevitability that scientists look for in a fundamental theory. And string theory cannot yet account for the most important recent discovery in physics—that only 4% of the observable universe is made of ordinary matter

---

[86]Lee Smolin, *The Trouble with Physics: The Rise of String Theory, the Fall of a Science, and What Comes Next* (Boston: Mariner Books, 2007), p. xiv.

(the quarks and leptons described by the standard model), the rest being composed of so-called dark energy and dark matter whose nature and laws we have not yet begun to understand.

For all of its potential failings as a physical theory, however, string theory has been an enormous boon to mathematics. The very simple starting point of strings vibrating and moving in ten dimensions has led over the past thirty years to the development of an incredibly complex set of mathematical tools—group theory, differential geometry, homology, homotopy, fiber bundles, and supersymmetry, to name only a few.[87] String theorists have adopted mathematical methods that physicists had never used before, and they have created their own methods that mathematicians had never seen before. Edward Witten, one of the founders of string theory even won the Fields Medal, the most prestigious award granted to a mathematician (the award that John Lighton Synge had a hand in establishing). Ironically, the closer that physicists get to a 'simple' underlying theory, the more complex the mathematics becomes. In fact, string theory has been driven more by advances in mathematics than by advances in experimental data. In the absence of experimental guidance, physicists and mathematicians follow symmetry principles—they assume (like Hamilton) that the fundamental structure of the universe must match our own internal sense of beauty, harmony, and simplicity.

For this reason, the string theorists at Trinity are the closest modern heirs to the Hamiltonian tradition. Samson Shatashvili, the University Chair of Natural Philosophy at Trinity and an international leader in the mathematical methods of string theory, explains, "About one third of my work is on hyperkähler manifolds, which are very closely related to quaternions (in fact its three holomorphic structures are called i, j, and k after Hamilton's notation). One of my most cited papers is titled 'Algebraic and

---

[87]superstringtheory.com/math/math2.html.

Hamiltonian Methods'. In a sense we are continuing Hamilton and MacCullagh's tradition in dynamical systems."[88] With five positions in the School of Mathematics, the string theorists at Trinity represent one of the department's largest and most internationally renowned research groups (about the same size as the lattice QCD group).

Shatashvili studied at the Leningrad Department of the Steklov Institute of Mathematics and was a member of L.D. Fadeev's Leningrad School of theoretical and mathematical physics. In 1990 he left Russia for a position at the Institute for Advanced Studies in Princeton and then in 1994 he was hired as a professor by the Yale physics department. He came to Trinity in 2002 with a mandate to create a world-class School of Mathematics out of the pre-existing Department of Pure and Applied Mathematics. Shatashvili became the Head of the School and ultimately succeeded in bringing 18 new permanent mathematicians. Together with his colleagues, he has made Trinity's School of Mathematics one of the top departments in the world for string theory. In 2011, the department was ranked 15th in the world, and in 2014 it was ranked number one in the world for citations.[89] In 2014 two of Trinity's mathematicians (Shatashvili and Gerasimov) were invited to speak at the quadrennial International Congress of Mathematicians (a rare honor, particularly for two speakers from the same university).

Like Schrödinger before him, Shatashvili is part of a proud tradition of foreign scientists drawn to Ireland. In fact, the School of Mathematics has diversified to the point where only a handful of its faculty are Irish nationals. The Hamiltonian tradition still serves as a unifying theme bringing talent to Ireland. Shatashvili was instrumental in creating a Hamilton Mathematics Institute at

[88]Interview with Samson Shatashvili, University Chair of Natural Philosophy, Trinity College Dublin (January 14, 2013).

[89]QS World University Rankings by Subject: Mathematics (2011), www.topuniversities.com/university-rankings/university-subject-rankings/2011/mathematics (Accessed on April 30, 2014).

Trinity in 2003 (it launched officially in 2005) and he still serves
as its director. (NUI Maynooth had already created a Hamilton
Institute in 2001 focused not on pure mathematics but on providing "a bridge between mathematics and its applications in ICT and
biology."[90])

The fate of the Hamilton Mathematics Institute, however, illustrates how much the context for science in Ireland has changed
in the early twenty-first century. While de Valera was able to find
funding for the Dublin Institute for Advanced Studies at time when
Ireland faced extreme poverty, the Hamilton Institute has struggled
even as Ireland became one of the wealthiest countries in the world.
Science Foundation Ireland refused to fund the Hamilton Institute,
arguing that it fell outside of their mandate. Trinity was able to
leverage private donors to support the Institute at first, but private funding dried up in 2011 a few years after the financial crisis
hit. The Institute is now in a kind of suspended animation, but
Shatashvili is optimistic that it will soon be able to operate at the
level of 2005–2011 again.

The lack of funding has not stopped the work of the string theorists. They don't require supercomputers or nanofabrication devices for their work. But as de Valera realized, while theoretical
physics may be cheap, money is critical to recruit the world's best
talent and to train the next generation of students. The failing
Irish economy and cutbacks to higher education funding are making it harder to attract and retain top professors who face increased
taxes, declining salaries, and brighter opportunities abroad. The
continued success of string theory in Dublin will depend on the
ability of the mathematicians to make the case to faculty, students,
administrators, donors, and government agencies that this type of
fundamental research is important for the future of Ireland.

---

[90]www.hamilton.ie. (Accessed January 26, 2014.)

9.6 The Science of Business and the Business of Science

The Hamiltonian tradition now sits somewhat uncomfortably within Trinity College and Ireland. The idealist vision of science that inspired Hamilton, Schrödinger, and de Valera—the search for underlying beauty and fundamental meaning—contrasts with the current focus on application, commercialization, and innovation. In the Celtic Tiger, science and mathematics are big business. The Irish economy is increasingly built on science and mathematics-based industry, and at the same time, the research enterprise itself has come to resemble a multinational corporation. The latest research facilities at Trinity have the same drug development facilities, nanofabrication equipment, clean rooms, and high performance computers as the most sophisticated corporations. Scientists are now supported by armies of technical and administrative support staff. Research teams involve hundreds or thousands of individuals working across dozens of universities, national laboratories, and corporations on a global scale. Supporting this enterprise depends on a constant flow of funding from government agencies and corporate research contracts. As a result, scientists spend much of their time competing for grants and commercializing their research. The new systems for measuring academic productivity—counting publications, tracking grant funds, calculating impact factors—apply a corporate standard of output to academic work.

Star scientists themselves are now like multinational corporations—globally mobile and highly sought after. They can do their work from any major university, and they are courted with offers of facilities, funding, graduate students, and reduced teaching loads (not unlike the tax breaks offered to multinationals). A single top researcher can rocket a department instantly to the top of the rankings, bring in tens of millions of euros in grants, and attract major donors or corporate partners. Recruiting and retaining these star researchers has become a major focus for top research universities.

Perhaps most importantly, science and business increasingly now share similar problems. In the nineteenth century, Hamilton's mathematical work had nothing to do with the problems of industry. While his colleagues embraced the study of energy, heat, and electromagnetism in the hopes of contributing to industrial progress, Hamilton sought solutions to philosophical and aesthetic problems. And yet Hamilton's mathematics is now crucial to solving a wide range of industrial problems. Quaternions are used in computer graphics and satellite control systems. The Hamiltonian function is used in weather prediction, protein folding, circuit design, and network analysis. As Carlyle and Burke lamented in the early nineteenth century, we now live in an age of calculators. Today's mathematicians (or 'mathematical science practitioners' as recent reports tend to call them) resemble supply chain managers more than poets. While the search for beauty and symmetry is still the overriding concern in areas such as string theory, most researchers are more like the lattice QCD researchers—focused on optimization, approximation, and faster and cheaper solutions to technical problems. The algorithms they discover are major accomplishments with important applications (and they move us closer to that long dreamed of unified theory), but no one is likely to carve them into a bridge or compare them to poetry.

While Hamilton spent most of his time with poets and novelists or reading German philosophy, in the current environment there is little time for philosophical reflection. In a publish or perish environment, where competition for funding determines which areas of research are possible or not, practitioners focus on writing papers and winning grants rather than on stopping to think about what it all means.

Scientists today are significantly more productive than those in the nineteenth century—they write more papers (though Hamilton's collected works run to four large volumes), they patent more technologies, and they consult with industry and even launch their own companies. But no Irish scientist has won a Nobel prize since E.T.S. Walton, and even his research was done in the 1930s at Cambridge University. Hamilton's work would never be funded by SFI today. And yet it is important to remember that Hamilton's freedom to pursue 'curiosity-driven research' depended on a very specific social, economic, and political structure, one that the Irish people today would find completely unacceptable. That takes nothing away from his intellectual accomplishment, but it forces us to recognize that asking what kind of science we want to promote cannot be separate from the question of what kind of society we want to live in.

Bibliography

Thoughts on the Present State of the College of Dublin; Addressed to the Gentlemen of the University (Dublin, 1782)

Strictures on the Bishop of Cloyne's Present State of the Church of Ireland (London, 1787)

Dublin Problems: Being a Collection of Questions Proposed to the Candidates for the Gold Medal at the General Examinations from 1816 to 1822 Inclusive, Which Is Succeeded by an Account of the Fellowship Examination in 1823 (London, 1823)

"Review: *A Short View of the Principles of the Differential Calculus* by the Rev. Arthur Browne, Fellow of St. John's College, Cambridge (Cambridge, 1824)," *Dublin Philosophical Journal* (1825) 1: 203-11.

"The Dublin University," *Blackwood's Edinburgh Magazine* (1829) 26: 153-77.

"Dr. Lloyd on Mechanical Philosophy," *The Quarterly Review* (1829) 39: 432-51.

"The Universities of Oxford and Cambridge," *Westminster Review* (1831) 15: 56-69.

"Edmund Burke," *Blackwood's Edinburgh Magazine* (1833) 33: 277-97.

"The University of Dublin," *Quarterly Journal of Education* (1833) 6: 5-27, 201-237.

"The Dublin University Calendar for 1833," *Dublin University Magazine* (1833) 1: 105-6.

"Academical Reform: The Dublin University System of Education Considered in Relation to Its Practicable and Probable Reform," *Dublin University Magazine* (1834) 3: 81-96.

"Obituary of John Brinkley," *Gentleman's Magazine* (1835) 2: 547.

"State of the Irish Church," *Edinburgh Review* (1835) 61: 490-525.

"The Irish and English Universities," *Dublin Review* (1836) 1: 68-100.

"Trinity College Dublin," *Dublin Review* (1838) 4: 281-307.

"The Late Provost," *Dublin University Magazine* (1838) 11: 111-21.

"Award of the Copley Medal to James MacCullagh," *Proceedings of the Royal Society* (1842) 4: 419-22.

"The Scholarship Question," *Dublin Review* (1847) 23: 228-51.

"Obituary of James MacCullagh," *Proceedings of the Royal Society* (1847) 5: 712-18.

University of Dublin Military Class: The Conditions of Admission, and the Examination Papers Set at the Entrance Examination in October 1857 (Dublin, 1857)

University Education in Ireland in the Year 1860 (Dublin, 1861)

"The British Association–Breaking Down the Atom," *The Times* (September 12, 1933)

"Irish Culture," *The Irish Times* (1939)

"Schrödinger," *Time* (April 5, 1943)

"The Hamilton Postage Stamp: An Announcement by the Irish Minister of Posts and Telegraphs," pp. 81-82 in *A Collection of Papers in Memory of Sir William Rowan Hamilton* (New York: Scripta Mathematica, 1945)

"Science: Einstein Stopped Here," *Time* (February 10, 1947)

"Letters to the Editor," *Time* (March 3, 1947)

"Havok Plays Master Chief at Innovation Awards," *Silicon Republic* (December 20, 2007)

"Data, Data Everywhere: A Special Report on Managing Information," *The Economist* (February 27, 2010)

"U2 in SL Live Concert in Second Life (2008.3.29)," www.youtube.com/watch?v=K4LE6VBAPd8 (Accessed December 11, 2013)

Abrams, M.H. *The Mirror and the Lamp: Romantic Theory and the Critical Tradition* (Oxford: Oxford University Press, 1953)

Adelman, J. *Communities of Science in Nineteenth-Century Ireland* (London: Pickering & Chatto, 2009)

Airy, G.B. *Mathematical Tracts on the Lunar and Planetary Theories, the Figure of the Earth, Precession and Nutation, the Calculus of Variations and the Undulatory Theory of Optics Designed for the Use of Students in the University*, 2nd Ed. (Cambridge, 1831)

Andrews, J.H. *Plantation Acres: An Historical Study of the Irish Land Surveyor and His Maps* (Omagh: Ulster Historical Foundation, 1985)

Andrews, J.H. *Shapes of Ireland: Maps and Their Makers, 1564-1839* (Dublin: Geography Publications, 1997)

Arago, F. "Sur la vie et les ouvrages de M. John Brinkley, correspondant de l'Académie des Sciences," *Comptes rendus hebdomadaires des séances de L'Académie des Sciences* (1835) 1: 212-25.

Ashe, S.G. "A New and Easy Way of Demonstrating Some Propositions in Euclid," *Philosophical Transactions* (1684) 14: 672-76.

Ashe, S.G. *A Sermon Preached in Trinity College Chapell Before the University of Dublin, January 9, 1693-94* (Dublin, 1694)

Ashe, S.G. "St. George Ashe's Speech to Lord Clarendon, 25 January 1685/86," pp. 275-78 in I. Ehrenpreis (ed.) *Swift: The Man, His Works, and the Age* (London: Methuen, 1962), Vol. 1.

Ashworth, W.J. "Memory, Efficiency, and Symbolic Analysis: Charles Babbage, John Herschel, and the Industrial Mind," *Isis* (1996) 87: 629-653.

Aste, T. and D. Weaire, *The Pursuit of Perfect Packing*, 2nd Ed (New York: Taylor & Francis, 2008)

Atherton, M. "Corpuscles, Mechanism and Essentialism in Berkeley and Locke," *Journal of the History of Philosophy* (1991) 29: 47-67.

Attis, D.A. "The Social Context of W.R. Hamilton's Prediction of Conical Refraction," pp. 19-36 in P. J. Bowler and N. Whyte (eds.), *Science and Society in Ireland: The Social Context of Science and Technology in Ireland, 1800-1950* (Belfast: Institute of Irish Studies-Queen's University Belfast, 1997)

Attis, D.A., P. Kelly and D. Weaire, "Helsham and the Rise of Newtonian Physics," pp. 1-18 in D. Weaire, P. Kelly and D. A. Attis (eds.), *Richard Helsham's Lectures on Natural Philosophy [1739]* (Dublin: Trinity College Dublin Physics Department, Institute of Physics Publishing, Verlag MIT, 1999)

Attis, D.A. "John Sealy Edward Townsend," pp. 202-203 in Charles Mollan, Brendan Finucane, and William Davis (eds.), *Irish Innovators in Science & Technology* (Dublin: Royal Irish Academy and Enterprise Ireland, 2002)

Attis, D.A. and C. Mollan (eds.) *Science and Irish Culture Volume One: Why the History of Science Matters in Ireland* (Dublin: Royal Dublin Society, 2004)

Attis, D.A. "More than a Maxwellian: Fitzgerald and Technology," *European Review* (2007) 15: 561-73.

Babbage, C. *The Ninth Bridgewater Treatise: A Fragment, Second Edition [1838]*, Ed. by M. Campbell-Kelly (New York, 1989)

Bacon, F. "The New Atlantis," in J. Weinberger (ed.) *The Great Instauration and the New Atlantis* (Arlington Heights, IL: Harlan Davidson, 1980)

Bacon, F. *Novum Organum*, Trans. by Peter Urbach and John Gibson (Chicago: Open Court, 1994)

Bailie, W. "Rev. Samuel Barber, 1738-1811: National Volunteer and United Irishman," in J. Haire (ed.) *Challenge and Conflict: Essays in Presbyterian Doctrine* (Antrim, 1981)

Barber, S. *Remarks on a Pamphlet, Entitled the Present State of the Church of Ireland* (Dublin, 1787)

Barber, S. *A Reply to the Revd. Mr. Burrowes's Remarks* (Dublin, 1787)

Barnard, T.C. "Miles Symner and the New Learning in Seventeenth-Century Ireland," *Journal of the Royal Society of Antiquaries of Ireland* (1972) 102: 129-142.

Barnard, T.C. "The Hartlib Circle and the Origins of the Dublin Philosophical Society," *Irish Historical Studies* (1974) 19: 56-71.

Barnard, T.C. *Cromwellian Ireland* (London: Oxford University Press, 1975)

Barnard, T.C. "Sir William Petty, Irish Landowner," pp. 201-17 in H. Lloyd-Jones, V. Pearl and B. Worden (eds.), *History & Imagination: Essays in Honour of H.R. Trevor-Roper* (London: Duckworth, 1981)

Barnard, T.C. "Gardening, Diet and 'Improvement' in Later Seventeenth-Century Ireland," *Journal of Garden History* (1990) 10: 71-85.

Barnard, T.C. "The Hartlib Circle and the Cult and Culture of Improvement in Ireland," pp. 281-97 in M. Greengrass, M. Leslie and T. Raylor (eds.), *Samuel Hartlib and Universal Reformation* (Cambridge: Cambridge University Press, 1994)

Barr, C. "The Failure of Newman's Catholic University of Ireland," *Archivium Hibernicum* (2001) 55: 126-39.

Barrington, S.J. *Personal Sketches of His Own Times* (London, 1827)

Barrington, S.J. *Historic Memoirs of Ireland*, 2nd Ed. (London, 1835)

Barry, N. "Irish Contributions to Wireless Communications," *Technology Ireland* (May 1994), p. 23-26.

Becher, H.W. "William Whewell and Cambridge Mathematics," *Historical Studies in the Physical Sciences* (1980) 11: 1-48.

Becher, H.W. "Woodhouse, Babbage, Peacock, and Modern Algebra," *Historia Mathematica* (1980) 7: 389-400.

Becher, H. "Radicals, Whigs and Conservatives: The Middle and Lower Classes in the Analytical Revolution in Cambridge in the Age of Aristocracy," *British Journal for the History of Science* (1995) 28: 405-26.

Beckett, J.C. *The Anglo-Irish Tradition* (London: Faber and Faber, 1976)

Belhoste, B., A.D. Dalmedico and A. Picon (eds.), *La Formation Polytechnicienne, 1794-1994* (Paris: Dunod, 1994)

Bell, D. "Ireland Without Frontiers? The Challenge of the Communications Revolution," pp. 219-30 in R. Kearney (ed.) *Across the Frontiers: Ireland in the 1990s* (Dublin: Wolfhound Press, 1988)

Bence-Jones, M. *Twilight of the Ascendency* (London: Constable, 1987)

Benjamin, M. "Medicine, Morality and the Politics of Berkeley's Tar-Water," pp. 165-93 in A. Cunningham and R. French (eds.), *The Medical Enlightenment of the Eighteenth Century* (Cambridge: Cambridge University Press, 1990)

Bennett, J.A. "The Mechanics' Philosophy and the Mechanical Philosophy," *History of Science* (1986) 24: 1-28.

Bennett, J. *Church, State and Astronomy in Ireland: 200 Years of Armagh Observatory* (Armagh: Armagh Observatory, 1990)

Bennett, J. "The Challenge of Practical Mathematics," pp. 176-90 in S. Pumfrey, P. Rossi and M. Slawinski (eds.), *Science, Culture and Popular Belief in Renaissance Europe* (Manchester: Manchester University Press, 1991)

Bennett, J.A. "Geometry and Surveying in Early-Seventeenth-Century England," *Annals of Science* (1991) 48: 345-54.

Bennett, J. "Science and Social Policy in Ireland in the mid-Nineteenth Century," pp. 37-47 in P. J. Bowler and N. Whyte (eds.), *Science and Society in Ireland: The Social Context of Science and Technology in Ireland, 1800-1950* (Belfast: The Institute of Irish Studies-Queen's University of Belfast, 1997)

Bennett, J. "MacCullagh's Ireland: The Institutional and Cultural Space for Geometry and Physics," *The European Physics Journal H* (2010) 35: 123–132.

Berkeley, G. "De Motu," pp. 253-76 in M. Ayers (ed.) *George Berkeley: Philosophical Works Including the Works on Vision* (London: Everyman, 1975)

Berkeley, G. "A Treatise Concerning the Principles of Human Knowledge," pp. 71-154 in M. Ayers (ed.) *George Berkeley: Philosophical Works Including the Works on Vision* (London: Everyman, 1975)

Berkeley, G. "Three Dialogues Between Hylas and Philonous," pp. 155-252 in M. Ayers (ed.) *George Berkeley: Philosophical Works Including the Works on Vision* (London: Everyman, 1975)

Berkeley, G. "An Essay Towards a New Theory of Vision," pp. 1-70 in M. Ayers (ed.) *George Berkeley: Philosophical Works Including the Works on Vision* (London: Everyman, 1975)

Berkeley, G. "Siris: A Chain of Philosophical Reflexions and Inquiries," pp. 1-164 in A. A. Luce and T. E. Jessop (eds.), *The Works of George Berkeley, Bishop of Cloyne* (Nendeln: Kraus Reprint, 1979), Vol. 5.

Berkeley, G. "A Defence of Free-Thinking in Mathematics," pp. 103-142 in A. A. Luce and T. E. Jessop (eds.), *The Works of George Berkeley, Bishop of Cloyne* (Nendeln: Kraus Reprint, 1979), Vol. 4.

Berkeley, G. "The Analyst," pp. 53-103 in A. A. Luce and T. E. Jessop (eds.), *The Works of George Berkeley, Bishop of Cloyne* (Nendeln: Kraus Reprint, 1979), Vol. 4.

Berman, M. *Social Change and Scientific Organization: The Royal Institution, 1799-1844* (Ithaca: Cornell University Press, 1978)

Berman, D. "Enlightenment and Counter-Enlightenment in Irish Philosophy," *Archiv für Geschichte der Philosophie* (1982) 64: 148-65.

Berman, D. "The Culmination and Causation of Irish Philosophy," *Archiv für Geschichte der Philosophie* (1982) 64: 257-79.

Berman, D. "The Irish Counter-Enlightenment," pp. 119-40 in R. Kearney (ed.) *The Irish Mind* (Dublin: Wolfhound Press, 1984)

Berman, D. *George Berkeley: Idealism and the Man* (Oxford: Clarendon, 1994)

Berry, H.F. *A History of the Royal Dublin Society* (London: Longmans, Green and Co. 1915)

Bleaney, B. "Two Oxford Science Professors: F. Soddy and J.S.E. Townsend," *Notes and Records of the Royal Society of London* (2002) 56: 83-88.

Bloor, D. "Hamilton and Peacock on the Essence of Algebra," in H. Mehrtens, H. Bos and I. Schneider (eds.), *Social History of Nineteenth Century Mathematics* (Boston: Birkhauser, 1981)

Bloor, D. *Knowledge and Social Imagery*, 2nd Ed. (Chicago: University of Chicago Press, 1991)

Boate, G. *Irelands Naturall History* (London, 1652)

Boole, G. *The Mathematical Analysis of Logic, Being an Essay Towards a Calculus of Deductive Reasoning* (Cambridge, 1847)

Boole, G. *An Investigation of the Laws of Thought, On Which Are Founded the Mathematical Theories of Logic and Probabilities* (London, 1854)

Booth, J. *How to Learn and What to Learn: Two Lectures Advocating the System of Examinations Established by the Society of Arts* (London, 1856)

Boran, E. "The Libraries of Luke Challoner and James Ussher, 1595-1608," pp. 75-115 in H. Robinson-Hammerstein (ed.) *European Universities in the Age of Reformation and Counter Reformation* (Dublin: Four Courts, 1998)

Bos, H.J.M. "Mathematics and Rational Mechanics," pp. 327-55 in G. S. Rousseau and R. Porter (eds.), *The Ferment of Knowledge: Studies in the Historiography of Eighteenth-Century Science* (Cambridge: Cambridge University Press, 1980)

Bowler, P.J. and N. Whyte (eds.), *Science and Society in Ireland: The Social Context of Science and Technology in Ireland, 1800-1950* (Belfast: The Institute of Irish Studies-Queen's University of Belfast, 1997)

Bowler, P.J. *Reconciling Science and Religion: The Debate in Early Twentieth-Century Britain* (Chicago: University of Chicago Press, 2001)

Bowen, K. *Protestants in a Catholic State: Ireland's Privileged Minority* (Kingston: McGill-Queen's University Press, 1981)

Bowers, S. "Apple's Multi-Billion Dollar, Low-Tax Profit Hub: Knocknaheeny, Ireland," *The Guardian* (May 29, 2013)

Boyce, D.G. *Nationalism in Ireland*, 2nd ed. (New York: Routledge, 1991)

Boyton, C. *Observations on the Church;- Its Relation to the State;- The Authority of Its Governors;- and the Duties and Obligations of Its Ministers: Being A Sermon, Preached in the Cathedral of Derry, Before the Lord Primate... On the Occasion of the Triennial Visitation, September 14, 1838* (Dublin, 1838)

Bradley, D.J. "University-Industry Cooperation in Action," pp. 20-22 in *Innovation Report 1981* (Dublin: Confederation of Irish Industry, 1982)

Brewster, D. "Review of *Reflexions on the Decline of Science in England, and on some of its Causes* by Charles Babbage," *Quarterly Review* (1830) 43: 305-42.

Brewster, D. "Report on the Recent Progress of Optics," pp. 308-22 in *Report of the First and Second Meetings of the BAAS; At York in 1831, and at Oxford in 1832* (London, 1833)

Brinkley, J. "Investigations in Physical Astronomy, Principally Relative to the Mean Motion of the Lunar Perigee," *Transactions of the Royal Irish Academy* (1818) 13: 25-52.

Brinkley, J. *Elements of Astronomy*, 2nd Ed. (Dublin, 1819)

Brinkley, J. *Synopsis of Astronomical Lectures, To Commence October 29, 1799, at the Philosophy School, Trinity College, Dublin* (Dublin, n.d.)

Brinton, C. *The Political Ideas of the English Romanticists* (Oxford: Oxford University Press, 1926)

Brooke, J.H. "Natural Theology and the Plurality of Worlds: Observations on the Brewster-Whewell Debate," *Annals of Science* (1977) 34: 221-86.

Brown, T.M. "Medicine in the Shadow of the Principia," *Journal of the History of Ideas* (1987) 48: 629-48.

Browne, M. "448 Computers Identify Particle Called Glueball," *New York Times* (December 19, 1995)

Brynn, E. *The Church of Ireland in the Age of Catholic Emancipation* (New York: Garland, 1982)

Buchdahl, H.A. *An Introduction to Hamiltonian Optics* (New York: Dover, 1993)

Buchwald, J.Z. "The Quantitative Ether in the First Half of the Nineteenth Century," pp. 215-38 in G. N. Cantor and M. J. S. Hodge (eds.), *Conceptions of Ether: Studies in the History of Ether Theories, 1740-1900* (Cambridge: Cambridge University Press, 1981)

Buchwald, J.Z. *From Maxwell to Microphysics: Aspects of Electromagnetic Theory in the Last Quarter of the Nineteenth Century* (Chicago: University of Chicago Press, 1985)

Buchwald, J.Z. *Rise of the Wave Theory of Light* (Chicago: University of Chicago Press, 1989)

Buchwald, J.Z. and Andrew Warwick (eds.), *Histories of the Electron: The Birth of Microphysics* (Cambridge: MIT Press, 2001)

Buck, P. "Seventeenth-Century Political Arithmetic: Civil Strife and Vital Statistics," *Isis* (1977) 68: 67-84.

Buick, K.R. *James Ussher, Archbishop of Armagh* (Cardiff: University of Wales Press, 1967)

Burchfield, J.D. *Lord Kelvin and the Age of the Earth* (Chicago: University of Chicago Press, 1990)

Burke, E. *Reflections on the Revolution in France [1795]* (New York: Anchor Books, 1973)

Burnett, J.E. and A.D. Morrison-Low, *'Vulgar and Mechanick': The Scientific Instrument Trade in Ireland, 1650-1921* (Edinburgh and Dublin: National Museums of Scotland and Royal Dublin Society, 1989)

Burrowes, R. "Preface," *Transactions of the Royal Irish Academy* (1787) 1: ix-xvii.

Burrowes, R. *A Letter to the Rev. Samuel Barber, Minister of the Presbyterian Congregation of Rathfryland, Containing a Refutation of Certain Dangerous Doctrines Advanced in His Remarks on the Bishop of Cloyne's Present State of the Church of Ireland* (Dublin, 1787)

Burrowes, R. *Observations on the Course of Science Taught At Present in Trinity College, Dublin, With Some Improvements Suggested Therein* (Dublin, 1792)

Butler, M. *Romantics, Rebels and Reactionaries: English Literature and its Background 1760-1830* (Oxford: Oxford University Press, 1981)

Cahan, D. "The Institutional Revolution in German Physics, 1865-1914," *Historical Studies in the Physical and Biological Sciences* (1985) 15: 1-65.

Cahan, D. *An Institute for an Empire: The Physikalisch-Technische Reichsanstalt, 1871-1918* (Cambridge: Cambridge University Press, 1989)

Cahill, T. *How the Irish Saved Civilization* (New York: Doubleday, 1995)

Cajori, F. *A History of the Conceptions of Limits and Fluxions in Great Britain from Newton to Woodhouse* (Chicago: Open Court, 1919)

Cajori, F. *Mathematics in Liberal Education: A Critical Examination of the Judgments of Prominent Men of the Ages* (Boston: Christopher, 1928)

Caneva, K.L. "From Galvanism to Electrodynamics: The Transformation of German Physics and Its Social Context," *Historical Studies in the Physical Sciences* (1978) 9: 63-160.

Canny, N. *The Upstart Earl: A Study of the Social and Mental World of Richard Boyle, First Earl of Cork, 1566-1643* (Cambridge: Cambridge University Press, 1982)

Cantor, G. "Henry Brougham and the Scottish Methodological Tradition," *Studies in History and Philosophy of Science* (1971) 2: 69-89.

Cantor, G. "The Reception of the Wave Theory of Light in Britain: A Case Study Illustrating the Role of Methodology in Scientific Debate," *Historical Studies in the Physical Sciences* (1975) 6: 109-132.

Cantor, G. "Berkeley, Reid, and the Mathematization of Mid-Eighteenth-Century Optics," *Journal of the History of Ideas* (1977) 38: 429-48.

Cantor, G. "Brewster on the Nature of Light," pp. 67-76 in A. D. Morrison-Low and J. R. R. Christie (eds.), *'Martyr of Science': Sir David Brewster, 1781-1868,*

Proceedings of a Bicentennary Symposium held at the Royal Scottish Museum on 21 Novemeber 1981 (Edinburgh: Royal Scottish Museum, 1981)

Cantor, G. *Optics After Newton: Theories of Light in Britain and Ireland, 1704-1840* (Manchester: Manchester University Press, 1983)

Cantor, G. "Berkeley's *The Analyst* Revisited," *Isis* (1984) 75: 668-83.

Cardwell, D.S.L. *The Organization of Science in England* (London: Heinemann, 1972)

Carroll, P.K. and S. O'Connor, "Walter N. Hartley: Pioneering Spectroscopist," *Irish Chemical News* (1998) 12: 41-46.

Carter, C. "Magnetic Fever: Global Imperialism and Empiricism in the Nineteenth Century," *Transactions of the American Philosophical Society* (2009) 99: i-168.

Carlyle, T. "Signs of the Times [1829]," pp. 31-54 in G. B. Tennyson (ed.) *A Carlyle Reader* (Cambridge: Cambridge University Press, 1984)

Cathcart, B. *The Fly in the Cathedral: How a Group of Cambridge Scientists Won the International Race to Split the Atom* (New York: Farrar, Straus and Giroux, 2004)

Cawood, J. "Terrestrial Magnetism and the Development of International Collaboration in the Early Nineteenth Century," *Annals of Science* (1977) 34: 551-87.

Cawood, J. "The Magnetic Crusade: Science and Politics in Early Victorian Britain," *Isis* (1979) 70: 493-518.

Chambers, L. *Michael Moore, c. 1639-1726: Provost of Trinity, Rector of Paris* (Dublin: Four Courts Press, 2005).

Chen, X. "Young and Lloyd on the Particle Theory of Light: A Response to Achinstein," *Studies in History and Philosophy of Science* (1990) 21: 665-676.

Chen, X. and P. Barker, "Cognitive Appraisal and Power: David Brewster, Henry Brougham, and the Tactics of the Emission-Undulatory Controversy During the Early 1850s," *Studies in History and Philosophy of Science* (1992) 23: 75-101.

Chesney, H.C.G. "Enlightenment and Education," pp. 367-86 in J. W. Foster and H. C. G. Chesney (eds.), *Nature in Ireland: A Scientific and Cultural History* (Dublin: Lilliput Press, 1997)

Chillingworth, H.R. "TCD at the Beginning of the Century," *Hermathena* (1943) 61: 34-45.

Chitnis, A. *The Scottish Enlightenment: A Social History* (London: Croom Helm, 1976)

Clarke, D. "An Outline of the History of Science in Ireland," *Studies: An Irish Quarterly Review* (1973) 62: 287-302.

Cobban, A. *Edmund Burke and the Revolt Against the Eighteenth Century: A Study of the Political and Social Thinking of Burke, Wordsworth, Coleridge and Southey* (London: George Allen & Unwin, 1929)

Cocker, W. "A History of the University Chemical Laboratory, Trinity College, Dublin, 1711-1946," *Hermathena* (1978) 124: 58-76.

Cockroft, J.D. and E.T.S. Walton, "Disintegration of Lithium by Swift Protons," *Nature* (1932) 129: 242.

Coey, J.M.D. "Physics and New Materials," *Hermathena* (1982) 133: 47-57.

Cohen, I.B. (ed.) *Puritanism and the Rise of Modern Science: The Merton Thesis* (New Brunswick: Rutgers University Press, 1990)

Cohen, D.J. *Symbols of Heaven, Symbols of Man: Pure Mathematics and Victorian Religion* (PhD Thesis: Yale University, 1999)

Colburn, Z. *Memoirs of Zerah Colburn by Himself* (Springfield: G. and C. Merriam, 1833)

Coleridge, S.T. "The Statesman's Manual or The Bible the Best Guide to Political Skill and Foresight: A Lay Sermon Addressed to the Higher Classes of Society [1816]," pp. 3-114 in R. J. White (ed.) *Coleridge's Lay Sermons* (London: Routledge and Kegan Paul, 1972)

Coleridge, S.T. *On the Constitution of the Church and State, According to the Idea of Each [1829]*, Ed. by J. Colmer (London: Routledge and Kegan Paul, 1976)

Coleridge, S.T. *Biographia Literaria: or Biographical Sketches of My Literary Life and Opinions [1817]*, Ed. by J. Engell and W. J. Bate (London: Routledge and Kegan Paul, 1983), 2 Vols.

Colmer, J. *Coleridge: Critic of Society* (Oxford: Clarendon, 1959)

Colthurst, J.R. "The Influence of Irish Mathematicians on Modern Theoretical Physics," *The Mathematical Gazette* (1943) 27: 166-70.

Columbia University Department of Physics, "QCDOC Architecture," `phys.columbia.edu/~cqft/qcdoc/qcdoc.htm` (Accessed December 11, 2013)

Conway, A.W. "The Influence of the Work of Sir William Rowan Hamilton on Modern Mathematical Thought," *Scientific Proceedings of the Royal Dublin Society* (1931) 20: 125-128.

Corish, P.J. "The Cromwellian Conquest, 1649-53," pp. 336-52 in T. W. Moody, F. X. Martin and F. J. Byrne (eds.), *A New History of Ireland, Vol. III: Early Modern Ireland, 1534-1691* (Oxford: Clarendon, 1976)

Counihan, M.J. "Ireland and the Scientific Tradition," pp. 28-43 in P. O'Sullivan (ed.) *The Creative Migrant* (Leicester: Leicester University Press, 1994)

Cox, R.C. *Engineering at Trinity* (Dublin: School of Engineering, 1993)

Crosland, M. "The Image of Science as a Threat: Burke versus Priestley and the 'Philosophic Revolution," *British Journal for the History of Science* (1987) 20: 277-307.

Crowe, M.J. *A History of Vector Analysis* (South Bend: Notre Dame University Press, 1967)

Culverwell, E.P. *Mr. Bryce's Speech on the Proposed Reconstruction of the University of Dublin: Annotated Edition Issued by the Dublin University Defense Committee* (Dublin, 1907)

Cunningham, A. and N. Jardine (eds.), *Romanticism and the Sciences* (Cambridge: Cambridge University Press, 1990)

Curtis, L.P. "The Anglo-Irish Predicament," *Twentieth Century Studies* (1970) 4: 37-63.

Daly, M.E. *Industrial Development and Irish National Identity, 1922-1939* (Syracuse: Syracuse University Press, 1992)

Daly, M.E. *The Spirit of Earnest Inquiry: The Statistical and Social Inquiry Society of Ireland, 1847-1997* (Dublin: Statistical and Social Inquiry Society of Ireland, 1997)

Darrigol, O. "James MacCullagh's Ether: An Optical Route to Maxwell's Equations?" *European Physics Journal H* (2010) 35: 133-172.

Daston, L. *Classical Probability in the Enlightenment* (Princeton: Princeton University Press, 1988)

Daston, L. "Enlightenment Calculations," *Critical Inquiry* (1994) 21: 182-202.

Davie, G.E. *The Democratic Intellect: Scotland and Her Universities in the Ninenteenth Century* (Edinburgh: Edinburgh University Press, 1961)

Davis, P.J. and R. Hersh, *The Mathematical Experience* (Boston: Houghton Mifflin, 1982)

Davis, P.J. and R. Hersh, *Descartes' Dream: The World According to Mathematics* (Boston: Harcourt Brace Jovanovich, 1986)

DeArce, M. "The Parallel Lives of Joseph Allen Galbraith (1818-90) and Samuel Haughton (1821-97): Religion, Friendship, Scholarship and Politics in Victorian Ireland," *Proceedings of the Royal Irish Academy* (2011) 112C: 1-27.

De Morgan, A. "Review of Peacock's Treatise of Algebra," *Quarterly Journal of Education* (1835) 9: 293-311.

De Morgan, A. "On the Symbols of Logic, the Theory of the Syllogism, and in particular of the Copula [1850]," pp. 22-68 in P. Heath (ed.) *On the Syllogism and Other Logical Writings* (New Haven: Yale University Press, 1966)

Dear, P. *Discipline and Experience: The Mathematical Way in the Scientific Revolution* (Chicago: University of Chicago Press, 1995)

Department of Enterprise, Trade and Employment, *Technology Foresight Fund: Investing in Future Competitiveness,* (March 2000)

Department of Enterprise, Trade and Employment, *Science, Technology and Innovation: The White Paper* (Dublin: Stationery Office, 1996)

Department of Jobs, Enterprise and Innovation, *2013 Action Plan for Jobs* (February 2013)

Dewey, C.J. "The Education of a Ruling Caste: The Indian Civil Service in the Era of Competitive Examination," *English Historical Review* (1973) 88: 262-285.

Dixon, F.E. "Richard Kirwan, the Dublin Philosopher," *Dublin Historical Record* (1971) 24: 52-64.

Dixon, R.V. *A Treatise on Heat* (Dublin, 1849)

Dobbs, B.J.T. and M.C. Jacob, *Newton and the Culture of Newtonianism* (Atlantic Highlands, NJ: Humanties Press International, 1995)

Dodd, G. "Wordsworth and Hamilton," *Nature* (1970) 228: 1261-1263.

Donegan, J.F., D. Weaire, P. Florides (eds.), *Hutchie: The Life and Works of Edward Hutchinson Synge, 1890-1957* (Atascadero: Living Edition, 2012)

Donovan, M. "Biographical Account of the Late Richard Kirwan," *Proceedings of the Royal Irish Academy* (1850) 4: lxxxi-cxviii.

Doyle, E. "Irishman Creates A Virtual Dublin on the Website Second Life," Prosperity Blog, blog.prosperity.ie/2007/11/12/ irishman-creates-a-virtual-dublin-on-thewebsite-second-life/

Dubbey, J.M. "The Introduction of the Differential Calculus to Great Britain," *Annals of Science* (1963) 19: 37-48.

Dubbey, J.M. "Babbage, Peacock and Modern Algebra," *Historia Mathematica* (1977) 4: 295-302.

Dublin Institute for Advanced Studies, "History of the School of Theoretical Physics," (www.dias.ie)

Dublin University Commission, *Report of Her Majesty's Commissioners Appointed to Inquire Into the State, Discipline, Studies, and Revenues of the University of Dublin, and of Trinity College* (Dublin, 1853)

Duddy, T. *A History of Irish Thought* (London: Routledge, 2002)

Dunbar, N. *Inventing Money: The Story of Long-Term Capital Management and the Legends Behind It* (New York: John Wiley & Sons, 2000)

Eagleton, T. *Scholars & Rebels in Nineteenth-Century Ireland* (Oxford: Blackwell, 1999)

Eaton, R. *An Abridgement of Astronomy and Natural Philosophy, As Read in the Undergraduate Course at TCD* (Dublin, 1797)

Eddington, A. S. "Joseph Larmor, 1857-1942," *Obituary Notices of Fellows of the Royal Society* (1942) 4: 197-226.

Edgeworth, R.L. and M. Edgeworth, *Memoirs of Richard Lovell Edgeworth Begun By Himself and Concluded By His Daughter, Maria Edgeworth* (London, 1820)

Elrington, T. *A Sermon Preached in the Chapel of Trinity College, Dublin on Sunday, December the 7th, 1800 Upon the Death of the Right Rev. Matthew [Young] Lord Bishop of Clonfert* (Dublin, 1800)

Enros, P.C. *The Analytical Society: Mathematics at Cambridge University in the Early Nineteenth Century* (PhD Thesis: University of Toronto, 1979)

Eolas, *National Research and Development Needs in the Information Technology Area* (October 1993)

Eolas, *National Framework of Research Needs: Human Capital and Mobility* (November 1993)

Eumenes, *An Address to a Young Student, on His Entrance into College* (Dublin, 1798)

European Science Foundation, *Mathematics and Industry* (2011)

Eve, A. S. *Rutherford: Being the Life and Letters of the Rt. Hon. Lord Rutherford, O. M.* (New York: Macmillan, 1939)

Falconer, I. "Corpuscles, Electrons and Cathode Rays: J.J. Thomson and the 'Discovery of the Electron'," *British Journal for the History of Science* (1987) 20: 205-242.

Feingold, M. *The Mathematicians' Apprenticeship: Science, Universities and Society in England, 1560-1640* (Cambridge: Cambridge University Press, 1984)

Feynman, R. *QED: The Strange Theory of Light and Matter* (Princeton: Princeton University Press, 1985)

Fisch, M. *William Whewell, Philosopher of Science* (Oxford: Clarendon, 1991)

Fisch, M. and S. Schaffer (eds.), *William Whewell: A Composite Portrait* (Oxford: Clarendon, 1991)

Fisch, M. "'The Emergency Which Has Arrived': The Problematic History of Nineteenth-Century British Algebra—A Programmatic Outline," *British Journal for the History of Science* (1994) 27: 247-276.

Fitzgerald, G.F. "On the Electromagnetic Theory of the Reflection and Refraction of Light," pp. 45-74 in J. Larmor (ed.) *The Scientific Writings of George Francis Fitzgerald* (Dublin: Hodges, Figgis and Co. 1902) [Originally published in *Philosophical Transactions of the Royal Society* (1880)]

Fitzgerald, G.F. "Experimental Science in Schools and Universities," pp. 191-96 in J. Larmor (ed.) *The Scientific Writings of George Francis Fitzgerald* (Dublin: Hodges, Figgis and Co. 1902) [Originally published in *Nature* (1886)]

Fitzgerald, G.F. "Address to the Mathematical and Physical Section of the British Association [1888]," pp. 229-40 in J. Larmor (ed.) *The Scientific Writings of George Francis Fitzgerald* (Dublin: Hodges, Figgis and Co. 1902)

Fitzgerald, G.F. "Engineering Schools," pp. 224-28 in J. Larmor (ed.) *The Scientific Writings of George Francis Fitzgerald* (Dublin: Hodges, Figgis and Co. 1902) [Originally published in *Nature* (August 2, 1888)]

Fitzgerald, G.F. "Electromagnetic Radiation," pp. 266-76 in J. Larmor (ed.) *The Scientific Writings of George Francis Fitzgerald* (Dublin: Hodges, Figgis and Co. 1902) [Originally a lecture to the Royal Institution in 1890]

Fitzgerald, G.F. "Helmholtz Memorial Lecture," pp. 340-77 in J. Larmor (ed.) *The Scientific Writings of George Francis Fitzgerald* (Dublin: Hodges, Figgis and Co. 1902) [Originally delivered before the Chemical Society of London in 1896.]

Fitzgerald, G.F. "The Meaning and Possibilities of Wireless Telegraphy," *Daily Express* (July 21, 1898)

Fitzgerald, G.F. "The Applications of Science: A Lesson from the Nineteenth Century," pp. 487-499 in J. Larmor (ed.) *The Scientific Writings of George Francis Fitzgerald* (Dublin: Hodges, Figgis and Co. 1902) [Inaugural address to the Dublin Section of the Institution of Electrical Engineers (1900)]

Fitzgerald, R.P. "Science and Politics in Swift's Voyage to Laputa," *Journal of English and Germanic Philology* (1988) 87: 213-29.

Fitzmaurice, G.P. *The Life of Sir William Petty, 1623-1687* (London: John Murray, 1895)

Fitzpatrick, D. "Eamon de Valera at Trinity College," *Hermathena* (1982) 133: 7-14.

Flannery, S. *In Code: A Mathematical Journey* (London: Profile Books, 2000)

Florides, P. "John Synge, 1897-1995," pp. 208-19 in Mark McCartney and Andrew Whitaker (eds.) *Physicists of Ireland: Passion and Precision* (Bristol: IOP Publishing, 2003)

Florides, P. "John Lighton Synge, 1897-1995," *Biographical Memoirs of Fellows of the Royal Society* (2008) 54: 402-24.

Flood, R. "Taking Root: Mathematics in Victorian Ireland," pp. 103-119 in Raymond Flood, Adrian Rice and Robin Wilson (eds.) *Mathematics in Victorian Britain* (Oxford: Oxford University Press, 2011)

Foden, F. "James Booth: The Father of Technical Examinations," pp. 367-75 in N. McMillan (ed.) *Prometheus's Fire: A History of Scientific and Technological Education in Ireland* (Carlow: Tyndall Publications, 2000)

Ford, A. "Dependent or Independent? The Church of Ireland and Its Colonial Context, 1536-1649," *The Seventeenth Century* (1995) 10: 163-87.

Ford, T.H. *Henry Brougham and His World: A Biography* (Chichester: Barry Rose, 1995)

Forfás, *Technology Foresight Ireland: An ICSTI Overview* (April 1999)

Forfás, *Report of the Research Prioritisation Steering Group* (November 2011)

Foster, R.F. *Modern Ireland, 1600-1972* (New York: Penguin Book, 1988)

Foster, J.W. "Natural History, Science and Irish Culture," *The Irish Review* (1990) 9: 61-69.

Foster, J.W. and H.C.G. Chesney (eds.), *Nature in Ireland: A Scientific and Cultural History* (Dublin: Lilliput Press, 1997)

Fountain, H. "Bubble Question Frames Design for Olympic Aquatics Center," *New York Times* (August 5, 2008)

Fox, J.W. "From Lardner to Massey: A History of Physics, Space Science and Astronomy at University College London, 1826-1975," www.phys.ucl.ac.uk/department/history/BFox1.html#Fox140 (Accessed December 11, 2013)

Fox, R. and A. Guagnini (eds.), *Education, Technology and Industrial Performance in Europe, 1850-1939* (Cambridge: Cambridge University Press, 1993)

Frankel, E. "The Search for a Corpuscular Theory of Double Refraction: Malus, Laplace and the Prize Competition of 1808," *Centaurus* (1974) 18: 223-245.

Frängsmyr, T., J.L. Heilbron and R.E. Rider (eds.), *The Quantifying Spirit in the 18th Century* (Berkeley: University of California Press, 1990)

Fraser, A.C. *Life and Letters of George Berkeley* (Oxford: Clarendon, 1871)

Garber, D. "Locke, Berkeley, and Corpuscular Scepticism," pp. 174-93 in C. Turbayne (ed.) *Berkeley: Critical and Interpretive Essays* (Minneapolis: University of Minnesota Press, 1982)

Gascoigne, J. "Mathematics and Meritocracy: The Emergence of the Cambridge Mathematical Tripos," *Social Studies of Science* (1984) 14: 547-84.

Gascoigne, J. *Cambridge in the Age of Enlightenment: Science, Religion and Politics from the Restoration to the French Revolution* (Cambridge: Cambridge University Press, 1989)

Gårding, L. "History of the Mathematics of Double Refraction," *Archive for History of Exact Sciences* (1989) 40: 355-385.

Gell-Mann, M. *The Quark and the Jaguar: Adventures in the Simple and the Complex* (New York: Henry Holt, 1994)

Gellai, B. *The Intrinsic Nature of Things: The Life and Science of Cornelius Lanczos* (American Mathematical Society, 2010)

Gibbons, L. "Coming Out of Hibernization? The Myth of Modernization in Irish Culture," in L. Gibbons (ed.) *Transformations in Irish Culture* (Cork: Cork University Press and Field Day, 1996)

Gigerenzer, G. (ed.) *The Empire of Chance: How Probability Changed Science and Everyday Life* (Cambridge: Cambridge University Press, 1989)

Gillispie, C.C. *Pierre-Simon Laplace, 1749-1827: A Life in Exact Science* (Princeton: Princeton University Press, 1997)

Gooday, G. "Precision Measurement and the Genesis of Physics Teaching Laboratories in Victorian Britain," *The British Journal for the History of Science* (1990) 23: 25-51.

Gooday, G. and D.J. Mitchell, "Rethinking 'Classical Physics'," pp. 721-64 in Jed Z. Buchwald and Robert Fox (eds.), *The Oxford Handbook of the History of Physics* (Oxford: Oxford University Press, 2013)

Gorgett, G. and G. Sheridan (eds.), *Ireland and the French Enlightenment* (London: Macmillan, 1999)

Grattan-Guinness, I. "Mathematics and Mathematical Physics from Cambridge, 1815–40: A Survey of the Achievements and of the French Influences," pp. 84-111 in P. M. Harman (ed.) *Wranglers and Physicists: Studies on Cambridge Physics in the Nineteenth Century* (Manchester: Manchester University Press, 1985)

Grattan-Guinness, I. "Before Bowditch: Henry Harte's Translation of Books 1 and 2 of Laplace's *Mecanique Celeste*," *NTM* (2 1987) 24: 53-55.

Grattan-Guinness, I. "Work for the Workers: Advances in Engineering Mechanics and Instruction in France, 1800-1830," *Annals of Science* (1984) 41: 1-33.

Grattan-Guinness, I. "Mathematical Research and Instruction in Ireland, 1782-1840," pp. 11-30 in J. Nudds, N. McMillan, D. Weaire and S. M. Lawlor (eds.), *Science in Ireland 1800-1930: Tradition and Reform, Proceedings of an International Symposium held at Trinity College Dublin March 1986* (Dublin: Trinity College Dublin, 1988)

Grattan-Guinness, I. *Convolutions in French Mathematics, 1800-1840: From the Calculus and Mechanics to Mathematical Analysis and Mathematical Physics* (Boston: Birkhäuser, 1990)

Graves, R.P. *Life of Sir William Rowan Hamilton* (Dublin: Hodges, Figgis, 1882-89), 3 Vols.

Greengrass, M., M. Leslie and T. Raylor (eds.), *Samuel Hartlib and Universal Reformation* (Cambridge: Cambridge University Press, 1994)

Guerrini, A. "James Keill, George Cheyne, and Newtonian Physiology," *Journal of the History of Biology* (1985) 18: 247-66.

Guicciardini, N. *The Development of the Newtonian Calculus in Britain, 1700-1800* (Cambridge: Cambridge University Press, 1989)

Hacking, I. *The Emergence of Probability* (Cambridge: Cambridge University Press, 1975)

Hacking, I. *The Taming of Chance* (Cambridge: Cambridge University Press, 1990)

Hackney, P. "Edward Sabine," pp. 331-36 in J. W. Foster and H. C. G. Chesney (eds.), *Nature in Ireland: A Scientific and Cultural History* (Dublin: Lilliput, 1997)

Hales, W. *The Inspector, Or Select Literary Intelligence for The Vulgar A.D. 1798, But Correct A.D. 1801, The First Year of the XIXth Century* (London, 1799)

Hales, W. *Analysis Fluxionum* (London, 1800)

Hales, W. *A New Analysis of Chronology, In Which An Attempt Is Made to Explain the History and Antiquities of the Primitive Nations of the World and the Prophecies Relating to Them, On Principles Tending to Remove the Imperfection and Discordance of Preceding Systems* (London, 1809), 3 Vols.

Hales, W. *A Synopsis of the Signs of the Times, Past, Present, and Future; Humbly Attempted to Be Traced From the Chronological Prophecies, In the Original Scriptures* (Dublin, 1817)

Halévy, E. *The Growth of Philosophic Radicalism [1928]*, Trans. by Mary Morris (London: Faber and Faber, 1972)

Hall, A.R. *All Was Light: An Introduction to Newton's Opticks* (Oxford: Clarendon, 1993)

Hamilton, H. "Four Introductory Lectures on Natural Philosophy," in A. Hamilton (ed.) *The Works of the Right Rev. Hugh Hamilton, D.D. Late Bishop of Ossory; Collected and Published With Some Alterations and Additions From His Manuscripts* (London, 1809)

Hamilton, W. "On the State of the English Universities, with more especial reference to Oxford," pp. 386-434 in W. Hamilton (ed.) *Discussions on Philosophy and Literature, Education and University Reform* (London, 1852)

Hamilton, W.R. "Theory of Systems of Rays. Part First [1827]," *Transactions of the Royal Irish Academy* (1828) 15: 69-174. [*Mathematical Papers* 1: 1-88]

Hamilton, W.R. "Supplement to an Essay on the Theory of Systems of Rays [1830]," *Transactions of the Royal Irish Academy* (1830) 16: 1-61. [*Mathematical Papers* 1: 107-144]

Hamilton, W.R. "Second Supplement to an Essay on the Theory of Systems of Rays [1830]," *Transactions of the Royal Irish Academy* (1831) 16: 93-125. [*Mathematical Papers* 1: 145-163]

Hamilton, W.R. "On a General Method in Dynamics," *Philosophical Transactions of the Royal Society of London* (1834) 24: 247-308. [*Mathematical Papers*, Vol. II]

Hamilton, W.R. "Second Essay on a General Method in Dynamics," *Philosophical Transactions of the Royal Society of London* (1835) 25: 95-144. [*Mathematical Papers*, Vol. II]

Hamilton, W.R. "Third Supplement to an Essay on the Theory of Systems of Rays [1832]," *Transactions of the Royal Irish Academy* (1837) 17: 1-144. [*Mathematical Papers* 1: 164-329]

Hamilton, W.R. "Letter to Graves on Quaternions; or on a New System of Imaginaries," *Philosophical Magazine* (1844) 25: 489-95. [*Mathematical Papers* 3: 106-110]

Hamilton, W.R. "On a New Species of Imaginary Quantities Connected with the Theory of Quaternions," *Proceedings of the Royal Irish Academy* (1844) 2: 424-34. [*Mathematical Papers* 3: 111-16]

Hamilton, W.R. *Lectures on Quaternions* (Dublin, 1853)

Hamilton, W.R. "Theory of Systems of Rays. Part Second [1827]," pp. 88-106 in A. W. Conway and J. L. Synge (eds.), *The Mathematical Papers of Sir William Rowan Hamilton. Vol. I: Geometrical Optics* (Cambridge: Cambridge University Press, 1931)

Hamilton, W.R. "On Caustics. Part First [1824]," pp. 345-63 in A. W. Conway and J. L. Synge (eds.), *The Mathematical Papers of Sir William Rowan Hamilton. Vol. I: Geometrical Optics* (Cambridge: Cambridge University Press, 1931)

Hamilton, W.R. "On a General Method of expressing the Paths of Light, and of the Planets, by the Coefficients of a Characteristic Function [1833]," pp. in A. W. Conway and A. J. McConnell (eds.), *The Mathematical Papers of Sir William Rowan Hamilton. Volume II: Dynamics* (Cambridge: Cambridge University Press, 1940)

Hamilton, W.R. "Theory of Conjugate Functions, or Algebraic Couples; With a Preliminary and Elementary Essay on Algebra as the Science of Pure Time [1837]," pp. 3-96 in H. Halberstam and R. E. Ingram (eds.), *The Mathematical Papers of Sir William Rowan Hamilton, Vol. 3: Algebra* (Cambridge: Cambridge University Press, 1967)

Hankins, T.L. "Algebra as Pure Time: William Rowan Hamilton and the Foundation of Algebra," pp. 327-59 in P. K. Machamer and R. G. Turnbull (eds.), *Motion and Time, Space and Matter* (Columbus: Ohio State University Press, 1976)

Hankins, T.L. "Triplets and Triads: Sir William Rowan Hamilton on the Metaphysics of Mathematics," *Isis* (1977) 68: 175-193.

Hankins, T.L. *Sir William Rowan Hamilton* (Baltimore: Johns Hopkins University Press, 1980)

Hankins, T.L. *Science and the Enlightenment* (Cambridge: Cambridge University Press, 1985)

Hardiman, T.P. "Development of an Indigenous Innovative Technology Base in Industry, Agriculture, and Society," pp. 9-12 in *Innovation Report 1981* (Dublin: Confederation of Irish Industry, 1982)

Hardiman, T.P. and E.P. O'Neill, "The Role of Science and Innovation," pp. 253-84 in F. Ó Muircheartaigh (ed.) *Ireland in the Coming Times: Essays to Celebrate T.K. Whitaker's 80 Years* (Dublin: Institute of Public Administration, 1997)

Harman, P.M. (ed.) *Wranglers and Physicists: Studies on Cambridge Physics in the Nineteenth Century* (Manchester: Manchester University Press, 1985)

Harman, P.M. *The Natural Philosophy of James Clerk Maxwell* (Cambridge: Cambridge University Press, 1998)

Harris, R.W. *Romanticism and the Social Order* (London: Blandford Press, 1969)

Hayes, B. "Getting Your Quarks in a Row," *American Scientist* (November-December 2008) 96: 450-454.

Hays, J.N. "Science and Brougham's Society," *Annals of Science* (1964) 20: 227-41.

Hays, J.N. "The Rise and Fall of Dionysius Lardner," *Annals of Science* (1981) 38: 527-42.

Heilbron, J.L. "A Mathematicians' Mutiny with Morals," pp. 81-129 in P. Horwich (ed.) *World Changes: Thomas Kuhn and the Nature of Science* (Cambridge: MIT Press, 1993)

Heimann, P. "Nature is a Perpetual Worker, Newton's Aether and Eighteenth Century Natural Philosophy," *Ambix* (1973) 20: 1-25.

Hendry, J. "The Evolution of William Rowan Hamilton's View of Algebra as the Science of Pure Time," *Studies in the History and Philosophy of Science* (1984) 15: 63-81.

hEocha, C.O. "University College Galway: A Case Study of Response to Industrial Innovation," pp. 24-27 in *Innovation Report 1981* (Dublin: Confederation of Irish Industry, 1982)

Herries Davies, G.L. "Irish Thought in Science," pp. 294-310 in R. Kearney (ed.) *The Irish Mind: Exploring Intellectual Traditions* (Dublin: Wolfhound Press, 1985)

Herschel, J.F.W. "On the Aberrations of Compound Lenses and Object-Glasses," *Philosophical Transactions* (1821) 11: 222-267.

Herschel, J.F.W. *A Preliminary Discourse on the Study of Natural Philosophy [1830]* (Chicago: University of Chicago Press, 1987)

Hill, C. *Intellectual Origins of the English Revolution* (Oxford: Clarendon, 1965)

Hill, J. "The Meaning and Significance of 'Protestant Ascendancy', 1787-1840," pp. 1-22 in *Ireland After the Union: Proceedings of the Second Joint Meeting of the Royal Irish Academy and the British Academy, London, 1986* (New York: Oxford University Press, 1989)

Hoddeson, L. "The Emergence of Basic Research in the Bell Telephone System, 1875-1915," *Technology and Culture* (1981) 22: 512-44.

Hogan, E. "Robert Adrain: American Mathematician," *Historia Mathematica* (1977) 4: 157-72.

Holland, C.H. (ed.) *Trinity College Dublin & The Idea of a University* (Dublin: TCD Press, 1991)

Hong, S. "Marconi and the Maxwellians: The Origins of Wireless Telegraphy Revisited," *Technology and Culture* (1994) 35: 717-49.

Hopkins, R.H. "The Personation of Hobbism in Swift's *Tale of a Tub* and *Mechanical Operation of the Spirit*," *Philological Quarterly* (1966) 45: 372-78.

Hoppen, K.T. *The Common Scientist in the Seventeenth Century: A Study of the Dublin Philosophical Society, 1683-1708* (London: Routledge & Kegan Paul, 1970)

Hoppen, K.T. "Correspondence: The Hartlib Circle and the Origins of the Dublin Philosophical Society," *Irish Historical Studies* (1976) 20: 40-48.

Houston, K. (ed.) *Creators of Mathematics: The Irish Connection* (Dublin: University College Dublin Press, 2000)

Huch, R.K. *Henry, Lord Brougham: The Later Years, 1830-1868* (Lewiston, NY: Edwin Mellen, 1993)

Hunt, B. "'Practice vs. Theory': The British Electrical Debate, 1888-1891," *Isis* (1983) 74: 341-55.

Hunt, B. "'How My Model Was Right': G.F. Fitzgerald and the Reform of Maxwell's Theory," pp. 299-321 in R. Kargon and P. Achinstein (eds.), *Kelvin's Baltimore Lectures and Modern Theoretical Physics: Historical and Philosophical Perspectives* (Cambridge: MIT Press, 1987)

Hunt, B. "The Origins of the FitzGerald Contraction," *British Journal for the History of Science* (1988) 21: 67-76.

Hunt, B. *The Maxwellians* (Ithaca: Cornell University Press, 1990)

Hunt, B. "'Our Friend of Brilliant Ideas': G.F. Fitzgerald and the Maxwellian Circle," *European Review* (2007) 15: 531-44.

Hurst, M. *Maria Edgeworth and the Public Scene: Intellect, Fine Feeling and Landlordism in the Age of Reform* (London: Macmillan, 1969)

Hutchinson, J. *The Dynamics of Cultural Nationalism: The Gaelic Revival and the Creation of the Irish Nation State* (London: Allen & Unwin, 1987)

IBM, "Blue Gene Overview", /www-03.ibm.com/ibm/history/ibm100/us/en/icons/bluegene/ (Accessed December 11, 2013)

IDA, "IDA Ireland launches Technology & Innovation Showcase in Second Life," (01/04/09) www.idaireland.com/news-media/press-releases/ida-ireland-launches-tech/ (Accessed December 11, 2013)

Inkster, I. "Science and the Mechanics Institutes, 1820-50," *Annals of Science* (1975) 32: 451-74.

Institute of Physics, "The Importance of Physics to the Irish Economy" (November 2012)

Israel, P. *From Machine Shop to Industrial Laboratory: Telegraphy and the Changing Context of American Invention, 1830-1920* (Baltimore: Johns Hopkins University Press, 1992)

Jackson, P.W. "A Man of Invention: John Joly (1857-1933): Engineer, Physicist and Geologist," pp. 89-96 in D. Scott (ed.) *Treasures of the Mind* (London: Sotheby, 1992)

Jacob, M. *The Newtonians and the English Revolution, 1687-1720* (Ithaca: Cornell University Press, 1976)

Jacob, J.R. *Robert Boyle and the English Revolution: A Study in Social and Intellectual Change* (New York: Burt Franklin & Co. 1977)

Jacob, J.R. and M. Jacob, "The Anglican Origins of Modern Science: The Metaphysical Foundations of the Whig Constitution," *Isis* (1980) 71: 251-67.

Jacob, M. *The Radical Enlightenment: Pantheists, Freemasons and Republicans* (London: George Allen and Unwin, 1981)

Jacobsen, J.K. *Chasing Progress in the Irish Republic: Ideology, Democracy and Dependent Development* (Cambridge: Cambridge University Press, 1994)

Jarrell, R.A. "Colonialism and the Truncation of Science in Ireland and French Canada During the Nineteenth Century," *HSTC Bulletin: Journal of the History of Canadian Science, Technology and Medicine* (1981) 5: 140-57.

Jarrell, R.A. "The Department of Science and Art and the Control of Irish Science, 1853-1905," *Irish Historical Studies* (1983) 23: 330-47.

Jarrell, R.A. "Some Aspects of the Evolution of Agricultural and Technical Education in Ireland, 1800-1950," pp. 101-17 in P. J. Bowler and N. Whyte (eds.), *Science and Society in Ireland: The Social Context of Science and Technology in Ireland, 1800-1950* (Belfast: The Institute of Irish Studies- Queen's University of Belfast, 1997)

Jarrell, R.A. "Technical Education and Colonialism in 19th Century Ireland," pp. 170-87 in N. McMillan (ed.) *Prometheus's Fire: A History of Scientific and Technological Education in Ireland* (Carlow: Tyndall Press, 2000)

Jesseph, D.M. *Berkeley's Philosophy of Mathematics* (Chicago: University of Chicago Press, 1993)

Johnston, R. "Science in a Post-Colonial Culture," *The Irish Review* (1989) 8: 70-76.

Johnston, S. "Mathematical Practitioners and Instruments in Elizabethan England," *Annals of Science* (1991) 48: 319-44.

Joly, J. *An Epitome of the Irish University Question* (Dublin: Hodges, Figgis & Company, 1907)

Joly, J. "George Johnstone Stoney," *Proceedings of the Royal Society* (1912) 86A: xx-xxxv.

Joly, J. "In Trinity College During the Sinn Fein Rebellion, by One of the Garrison," *Blackwood's Magazine* (1916) 200: 101-125.

Jones, G. "Catholicism, Nationalism and Science," *Irish Review* (1997) 20: 47-61.

Kargon, R. "William Petty's Mechanical Philosophy," *Isis* (1965) 56: 63-66.

Kargon, R.H. *Science in Victorian Manchester: Enterprise and Expertise* (Manchester: Manchester University Press, 1977)

Kearney, R. "George Berkeley: We Irish Think Otherwise," pp. 145-56 in R. Kearney (ed.) *Postnationalist Ireland: Politics, Culture, Philosophy* (London: Routledge, 1997)

Kelham, B.B. "The Royal College of Science for Ireland (1867-1926)," *Studies* (1967) 56: 297-309.

Keller, E.F. *Refiguring Life: Metaphors of Twentieth-Century Biology* (New York: Columbia University Press, 1995)

Kelly, J. "The Genesis of 'Protestant Ascendancy': The Rightboy Disturbances of the 1780s and their Impact upon Protestant Opinion," in G. O'Brien (ed.) *Parliament, Politics, and People: Essays in Eighteenth-Century Irish History* (Dublin: Irish Academic Press, 1989)

Kelly, J. "Eighteenth-Century Ascendancy: A Commentary," *Eighteenth-Century Ireland* (1990) 5: 173-87.

Kelly, P. "From Molyneux to Berkeley: The Dublin Philosophical Society and Its Legacy," pp. 23-31 in D. Scott (ed.) *Treasures of the Mind* (London: Sotheby's, 1992)

Kelly, J. *Prelude to Union: Anglo-Irish Politics in the 1780s* (Cork: Cork University Press, 1992)

Kennedy, D. "The Ulster Academies and the Teaching of Science (1785-1835)," *Irish Ecclesiastical Record* (1944) 63: 25-38.

Kennedy, M. "The *Encyclopédie* in 18th-Century Ireland." *Book Collector* (1996) 45: 201-13.

Kiernan, C. "Swift and Science," *The Historical Journal* (1971) 14: 722.

Kirkpatrick, T.P.C. *History of the Medical Teaching in Trinity College Dublin and of the School of Physic in Ireland* (Dublin: Hanna and Neale, 1912)

Knights, B. *The Idea of the Clerisy in the Nineteenth Century* (Cambridge: Cambridge University Press, 1978)

Koetsier, T. "Explanation in the Historiography of Mathematics: The Case of Hamilton's Quaternions," *Studies in History and Philosophy of Science* (1995) 26: 593-616.

Koppelman, E. "The Calculus of Operations and the Rise of Abstract Algebra," *Archive for History of Exact Sciences* (1971) 8: 155-242.

Kramnick, I. "Eighteenth-Century Science and Radical Social Theory: The Case of Joseph Priestley's Scientific Liberalism," *Journal of British Studies* (1986) 25: 1-30.

Krüger, L. (ed.) *The Probabilistic Revolution* (Cambridge: MIT Press, 1987)

Kuhn, T. *The Structure of Scientific Revolutions*, 2nd Ed. (Chicago: University of Chicago Press, 1970)

Kuhn, T. "Mathematical versus Experimental Traditions in the Development of Physical Science," pp. 31-65 in T. Kuhn (ed.) *The Essential Tension: Selected Studies in Scientific Tradition and Change* (Chicago: University of Chicago Press, 1977)

Kuhn, T.S. "Energy Conservation as an Example of Simultaneous Discovery," pp. 66-104 in T. S. Kuhn (ed.) *The Essential Tension: Selected Studies in Scientific Tradition and Change* (Chicago: University of Chicago Press, 1977)

Laita, L.M. "The Influence of Boole's Search for a Universal Method in Analysis on the Creation of His Logic," *Annals of Science* (1977) 34: 163-76.

Laita, L.M. "Influences on Boole's Logic: The Controversy Between William Hamilton and Augustus De Morgan," *Annals of Science* (1979) 36: 45-65.

Lamb, H. "Presidential Address to Section A of the BAAS" (1904)

Landy, S. "Francis Beaufort," pp. 327-30 in J. W. Foster and H. C. G. Chesney (eds.), *Nature in Ireland: A Scientific and Cultural History* (Dublin: Lilliput Press, 1997)

Lansdowne, M. *The Petty Papers: Some Unpublished Writings of Sir William Petty* (London: Constable & Company, 1927), 2 Vols.

Larcom, T.A. (ed.) *A History of the Survey of Ireland Commonly Called the Down Survey by Doctor William Petty, A.D. 1655-6* (Dublin, 1851)

Lardner, D. *The Elements of the Theory of Central Forces* (Dublin, 1820)

Larmor, J. "Least Action as the Fundamental Formulation in Dynamics and Physics," *Proceedings of the London Mathematical Society* (1884) 15: 158-84.

Larmor, J. "Address of the President of the Mathematical and Physical Section of the British Association for the Advancement of Science," *Science* (1900) 12: 417-36.

Larmor, J. (ed.) *The Scientific Writings of the Late George Francis Fitzgerald* (Dublin: Hodges, Figgis and Co. 1902)

Larmor, J. "Historical Note on Hamiltonian Action (1927)," pp. 640-41 in Sir Joseph Larmor, *Mathematical and Physical Papers* (Cambridge: Cambridge University Press, 1929)

Laudan, L.L. "Thomas Reid and the Newtonian Turn of British Methodological Thought," pp. 103-31 in R. E. Butts and J. W. Davis (eds.), *The Methodological Heritage of Newton* (Oxford: Basil Blackwell, 1970)

Laudan, L.L. "The Medium and Its Message: A Study of Some Philosophical Controversies About Ether," pp. 157-85 in G. N. Cantor and M. J. S. Hodge (eds.), *Conceptions of Ether: Studies in the History of Ether Theories, 1740-1900* (Cambridge: Cambridge University Press, 1981)

Lee, J.J. *Ireland, 1912-1985: Politics and Society* (Cambridge: Cambridge University Press, 1995)

Levere, T.H. *Poetry Realized in Nature: Samuel Taylor Coleridge and Early Nineteenth-Century Science* (Cambridge: Cambridge University Press, 1981)

Lewis, M. "When Irish Eyes Are Crying," *Vanity Fair* (March 2011)

Lievers, M. "The Molyneux Problem," *Journal of the History of Philosophy* (1992) 30: 399-415.

Lillington, K. "Super-Duper Computers," *Technology Ireland* (October 1998), pp. 18-21.

Lindberg, D.C. "Conceptions of the Scientific Revolution from Bacon to Butterfield: A Preliminary Sketch," pp. 1-26 in D. C. Lindberg and R. S. Westman (eds.), *Reappraisals of the Scientific Revolution* (Cambridge: Cambridge University Press, 1991)

Lloyd, B. *Analytic Geometry; or A Short Treatise on the Application of Algebra to Geometry, Intended Chiefly for the Use of Undergraduates in the University of Dublin* (Dublin, 1819)

Lloyd, H. "On the Phenomena presented by Light in its Passage along the Axes of Biaxal Crystals," *Philosophical Magazine* (1833) 37: 112-120.

Lloyd, H. "Further Experiments on the Phenomena presented by Light in its Passage along the Axes of Biaxal Crystals," *Philosophical Magazine* (1833) 37: 207-210.

Lloyd, H. "On Conical Refraction," pp. 370-73 in *Report of the Third Meeting of the BAAS; Held at Cambridge in 1833* (London, 1834)

Lloyd, H. "Report on the Progress and Present State of Physical Optics," pp. 295-413 in *Report of the Fourth Meeting of the BAAS, Held at Edinburgh in 1834* (London, 1835)

Lloyd, H. *Praelection on the Studies Connected with the School of Engineering. Delivered on the Occasion of the Opening of the School, November 15, 1841* (Dublin, 1841)

Lloyd, H. *Account of the Magnetical Observatory of Dublin and of the Instruments and Methods of Observation Employed There* (Dublin, 1842)

Luce, J.V. *Trinity College Dublin: The First 400 Years* (Dublin: Trinity College Dublin Press, 1992)

Lyons, F.S.L. *Culture and Anarchy in Ireland, 1890-1939* (Oxford: Clarendon, 1979)

Lysaght, S. "Themes in the Irish History of Science," *Irish Review* (1996) 19: 87-97.

MacDonnell, R. *A Letter to Dr. Phipps, SFTD, Registrar of Trinity College, Concerning the Undergraduate Exams in the University of Dublin* (Dublin, 1828)

MacHale, D. *George Boole: His Life and Work* (Dublin: Boole Press, 1985)

Mackenzie, D. *An Engine, Not a Camera: How Financial Models Shape Markets* (Cambridge: MIT Press, 2008)

McKinsey Global Institute, *Big Data: The Next Frontier for Innovation, Competition, and Productivity* (June 2011)

MacLeod, R. (ed.) *Days of Judgment: Science, Examinations and the Organization of Knowledge in Late Victorian England* (Driffield: Studies in Education, 1982)

MacLeod, R. "On Science and Colonialism," pp. 1-18 in P. J. Bowler and N. Whyte (eds.), *Science and Society in Ireland: The Social Context of Science and Technology in Ireland, 1800-1950* (Belfast: The Institute of Irish Studies-Queen's University of Belfast, 1997)

MacSharry, R. and P. White, *The Making of the Celtic Tiger: The Inside Story of Ireland's Boom Economy* (Cork: Mercier Press, 2000)

Mahoney, M.S. "Mathematics," pp. 145-78 in D. Lindberg (ed.) *Science in the Middle Ages* (Chicago: University of Chicago Press, 1978)

Mahoney, M.S. "Changing Canons of Mathematical and Physical Intelligibility in the Later 17th Century," *Historia Mathematica* (1984) 11: 417-23.

Mahoney, M.S. "The Mathematical Realm of Nature," pp. 702-55 in D. Garber and M. Ayers (eds.), *Cambridge History of Seventeenth-Century Philosophy* (Cambridge: Cambridge University Press, 1998), Vol. 1.

Marconi, E. "The Queen's Cup Regatta Kingstown 1898" in Maria Cristina Marconi and Elettra Marconi, *Marconi My Beloved* (Boston: Dante University of America Press, 1999),

Mathews, J. "William Rowan Hamilton's Paper of 1837 On the Arithmetization of Analysis," *Archive for History of Exact Sciences* (1978) 19: 177-200.

Maull, N.L. "Berkeley on the Limits of Mechanistic Explanation," pp. 95-107 in C. Turbayne (ed.) *Berkeley: Critical and Interpretive Essays* (Minneapolis: University of Minnesota Press, 1982)

Maxwell, C. *Dublin Under the Georges, 1714-1830* (London: Faber and Faber, 1956)

McBride, I. "William Drennan and the Dissenting Tradition," pp. 49-61 in D. Dickson, D. Keogh and K. Whelan (eds.), *The United Irishmen: Republicanism, Radicalism and Rebellion* (Dublin: Lilliput Press, 1993)

McBrierty, V.P. "The Strategy for Forging the Links between Academics and Industry," pp. 12-14 in *Innovation Report 1981* (Dublin: Confederation of Irish Industry, 1982)

McBrierty, V.J. and R.P. Kinsella, *Ireland and the Knowledge Economy: The New Techno-Academic Paradigm* (Dublin: Oak Tree Press, 1998)

McBrierty, V.J. *Ernest Thomas Sinton Walton, 1903-1995: The Irish Scientist* (Dublin: Trinity College Dublin Press, 2003)

McCarthy, R.B. *The Trinity College Estates 1800-1923: Corporate Management in an Age of Reform* (Dublin: Dundalgan Press, 1992)

McCartney, D. *UCD: A National Idea* (Dublin: Gill & McMillan, 1999)

McCartney, M. and A. Whitaker (eds.) *Physicists of Ireland: Passion and Precision* (Bristol: IOP Publishing, 2003)

McClellan, J.E. *Science Reorganized: Scientific Societies in the Eighteenth Century* (New York: Columbia University Press, 1985)

McConnell, A.J. "The Dublin Mathematical School of the First Half of the Nineteenth Century," *Proceedings of the Royal Irish Academy* (1944) 50A: 75-88.

McConville, M. *Ascendancy to Oblivion: The Story of the Anglo-Irish* (London: Quartet, 1986)

McCormack, W.J. *Ascendancy and Tradition in Anglo-Irish Literary History from 1789 to 1939* (Oxford: Clarendon, 1985)

McCormack, W.J. "Eighteenth-Century Ascendancy: Yeats and the Historians," *Eighteenth-Century Ireland* (1989) 4: 159-81.

McCormack, W.J. *The Dublin Paper War of 1786-1788: A Bibliographical and Critical Inquiry* (Dublin: Irish Academic Press, 1993)

McCormack, W.J. *From Burke to Beckett: Ascendancy, Tradition and Betrayal in Literary History* (Cork: Cork University Press, 1994)

McCue, D. "Samuel Garth, Physician and Man of Letters," *Bulletin of the New York Academy of Medicine* (1977) 53: 368-402.

McCullagh, J. "On the Double Refraction of Light in a Crystallized Medium According to the Principles of Fresnel," in J. H. Jellett and S. Haughton (eds.), *The Collected Works of James MacCullagh* (Dublin: Hodges, Figgis & Co., 1880)

McDowell, R.B. and D.A. Webb, *Trinity College Dublin, 1592-1952: An Academic History* (Cambridge: Cambridge University Press, 1982)

McDowell, R.B. "The Main Narrative," pp. 1-92 in T. O'Raifeartaigh (ed.) *The Royal Irish Academy: A Bicentennial History, 1785-1985* (Dublin: The Royal Irish Academy, 1985)

McDowell, R.B. "The Age of the United Irishmen: Reform and Reaction, 1789-94," pp. 289-338 in T. W. Moody and W. E. Vaughan (eds.), *A New History of Ireland: Eighteenth-Century Ireland, 1691-1800* (Oxford: Clarendon Press, 1986), Vol. 4.

McGuiness, P. "*Christianity Not Mysterious* and the Enlightenment," pp. 231-42 in P. McGuinness, A. Harrison and R. Kearney (eds.), *John Toland's Christianity Not Mysterious* (Dublin: Lilliput, 1997)

McGuinness, P. A. Harrison and R. Kearney (eds.), *John Toland's Christianity Not Mysterious* (Dublin: Lilliput, 1997)

McGuinness, P. "'Perpetual Flux': Newton, Toland, Science and the Status Quo," pp. 317-29 in P. McGuinness, A. Harrison and R. Kearney (eds.), *John Toland's Christianity Not Mysterious* (Dublin: Lilliput, 1997)

McGuinness, P. "John Toland and Irish Politics," pp. 261-92 in P. McGuinness, A. Harrison and R. Kearney (eds.), *John Toland's Christianity Not Mysterious* (Dublin: Lilliput, 1997)

McLaughlin, P.J. "Centenary of the Discovery of Quaternions," *Studies* (1943) 32: 441-56.

McLaughlin, J. "The 'New' Intelligentsia and the Reconstruction of the Irish Nation," *Irish Review* (1999) 24: 53-65.

McMillan, N.D. "The Analytical Reform of Irish Mathematics, 1800-1831," *Irish Mathematical Newsletter* (November 1984) 61-75.

McMillan, N.D. "Richard Helsham M.D. (1683-1737), Medical Man, Virtuoso and Educationalist: Author of the First Purpose-Written Student Textbook on Natural Philosophy in the Vernacular," *Science (Journal of the Irish Science Teachers Association)* (October 1988) 13-18.

McMillan, N. (ed.) *Prometheus's Fire: A History of Scientific and Technological Education in Ireland* (Carlow: Tyndall Publications, 2000)

McNulty, E. "Micro-Ethnography Study of the Virtual Dublin Community in Second Life," Eneas' E-learning and Digital Cultures Blog, digitalculture-ed.net/eneasm/micro-ethnography-study-of-the-virtual-dublin-community-in-second-life/ (Accessed December 11, 2013)

Meenan, J. and D. Clarke (eds.), *RDS: The Royal Dublin Society, 1731-1981* (Dublin: Gill & Macmillan, 1981)

Mendyck, S. "Gerard Boate and Irelands Naturall History," *Journal of the Royal Society of Antiquaries of Ireland* (1985) 115: 5-12.

Merchant, C. *The Death of Nature: Women, Ecology and the Scientific Revolution* (San Francisco: Harper & Row, 1980)

Merrill, D.D. *Augustus de Morgan and the Logic of Relations* (Dordrecht: Kluwer, 1990)

Merton, R. *Science, Technology and Society in Seventeenth Century England [1938]* (New York: Howard Fertig, 1970)

Mill, J. *Analysis of the Phenomena of the Human Mind*, 2nd Ed. (London, 1878)

Mill, J.S. *Autobiography [1873]* (New York: Bobbs-Merrill, 1957)

Mill, J.S. "Bentham [1838]," pp. 132-75 in A. Ryan (ed.) *Utilitarianism and Other Essays: J.S. Mill and Jeremy Bentham* (London: Penguin, 1987)

Mill, J.S. "Whewell on Moral Philosophy," pp. 228-70 in A. Ryan (ed.) *Utilitarianism and Other Essays: J.S. Mill and Jeremy Bentham* (London: Penguin, 1987)

Mill, J.S. "Coleridge [1840]," pp. 177-226 in A. Ryan (ed.) *Utilitarianism and Other Essays: J.S. Mill and Jeremy Bentham* (London: Penguin, 1987)

Miller, G. *The Policy of the Roman Catholic Question Discussed, In a Letter to the Right Honourable W.C. Plunket* (London, 1826)

Miller, J.T. *Ideology and Enlightenment: The Political and Social Thought of Samuel Taylor Coleridge* (London: Garland, 1987)

Mollan, C., W. Davis, and B. Finucane (eds.) *Some People and Places in Irish Science and Technology* (Dublin: Royal Dublin Society, 1985)

Mollan, C., W. Davis, and B. Finucane (eds.) *More People and Places in Irish Science and Technology* (Dublin: Royal Dublin Society, 1990)

Mollan, C. and J. Upton, *The Scientific Apparatus of Nicholas Callan and Other Historic Instruments* (Dublin: St. Patrick's College Maynooth and Samton, 1994)

Mollan, C., B. Finucane and W. Davis (eds.), *Irish Innovators in Science and Technology* (Dublin: Royal Irish Academy and Enterprise Ireland, 2002)

Mollan, C. "Ernest Thomas Sinton Walton (1903-1995)," pp. 1575-95 in Charles Mollan, *It's Part of What We Are: Some Irish Contributors to the Development of the Chemical and Physical Sciences, Vol. II* (Dublin: Royal Dublin Society, 2007)

Mollan, C. *It's Part of What We Are: Some Irish Contributors to the Development of the Chemical and Physical Sciences, 2 Vols.* (Dublin: Royal Dublin Society, 2007)

Molyneux, W. *Dioptrica Nova* (London, 1692)

Moody, T.W. and J.C. Beckett, *Queen's Belfast 1845-1949: The History of a University* (London: Faber and Faber, 1959)

Moody, T.W. "Early Modern Ireland," pp. xxxix-lxiii in T. W. Moody, F. X. Martin and F. J. Byrne (eds.), *A New History of Ireland, Vol. III: Early Modern Ireland, 1534-1691* (Oxford: Clarendon, 1976)

Mooney, T. and F. White, "The Gentry's Winter Season," pp. 1-16 in D. Dickson (ed.) *The Georgeous Mask: Dublin, 1700-1850* (Dublin: Trinity History Workshop, 1987)

Moore, W. *Schrödinger: Life and Thought* (Cambridge: Cambridge University Press, 1989)

Moran, D. "Nationalism, Religion and the Education Question," *Crane Bag* (1983) 7: 77-84.

Morrell, J.B. "Professors Robison and Playfair, and the Theophobia Gallica: Natural Philosophy, Religion and Politics in Edinburgh, 1789-1815," *Notes and Records of the Royal Society* (1971) 26: 43-63.

Morrell, J. and A. Thackray, *Gentlemen of Science: Early Years of the British Association for the Advancement of Science* (Oxford: Clarendon, 1981)

Morrell, J.B. "Brewster and the Early British Association for the Advancement of Science," pp. 25-29 in A. D. Morrison-Low and J. R. R. Christie (eds.), *'Martyr of Science': Sir David Brewster, 1781-1868, Proceedings of a Bicentennary Symposium held at the Royal Scottish Museum on 21 Novemeber 1981* (Edinburgh: Royal Scottish Museum, 1984)

Morse, E.W. *'Natural Philosophy, Hypotheses and Impiety,' Sir David Brewster Confronts the Undulatory Theory of Light* (PhD Thesis: University of California, Berkeley, 1972)

Moynahan, J. *Anglo-Irish: The Literary Imagination in a Hyphenated Culture* (Princeton: Princeton University Press, 1995)

Murdoch, J.E. and E.D. Sylla, "The Science of Motion," pp. 206-64 in D. Lindberg (ed.) *Science in the Middle Ages* (Chicago: University of Chicago Press, 1978)

Murnaghan, D. "History of Radium Therapy in Ireland: The 'Dublin Method' and the Irish Radium Institute," *Journal of the Irish Colleges of Physicians and Surgeons* (1988) 17: 174-6.

Mykkanen, J. "'To Methodize and Regulate Them': William Petty's Governmental Science of Statistics," *History of the Human Sciences* (1994) 7: 65-88.

Nagel, E. "Impossible Numbers: A Chapter in the History of Modern Logic," *Studies in the History of Ideas* (1935) 3: 429-74.

Nahin, P.J. *An Imaginary Tale: The Story of the Square Root of Negative One* (Princeton: Princeton University Press, 1998)

Nakane, M. "The Role of the Three-Body Problem in W.R. Hamilton's Construction of the Characteristic Function for Mechanics," *Historia Scientiarum* (1991) 1: 27-37.

Nasar, S. *A Beautiful Mind* (New York: Simon and Schuster, 1998)

National Board for Science and Technology, *Barriers to Research and Consultancy in the Higher Education Sector* (April 1986)

Neal, K. "The Rhetoric of Utility: Avoiding Occult Associations for Mathematics Through Profitability and Pleasure," *History of Science* (1999) 37: 151-78.

Newton, I. *Opticks [1730]*, 4th Ed. (New York: Dover, 1979)

Nicolson, M. and N.M. Mohler, "The Scientific Background to Swift's 'Voyage to Laputa'," *Annals of Science* (1937) 2: 299-334.

Nicolson, M. and N.M. Mohler, "Swift's 'Flying Island' in the Voyage to Laputa," *Annals of Science* (1937) 2: 405-30.

Nicolson, M. and G.S. Rousseau, "Bishop Berkeley and Tar-Water," pp. 102-37 in H. K. Miller, E. Rothstein and G. S. Rousseau (eds.), *The Augustan Milieu* (Oxford: Clarendon, 1970)

Nudds, J.R. N. McMillan, D. Weaire and S.M. Lawlor (eds.), *Science in Ireland 1800-1930: Tradition and Reform, Proceedings of an International Symposium held at Trinity College Dublin March 1986* (Dublin: Trinity College Dublin, 1988)

Nudds, J.R. "John Joly: Brilliant Polymath," pp. 162-78 in J. Nudds, N. McMillan, D. Weaire and S. M. Lawlor (eds.), *Science in Ireland 1800-1930: Tradition and Reform, Proceedings of an International Symposium held at Trinity College Dublin March 1986* (Dublin: Trinity College Dublin, 1988)

O'Connor, T.C. "John A. McClelland, 1870-1920," pp. 176-85 in Mark McCartney and Andrew Whitaker (eds.) *Physicists of Ireland: Passion and Precision* (Bristol: IOP Publishing, 2003)

O'Connor, T.C. "The Scientific Work of John A. McClelland: A Recently Discovered Manuscript," *Physics in Perspective* (2010) 12: 266-306.

O'Donnell, S. "Early American Science: The Irish Contribution," *Eire-Ireland* (1983) 18: 134-37.

O'Donnell, S. *William Rowan Hamilton: Portrait of a Prodigy* (Dublin: Boole Press, 1983)

O'Dwyer, D.P. et al. "Conical Diffraction of Linearly Polarised Light Controls the Angular Position of a Microscopic Object," *Optics Express* (2010) 18: 27319-27326.

O'Hara, J.G. "George Johnstone Stoney and the Concept of the Electron," *Notes and Records of the Royal Society of London* (1975) 29: 265-76.

O'Hara, J.G. *Humphrey Lloyd (1800-1881) and the Dublin Mathematical School of the Nineteenth Century* (PhD Thesis: University of Manchester, 1979)

O'Hara, J.G. "The Prediction and Discovery of Conical Refraction by William Rowan Hamilton and Humphrey Lloyd (1832-1833)," *Proceedings of the Royal Irish Academy* (1982) 82A: 231-57.

O'Hara, J.G. "Humphrey Lloyd: Ambassador of Irish Science and Technology," pp. 124-39 in in J. Nudds, N. McMillan, D. Weaire and S. M. Lawlor (eds.), *Science in Ireland 1800-1930: Tradition and Reform, Proceedings of an International Symposium held at Trinity College Dublin March 1986* (Dublin: Trinity College Dublin, 1988)

O'Hara, J.G. "A 'Horrible Conflict with Theory' in Heinrich Hertz's Experiments on Magnetic Waves," *European Review* (2007) 15: 545-59.

O'Hearn, D. *Inside the Celtic Tiger: The Irish Economy and the Asian Model* (London: Pluto Press, 1998)

O'Neill, E.P. (ed.) *What Did You Do Today, Professor? Fifteen Illuminating Responses from Trinity College Dublin* (Living Edition: Pöllauberg, Austria, 2008)

O'Neill, J. "Formalism, Hamilton and Complex Numbers," *Studies in the History and Philosophy of Science* (1986) 17: 351-372.

O'Raifeartaigh, T. (ed.) *The Royal Irish Academy: A Bicentennial History, 1785-1985* (Dublin: The Royal Irish Academy, 1985)

O'Raifeartaigh (ed.), L. *General Relativity: Papers in Honour of J.L. Synge* (Oxford: Clarendon Press, 1972)

O'Toole, F. *Ship of Fools: How Stupidity and Corruption Sank the Celtic Tiger* (New York: Public Affairs, 2010)

OECD, *Science and Irish Economic Development* (Dublin: Stationery Office, 1966)

Ogawa, T. "His Final Step to Hamilton's Discovery of the Characteristic Function," *Journal of the Society of Arts and Sciences, Chiba University* (1990) B-23: 45-62.

Øhrstrøm, P. "W.R. Hamilton's View of Algebra as the Science of Pure Time and His Revision of this View," *Historia Mathematica* (1985) 12: 45-55.

Olesko, K.M. *Physics as a Calling: Discipline and Practice in the Königsberg Seminar for Physics* (Ithaca: Cornell University Press, 1991)

Olson, R. "Scottish Philosophy and Mathematics, 1750-1830," *Journal of the History of Ideas* (1971) 32: 29-44.

Olson, R. *Scottish Philosophy and British Physics, 1750-1880: A Study in the Foundations of the Victorian Scientific Style* (Princeton: Princeton University Press, 1975)

Olson, R.G. "Tory-High Church Opposition to Science and Scientism in the Eighteenth Century: The Works of John Arbuthnott, Jonathan Swift, and Samuel Johnson," pp. 171-204 in J. G. Burke (ed.) *The Uses of Science in the Age of Newton* (Berkeley: University of California Press, 1983)

Outram, D. "Negating the Natural: Or Why Historians Deny Irish Science," *The Irish Review* (1986) 1: 45-49.

Peacock, G. *Treatise on Algebra* (Cambridge, 1830)

Peckham, M. "Dr. Lardner's Cabinet Cyclopedia," *Papers of the Bibliographical Society of America* (1951) 45: 37-58.

Perl, T. "The Ladies' Diary or Woman's Almanack, 1704-1841," *Historia Mathematica* (1979) 6: 36-53.

Pašeta, S. "Trinity College, Dublin, and the Education of Irish Catholics, 1873-1908," *Studia Hibernica* (1998/1999) 30: 7-20.

Pettit, S.F. "The Royal Cork Institution: A Reflection of the Cultural Life of a City," *Journal of the Cork Historical and Archaeological Society* (1976) 81: 70-90.

Petty, W. *Reflections Upon Some Persons and Things in Ireland* (London, 1660)

Petty, W. *A Discourse Made Before the Royal Society... Concerning the Use of Duplicate Proportion... Together With a New Hypothesis of Springing or Elastique Motions* (London, 1674)

Petty, W. "The Advice of W.P. to Mr. Samuel Hartlib for the Advancement of Some Particular Parts of Learning [1648]," *Harleian Miscellany* (1745) 6: 1-13.

Petty, W. "Political Arithmetick [1690]," pp. 233-313 in C. H. Hull (ed.) *The Economic Writings of Sir William Petty* (Fairfield, NJ: Augustus M. Kelley, 1986)

Petty, W. "A Treatise on Ireland," in C. H. Hull (ed.) *The Economic Writings of Sir William Petty* (Fairfield, NJ: Augustus M. Kelley, 1986)

Petty, W. "The Political Anatomy of Ireland [1691]," pp. 135-223 in C. H. Hull (ed.) *The Economic Writings of Sir William Petty* (Fairfield, NJ: Augustus M. Kelley, 1986)

Petty, W. "A Treatise on Taxation [1662]," in C. H. Hull (ed.) *The Economic Writings of Sir William Petty* (Fairfield, NJ: Augustus M. Kelley, 1986)

Phelan, W. "An Essay on the Subject Proposed by the Royal Irish Academy, 'Whether, and how far, the pursuits of Scientific, and Polite Literature, assist, or obstruct, each other'," *Transactions of the Royal Irish Academy* (1815) 12: 3-60.

Pickering, A. and A. Stephanides, "Constructing Quaternions: On the Analysis of Conceptual Practice," pp. 139-67 in A. Pickering (ed.) *Science as Practice and Culture* (Chicago: University of Chicago Press, 1992)

Pickering, A. "Concepts and the Mangle of Practice: Constructing Quaternions," *South Atlantic Quarterly* (1995) 94: 417-65.

Playfair, J. "On the Arithmetic of Impossible Quantities," *Philosophical Transactions* (1778) 68: 318-43.

Playfair, J. "La Place, *Traité de Méchanique Céleste*," *Edinburgh Review* (1808) 11: 249-84.

Poole, J.H.J. "John Joly," *Hermathena* (1958) 92: 66-73.

Poole, R. and J. Cash, *Views of the Most Remarkable Public Buildings, Monuments, and Other Edifices in the City of Dublin [1780]* (Shannon: Irish University Press, 1970)

Poovey, M. *A History of the Modern Fact: Problems of Knowledge in the Sciences of Wealth and Society* (Chicago: University of Chicago Press, 1998)

Porter, T.M. *The Rise of Statistical Thinking, 1820-1900* (Princeton: Princeton University Press, 1986)

Porter, T.M. *Trust in Numbers: The Pursuit of Objectivity in Science and Public Life* (Princeton: Princeton University Press, 1995)

Preston, T. "Radiating Phenomena in a Strong Magnetic Field, Part II: Magnetic Perturbations of the Spectral Lines," *Scientific Transactions of the Royal Dublin Society* (1899) 7: 7-22.

Priestley, J. *Experiments and Observations on Different Kinds of Air* (Birmingham, 1790)

Priestley, J. *Political Writings*, Ed. by P. N. Miller (Cambridge: Cambridge University Press, 1993)

Probyn, C.T. "Swift and the Physicians: Aspects of Satire and Status," *Medical History* (1974) 18: 249-61.

Purser, J. "A Note on the School of Engineering Since Its Foundation," *Hermathena* (1941) 58: 53-6.

Purser, M. *Jellett, O'Brien, Purser and Stokes: Seven Generations, Four Families* (Dublin: Prejmer Verlag, 2004)

Purser, M. "As We Saw It: The Context of the Life of George Francis Fitzgerald," *European Review* (2007) 15: 523-29.

Purser, M. *Beyond Buenos Aires: Stories of Peru and Chile (1)* (Dublin: Prejmer Verlag, 2011)

Purser, M. *La Rosa con El Clavel: Stories of Chile and Peru (2)* (Dublin: Prejmer Verlag, 2011)

Purser, M. *From Lvov to Lima: Stories of Argentina, Chile and Peru (3)* (Dublin: Prejmer Verlag, 2012)

Pycior, H. *The Role of Sir William Rowan Hamilton in the Development of British Modern Algebra* (PhD Thesis: Cornell University, 1976)

Pycior, H.M. "George Peacock and the British Origins of Symbolical Algebra," *Historia Mathematica* (1981) 8: 23-45.

Pycior, H.M. "Augustus de Morgan's Algebraic Work: The Three Stages," *Isis* (1983) 74: 211-266.

Pycior, H.M. "Internalism, Externalism, and Beyond: 19th-Century British Algebra," *Historia Mathematica* (1984) 11: 424-441.

Pycior, H.M. "Mathematics and Philosophy: Wallis, Hobbes, Barrow and Berkeley," *Journal of the History of Ideas* (1987) 48: 265-86.

Pycior, H. *Symbols, Impossible Numbers, and Geometric Entanglements: British Algebra Through the Commentaries on Newton's Universal Arithmetick* (Cambridge: Cambridge University Press, 1997)

Quinn, D.B. *The Elizabethans and the Irish* (Ithaca: Cornell University Press, 1966)

Reich, L. *The Making of American Industrial Research: Science and Business at GE and Bell, 1876-1926* (Cambridge: Cambridge University Press, 1985)

Reichel, C.P. and W. Anderson, *Trinity College Dublin and University Reform* (Dublin, 1858)

Reilly, J. and N. O'Flynn, "Richard Kirwan, An Irish Chemist of the Eighteenth Century," *Isis* (1930) 13: 298-319.

Reilly, S.M.P. *Aubrey De Vere: Victorian Observer*, 3rd Ed. (Dublin: Clonmore and Reynolds, 1956)

Renaker, D. "Swift's Liliputians as a Caricature of the Cartesians," *PMLA* (1979) 94: 936-44.

Rhodes, R. *The Making of the Atomic Bomb* (New York: Simon & Schuster, 1986)

Richards, J.L. "The Art and the Science of British Algebra: A Study in the Perception of Mathematical Truth," *Historia Mathematica* (1980) 7: 343-365.

Richards, J.L. "Boole and Mill: Differing Perspectives on Logical Psychologism," *History and Philosophy of Logic* (1980) 1: 19-36.

Richards, J.L. "Augustus de Morgan, The History of Mathematics, and the Foundations of Algebra," *Isis* (1987) 78: 7-30.

Richards, J.L. *Mathematical Visions: The Pursuit of Geometry in Victorian England* (London: Academic Press, 1988)

Richards, J.L. "Rigor and Clarity: Foundations of Mathematics in France and England, 1800-1840," *Science in Context* (1991) 4: 297-319.

Richards, J.L. "God, Truth and Mathematics in Nineteenth-Century England," pp. 51-78 in M. J. Nye (ed.) *The Invention of the Physical Sciences* (Dordrecht: Kluwer, 1992)

Richards, J.L. "The History of Mathematics and *L'esprit humain*: A Critical Reappraisal," *Osiris* (1995) 10: 122-35.

Rider, R.E. "Measures of Ideas, Rule of Language: Mathematics and Language in the 18th Century," pp. 113-40 in T. Frängsmyr, J. L. Heilbron and R. E. Rider (eds.), *The Quantifying Spirit in the 18th Century* (Berkeley: University of California Press, 1990)

Riehm, E.M. and F. Hoffman, *Turbulent Times in Mathematics: The Life of J.C. Fields and the History of the Fields Medal* (American Mathematical Society, 2011)

Robinson, B. *A Treatise of the Animal Oeconomy* (Dublin, 1732)

Robinson, B. *Dissertation on the Aether of Sir Isaac Newton* (Dublin, 1743)

Robinson, B. *Sir Isaac Newton's Account of the Aether, With Some Additions by way of Appendix* (Dublin, 1745)

Robinson, T.R. *A System of Mechanics* (Dublin, 1820)

Robinson, T.R. "Opening Address to Section A," in *Twenty-Seventh Report of the BAAS Held in Dublin in 1857* (London, 1858)

Robison, J. *Proofs of a Conspiracy Against All the Religions and Governments of Europe, Carried On in the Secret Meetings of Free Masons, Illuminati, and Reading Societies*, 4th Ed. (London, 1798)

Roncaglia, A. *Petty: The Origins of Political Economy* (Armonk, NY: M.E. Sharpe, Inc. 1985)

Rothblatt, S. *The Revolution of the Dons: Cambridge and Society in Victorian England* (London: Faber and Faber, 1968)

Russell, C.A. *Science and Social Change in Britain and Europe, 1700-1900* (New York: St. Martin's Press, 1983)

Ryan, S. "Quarks from Top to Bottom: Particle Physics at the Energy Frontier," pp. 73-88 in Eoin P. O'Neill (ed.), *What Did You Do Today, Professor? Fifteen Illuminating Responses from Trinity College Dublin* (Living Edition: Pöllauberg, Austria, 2008)

Sanders, C.R. *Coleridge and the Broad Church Movement* (Durham: Duke University Press, 1942)

Sarton, G. "The Discovery of Conical Refraction by William Rowan Hamilton and Humphrey Lloyd (1833)," *Isis* (1932) 17: 154-70.

Scaife, B.K.P. "James MacCullagh MRIA, FRS 1809-47," *Proceedings of the Royal Irish Academy* (1990) 90C: 67-106.

Scaife, W.G. "Technical Education and the Application of Technology in Ireland, 1800-1950," pp. 85-100 in P. J. Bowler and N. Whyte (eds.), *Science and Society in*

Ireland: The Social Context of Science and Technology in Ireland, 1800-1950 (Belfast: The Institute of Irish Studies-Queen's University of Belfast, 1997)

Schaffer, S. "Halley's Atheism and the End of the World," *Notes and Records of the Royal Society of London* (1977) 32: 17-40.

Schaffer, S. "Newtonianism," pp. 610-26 in R. C. Olby, G. N. Cantor, J. R. R. Christie and M. J. S. Hodge (eds.), *Companion to the History of Modern Science* (London: Routledge, 1990)

Schaffer, S. "States of Mind: Enlightenment and Natural Philosophy," pp. 233-90 in G. S. Rousseau (ed.) *The Languages of Psyche: Mind and Body in Enlightenment Thought* (Berkeley: University of California Press, 1990)

Schaffer, S. "Late Victorian Metrology and Its Instrumentation: A Manufactory of Ohms," pp. 23-56 in S. Cozzens and R. Bud (eds.), *Invisible Connections* (Bellingham: SPIE Optical Engineering Press, 1992)

Schaffer, S. "Babbage's Intelligence: Calculating Engines and the Factory System," *Critical Inquiry* (1994) 21: 203-27.

Schofield, R. *Mechanism and Materialism: British Natural Philosophy in an Age of Reason* (Princeton: Princeton University Press, 1970)

Schrödinger, E. *What Is Life? The Physical Aspect of the Living Cell* (Cambridge: Cambridge University Press, 2013)

Science, Technology and Innovation Advisory Council, *Making Knowledge Work for Us: A Strategic View of Science, Technology and Innovation in Ireland* (Dublin: Stationery Office, 1995)

Sedgwick, A. *A Discourse on the Studies of the University of Cambridge*, 5th Ed. (London, 1850)

Seth, S. "Quantum Physics," pp. 814-59 in Jed Z. Buchwald and Robert Fox (eds.), *The Oxford Handbook of the History of Physics* (Oxford: Oxford University Press, 2013)

Shapin, S. and B. Barnes, "Science, Nature and Control: Interpreting Mechanics Institutes," *Social Studies of Science* (1977) 7: 31-74.

Shapin, S. "Of Gods and Kings: Natural Philosophy and Politics in the Leibniz-Clarke Disputes," *Isis* (1981) 72: 187-215.

Shapin, S. and S. Schaffer, *Leviathan and the Air Pump: Hobbes, Boyle and the Experimental Life* (Princeton: Princeton University Press, 1985)

Shapin, S. "Robert Boyle and Mathematics: Reality, Representation, and Experimental Practice," *Science in Context* (1988) 2: 23-58.

Shelley, P.B. "A Defence of Poetry [1821]," pp. 478-508 in D. H. Reiman and S. B. Powers (eds.), *Shelley's Poetry and Prose* (New York: W.W. Norton & Company, 1977)

Sheridan, C. "Baltimore Technologies," *Technology Ireland* (January 1999), pp. 16-17.

Sheridan, C. "National Microelectronics Research Centre," *Technology Ireland* (October 1999), pp. 32-33.

Sherry, D. "The Logic of Impossible Quantities," *Studies in the History and Philosophy of Science* (1991) 22: 37-62.

Shinn, T. *Savior scientifique et pouvoir social: L'Ecole Polytechnique et les polytechniciens, 1794-1914* (Paris: Presse Fondation Nationale des Sciences Politiques, 1980)

Silke, J. "Irish Scholarship and the Renaissance, 1580-1673," *Studies in the Renaissance* (1973) 20: 169-206.

Silke, J.J. "The Irish Abroad, 1534-1691," pp. 587-633 in T. W. Moody, F. X. Martin and F. J. Byrne (eds.), *A New History of Ireland, Vol. III: Early Modern Ireland, 1534-1691* (Oxford: Clarendon, 1976)

Simms, J.G. and D.H. Kelly, *William Molyneux of Dublin, 1656-98*, (Dublin: Irish Academic Press, 1982)

Sinègre, L. "Les Quaternions et le Mouvement du Solide autour d'un Point Fixe Chez Hamilton," *Revue d'histoire des mathématiques* (1995) 1: 83-109.

Singer, J.H. "Memoir of the Late President," *Proceedings of the Royal Irish Academy* (1836-40) 1: 121-26.

Smith, C. and M.N. Wise, *Energy and Empire: A Biographical Study of Lord Kelvin* (Cambridge: Cambridge University Press, 1989)

Smith, C. *The Science of Energy: A Cultural History of Energy Physics in Victorian Britain* (Chicago: University of Chicago Press, 1998)

Smith, J. *Fact and Feeling: Baconian Science and the Nineteenth-Century Literary Imagination* (Madison: University of Wisconsin Press, 1994)

Smolin, L. *The Trouble with Physics: The Rise of String Theory, the Fall of a Science, and What Comes Next* (Boston: Mariner Books, 2007)

Spearman, T.D. "Humphrey Lloyd, 1800-1881," *Hermathena* (1981) 130: 37-52.

Spearman, T.D. "Mathematics and Theoretical Physics," pp. 201-39 in T. O'Raifeartaigh (ed.) *The Royal Irish Academy: A Bicentennial History, 1785-1985* (Dublin: The Royal Irish Academy, 1985)

Spearman, T.D. "James MacCullagh," pp. 41-59 in J. Nudds, N. McMillan, D. Weaire and S. M. Lawlor (eds.), *Science in Ireland 1800-1930: Tradition and Reform, Proceedings of an International Symposium held at Trinity College Dublin March 1986* (Dublin: Trinity College Dublin, 1988)

Spearman, T.D. "Four Hundred Years of Mathematics," pp. 280-493 in C. H. Holland (ed.) *Trinity College Dublin & and the Idea of a University* (Dublin: Trinity College Dublin Press, 1991)

Spearman, T.D. "James MacCullagh, 1809–1847," *The European Physics Journal H* (2010) 35: 113-122.

Spearman, T.D. "400 Years of Mathematics," www.maths.tcd.ie/about/400Hist/index.php (Accessed December 13, 2013)

Starkman, M.K. *Swift's Satire on Learning in 'A Tale of a Tub'* (Princeton: Princeton University Press, 1950)

Steffens, H.J. *The Development of Newtonian Optics in England* (New York: Science History Publications, 1977)

Stein, H. "'Subtler Forms of Matter' in the Period Following Maxwell," pp. 309-40 in G. N. Cantor and M. J. S. Hodge (eds.), *Conceptions of Ether: Studies in the History of Ether Theories, 1740-1900* (Cambridge: Cambridge University Press, 1981)

Stock, J. *Memoirs of George Berkeley, Late Bishop of Cloyne*, Second Edition (London 1784)

Stoney, G.J. "How Thought Presents Itself in Nature," *Proceedings of the Royal Institution* (1885) 11: 178-96.

Stoney, G.J. "Studies in Ontology: The First Step," *Proceedings of the Aristotelian Society* (1889) 1: 1-9.

Stoney, G.J. "Studies in Ontology, From the Standpoint of the Scientific Student of Nature," *Scientific Proceedings of the Royal Dublin Society* (1890) 6: 475-524.

Strauss, E. *Sir William Petty: Portrait of a Genius* (London: The Bodley Head, 1954)

Struik, D.J. *A Concise History of Mathematics*, 4th Ed. (New York: Dover, 1987)

Stubbs, J.W. *The History of the University of Dublin, From Its Foundation to the End of the Eighteenth Century* (Dublin, 1889)

Studiosus, *Alma's Defence, Or The Critic Refuted: A Poem With Explanatory Notes* (Dublin, 1790)

Styazhkin, N.I. *History of Mathematical Logic from Leibniz to Peano* (Cambridge: MIT Press, 1969)

Sullivan, R. *John Toland and the Deist Controversy* (Cambridge: Harvard University Press, 1982)

Susskind, L. *The Cosmic Landscape: String Theory and the Illusion of Intelligent Design* (New York: Back Bay Books, 2006)

Sweeney, P. *The Celtic Tiger: Ireland's Economic Miracle Explained* (Dublin: Oak Tree Press, 1998)

Swift, T. *Animadversions on the Fellows of Trinity College* (Dublin, 1794)

Swift, T. *Prison Pindarics; Or, a New Year's Gift from Newgate Humbly Presented to the Students of the University* (Dublin, 1795)

Swift, J. "Gulliver's Travels [1726]," pp. 21-278 in M. K. Starkman (ed.) *Gulliver's Travels and Other Writings* (New York: Bantam Books, 1981)

Synge, J.L. *Science: Sense and Nonsense* (New York: W.W. Norton & Company, 1950)

Synge, J.L. "Eamon de Valera," *Biographical Memoirs of Fellows of the Royal Society* (1976) 22: 634-653.

Taylor, F.S. "The Teaching of Science at Oxford in the Nineteenth Century," *Annals of Science* (1952) 8: 82-112.

Taylor, E.G.R. *The Mathematical Practitioners of Tudor and Stuart England* (Cambridge: Cambridge University Press, 1968)

Thackray, A. *Atoms and Powers: An Essay on Newtonian Matter-Theory and the Development of Chemistry* (Cambridge: Harvard University Press, 1970)

Theerman, P.H. "Dionysius Lardner's American Tour: A Case Study in Antebellum American Interest in Science, Technology and Nature," in P. H. Theerman and K. H. Parshall (eds.), *Experiencing Nature: Proceedings of a Conference in Honor of Allen G. Debus* (Dordrecht: Kluwer, 1997)

Thesing, G. "The Campus Whiz Kids," *Business & Finance: Ireland's Business Weekly* 1998), pp. 14-17.

Todhunter, I. *William Whewell: An Account of His Writings, with Selections from His Literary and Scientific Correspondence* (London, 1876)

Toland, J. *Letters to Serena [1704]* (New York: Garland, 1976)

Treacy, W.P. *Irish Scholars of the Penal Days: Glimpses of Their Labors on the Continent of Europe* (New York, 1889)

Trinity College Dublin, *School of Engineering: A Record of Past and Present Students* (Dublin, 1909)

Trinity College Dublin, "Minister for Education Launches IITAC's Visualisation and Supercomputing Facilities at TCD," (September 14, 2006)

Trinity College Dublin, "School of Mathematics Organises International Lattice Conference at TCD," *Trinity Research News* (November 2005)

Trinity College Dublin, "Towards a Future Higher Education Landscape" (July 2012)

Turbayne, C. "Berkeley and Molyneux," *Journal of the History of Ideas* (1955) 16: 339-55.

Turnbull, G.H. "Samuel Hartlib's Influence on the Early History of the Royal Society," *Notes and Records of the Royal Society* (1953) 10: 101-30.

Turner, R.S. "The Growth of Professorial Research in Prussia, 1818-1848—Causes and Context," *Historical Studies in the Physical Sciences* (1971) 3: 137-82.

Turner, A.J. "Mathematical Instruments and the Education of Gentlemen," *Annals of Science* (1973) 30: 51-88.

U.S. Lattice Quantum Chromodynamics, "Computing Hardware and Software: The Blue Gene/Q," `www.usqcd.org/BGQ.html` (Accessed December 11, 2013)

Vasallo, N. "Analysis versus Laws: Boole's Explanatory Psychologism versus His Explanatory Anti-Psychologism," *History and Philosophy of Logic* (1997) 18: 151-63.

Vaughn, W.E. *Landlords and Tenants in Ireland, 1848-1903* (Dublin: Economic and Social History of Ireland Press, 1984)

Walsh, E.M. "The Higher Education Business: Responding to the Market," pp. 66-75 in A. T. McKenna (ed.) *Higher Education: Relevance and Future* (Dublin: Higher Education Authority, 1985)

Walton, E.T.S. "The Need for More Attention to Science in Éire with some Suggestions for its Encouragement (1957)," in V. McBrierty, *Ernest Thomas Sinton Walton, 1903-1995: The Irish Scientist* (Dublin: Trinity College Dublin Press, 2003)

Ward, W. *Aubrey De Vere: A Memoir Based On His Unpublished Diairies and Correspondence* (London: Longmans, Green, and Co. 1904)

Warren, J. *A Treatise on the Geometrical Representation of the Square Roots of Negative Quantities* (Cambridge, 1823)

Warwick, A. "On the Role of the FitzGerald-Lorentz Contraction Hypothesis in the Development of Joseph Larmor's Electronic Theory of Matter," *Archive for History of Exact Sciences* (1991) 43: 29-91.

Warwick, A. "Cambridge Mathematics and Cavendish Physics: Cunningham, Campbell and Einstein's Relativity, 1905-1911, Part I: The Uses of Theory," *Studies in History and Philosophy of Science* (1992) 23: 625-656.

Warwick, A. "The Sturdy Protestants of Science: Larmor, Trouton, and the Earth's Motion Through the Ether," pp. 300-43 in J. Z. Buchwald (ed.) *Scientific Practice: Theories and Stories of Doing Physics* (Chicago: University of Chicago Press, 1995)

Warwick, A. "A Mathematical World on Paper: Written Examinations in Early 19th-century Cambridge," *Studies in History and Philosophy of Modern Physics* (1998) 29: 295-319.

Warwick, A. *Masters of Theory: Cambridge and the Rise of Mathematical Physics* (Chicago: University of Chicago Press, 2003).

Warwick, A. "'That Universal Aetherial Plenum': Joseph Larmor's Natural History of Physics," pp. 343-86 in Kevin C. Knox and Richard Noakes (eds.), *From Newton to Hawking: A History of Cambridge University's Lucasian Professors of Mathematics* (Cambridge: Cambridge University Press, 2003)

Watson, J.D. *The Double Helix: A Personal Account of the Discovery of the Structure of DNA* (New York: W.W. Norton, 1980)

Wayman, P.A. *Dunsink Observatory, 1785-1985: A Bicentennial History* (Dublin: The Royal Dublin Society and the Dublin Institute for Advanced Studies, 1987)

Wayman, P.A. "Rev. John Brinkley: A New Start," in J. Nudds, N. McMillan, D. Weaire and S. M. Lawlor (eds.), *Science in Ireland 1800-1930: Tradition and Reform, Proceedings of an International Symposium held at Trinity College Dublin March 1986* (Dublin: Trinity College Dublin, 1988)

Weaire, D. and S. O'Connor, "Unfulfilled Renown: Thomas Preston (1860-1900) and the Anomalous Zeeman Effect," *Annals of Science* (1987) 44: 617-44.

Weaire, D. "Experimental Physics at Trinity," pp. 270-79 in C. H. Holland (ed.) *Trinity College Dublin & the Idea of a University* (Dublin: Trinity College Dublin Press, 1991)

Weaire, D. P. Kelly and D.A. Attis (eds.), *Richard Helsham's Lectures on Natural Philosophy [1739]* (Dublin: Trinity College Dublin Physics Department, Institute of Physics Publishing, Verlag MIT, 1999)

Weaire, D. "Thomas Ranken Lyle" in Charles Mollan, Brendan Finucane and William Davis (eds.),*Irish Innovators in Science and Technology* (Dublin: Royal Irish Academy and Enterprise Ireland, 2002)

Weaire, D. and J.M.D. Coey, "Mentor and Constant Friend: The Life of George Francis Fitzgerald (1851-1901)," *European Review* (2007) 15: 513-21.

Weaire, D. "Sand and Foam," pp. 227-39 in Eoin P. O'Neill (ed.), *What Did You Do Today, Professor? Fifteen Illuminating Responses from Trinity College Dublin* (Living Edition: Pöllauberg, Austria, 2008)

Webster, C. (ed.) *Samuel Hartlib and the Advancement of Learning* (Cambridge: Cambridge University Press, 1970)

Webster, C. "New Light on the Invisible College," *Transactions of the Royal Historical Society* (1974) 24: 19-42.

Webster, C. *The Great Instauration: Science, Medicine and Reform, 1626-1660* (London: Duckworth, 1975)

Webster, C. "Benjamin Worsley: Engineering for Universal Reform From the Invisible College to the Navigation Act," pp. 213-35 in M. Greengrass, M. Leslie and T.

Raylor (eds.), *Samuel Hartlib and Universal Reformation* (Cambridge: Cambridge University Press, 1994)

Weiss, J.H. *The Making of Technological Man: The Social Origins of French Engineering Education* (Cambridge: Harvard University Press, 1982)

Whelan, K. "The United Irishmen, the Enlightenment and Popular Culture," pp. 269-96 in D. Dickson, D. Keogh and K. Whelan (eds.), *The United Irishmen: Republicanism, Radicalism and Rebellion* (Dublin: Lilliput Press, 1993)

Whelan, K. *The Tree of Liberty: Radicalism, Catholicism and the Construction of Irish Identity, 1760-1830* (Cork: Cork University Press, 1996)

Whewell, W. *On the Principles of English University Education (with "Thoughts on the Study of Mathematics as Part of a Liberal Education")* (London, 1837)

Whewell, W. *Of a Liberal Education in General and with a Particular Reference to the Leading Studies of the University of Cambridge* (Cambridge, 1845)

Whewell, W. *The Philosophy of the Inductive Sciences [1840]*, Ed. by G. Buchdahl and L. L. Laudan (London: Frank Cass, 1967)

Whittaker, E.T. *A History of the Theories of Aether and Electricity [1910]* (New York: Harper & Brothers, 1960)

Whittaker, E.T. "The Hamiltonian Revival," *The Mathematical Gazette* (1940) 24: 153-58.

Whittaker, E.T. "The Sequence of Ideas in the Discovery of Quaternions," *Proceedings of the Royal Irish Academy* (1944) 50A: 93-98.

Whittaker, E.T. "Arthur William Conway, 1875-1950," *Obituary Notices of the Fellows of the Royal Society* (1951) 7: 329-40.

Whyte, N. "'Lords of Ether and of Light': The Irish Astronomical Tradition of the Nineteenth Century," *Irish Review* (1995) 17/18: 127-41.

Whyte, N. *Science, Colonialism and Ireland* (Cork: Cork University Press, 1999)

Wickham, J. "An Intelligent Island?," pp. 81-88 in M. Peillon and E. Slater (eds.), *Encounters with Modern Ireland: A Sociological Chronicle, 1995-1996* (Dublin: Institute of Public Administration, 1998)

Wilczek, F. *The Lightness of Being: Mass, Ether, and the Unification of Forces* (NY: Basic Books, 2008)

Wilder, T. *Newton's Universal Arithmetick* (London, 1769)

Wilkins, D.R. (ed.), *Perplexingly Easy: Selected Correspondence between William Rowan Hamilton and Peter Guthrie Tait* (Dublin: Trinity College Dublin Press, 2005)

Williams, H. (ed.) *The Correspondence of Jonathan Swift* (Oxford: Clarendon, 1965)

Williams, P. "Passing on the Torch: Whewell's Philosophy and the Principles of English University Education," pp. 117-47 in M. Fisch and S. Schaffer (eds.), *William Whewell: A Composite Portrait* (Oxford: Clarendon, 1991)

Wilson, D.B. "Experimentalists Among the Mathematicians: Physics in the Cambridge Natural Sciences Tripos, 1851-1900," *Historical Studies in the Physical Sciences* (1982) 12: 325-71.

Winterbourne, A.T. "Algebra and Pure Time: Hamilton's Affinity with Kant," *Historia Mathematica* (1982) 9: 195-200.

Wise, G. "A New Role for Professional Scientists in Industry: Industrial Research at GE, 1900-16," *Technology and Culture* (1980) 21: 408-29.

Wise, M.N. "Mediating Machines," *Science in Context* (1988) 2: 81-117.

Wise, M.N. "Work and Waste: Political Economy and Natural Philosophy in Nineteenth Century Britain," *History of Science* (1989) 27: 263-301; (1989) 27: 391-449; (1990) 28: 221-261.

Wise, M.N. "Mediations: Enlightenment Balancing Acts, or the Technologies of Rationalism," pp. 207-56 in P. Horwich (ed.) *World Changes: Thomas Kuhn and the Nature of Science* (Cambridge: MIT Press, 1993)

Wise, M.N. *The Values of Precision* (Princeton: Princeton University Press, 1995)

Woodward, R. *The Present State of the Church of Ireland: Containing a Description of it's [sic] Precarious Situation; and the consequent Danger to the Public* (Dublin, 1787)

Wormell, D.E.W. "Latin Verses By William Thompson Spoken at the Opening in 1711 of the First Scientific Laboratory in Trinity College Dublin," *Hermathena* (1962) 96: 21-30.

Yavetz, I. *From Obscurity to Enigma: The Work of Oliver Heaviside, 1872-1891* (Boston: Birkhauser, 1995)

Yearley, S. "From One Dependency to Another: The Political Economy of Science Policy in the Irish Republic in the Second Half of the Twentieth Century," *Science, Technology, and Human Values* (1995) 20: 171-96.

Yearly, S. "The Social Construction of National Scientific Profiles: A Case Study of the Irish Republic," *Sociology of Science* (1987) 26: 191-210.

Yearly, S. "Colonial Science and Dependent Development: The Case of the Irish Experience," *Sociological Review* (1989) 37: 308-31.

Yeo, R. *Defining Science: William Whewell, Natural Knowledge, and Public Debate in Early Victorian Britain* (Cambridge: Cambridge University Press, 1993)

Young, J.R. *Three Lectures Addressed to the Students of Belfast College on Some of the Advantages of Mathematical Study* (London, 1845)

Young, M. *An Analysis of the Principles of Natural Philosophy* (Dublin, 1800)

Index

Accenture, 399
Adare, Viscount, 175, 184
Addison, Joseph, 62
Adrain, Robert, 138
Airy, George B., 178, 184, 190, 199, 202
algebra, 20, 21, 92, 117, 118, 123, 152, 160, 182, 220, 222, 223, 225, 228–236, 239–242, 245, 246, 248, 249, 253, 257, 300, 344, 385
alpha particles, 331, 335, 336, 338, 339, 390
analysis, 13, 20, 96, 122, 125, 145, 155, 156, 180, 204, 221, 222, 226, 229, 230, 236, 246, 247, 251, 253, 257, 314, 385, 389, 399–401, 408
Anglicanism, 17, 63, 81, 106
Arago, Francois, 126
Armagh Observatory, 16, 40, 110, 154, 163
Ashe, St. George, 70–73, 85, 91

astronomy, 14, 17, 25, 29, 38, 56, 69, 105, 106, 110, 112, 113, 119, 122–124, 126, 127, 129, 152, 177, 178, 185, 186, 210, 232, 277, 326, 343, 365
atheism, 79, 84, 94, 195
atom, 16, 57, 58, 88, 89, 284, 288, 289, 328, 329, 332, 335, 339, 342, 389, 390
Australia, 22, 272, 278

Babbage, Charles, 21, 156, 190, 193, 219, 227, 228, 253, 254
Bacon, Francis, 29–33, 35, 36, 57, 100, 184, 187, 194, 196
Baconianism, 31, 185, 205, 206, 320
Baker, Sean, 388
Baltimore Technologies, 368
Banks, Joseph, 50, 114, 115
Barber, Samuel, 135–137, 163
Beaufort, Francis, 175, 265, 266

beauty, 11, 20, 26, 59, 176, 177, 182, 185, 188, 205, 221, 270, 350, 351, 367, 404, 407, 408

Becquerel, Henri, 330

Belfast, 110, 116, 137, 138, 162, 168, 169, 173, 264, 268, 275, 285, 290, 291, 318, 319, 322, 394

Bentham, Jeremy, 129, 130, 207, 256

Benthamism, 165, 205–208

Beresford, Archbishop, 119, 149, 151, 163, 164

Berkeley, George, 8, 10, 15, 16, 21, 62–64, 67, 73, 82–96, 101, 102, 132, 155, 176, 177, 186, 188, 211, 219, 222, 224, 289, 290, 308, 317, 391

big data, 26, 366, 396, 399, 400

Black-Scholes equation, 397

Boate, Arnold, 35

Boate, Gerard, 35

Bohr, Niels, 341, 353

Boole, George, 21, 178, 220, 228, 246–253, 285

Booth, James, 272

Bott, Raoul, 357

Boyle, Robert, 14, 29, 34–36, 41, 57, 64, 83

Boyton, Charles, 213, 214

Brewster, David, 19, 190, 195–200, 202–204

Brinkley, John, 17, 105–107, 121, 123, 124, 126, 127, 140, 145, 152, 155, 156, 170, 178, 266

British Association for the Advancement of Science (BAAS), 154, 192, 197, 198, 200–204, 229, 264, 292, 293, 297

Brougham, Henry, 19, 162, 163, 190, 196, 197, 202–204

Browne, Arthur, 156

Browne, Paddy, 347, 349

Browne, Peter, 82, 85

Bryce, James, 319

Burke, Edmund, 82, 130, 131, 211, 254, 308, 382, 408

Burke, John, 326

Burrowes, Robert, 112–115, 117–120, 136, 139, 346

Butt, Isaac, 317

Cahill, Eamonn, 397, 398

calculation, 9, 19, 41, 50, 87, 97, 130, 141, 217, 219, 220, 239, 254–257, 278, 336, 389, 392–394

calculus, 8, 10, 11, 17, 20, 21, 45, 64, 73, 85, 86, 91–94, 96, 106, 121–123, 126, 132, 143, 153, 155, 156, 191,

219, 222, 224, 226–228, 245, 247–249, 255, 256, 276, 300

Cambridge University, 29, 32, 36, 38, 48, 120, 122, 130, 144, 146, 156, 160, 162, 163, 180, 182, 185, 206, 224, 227, 243, 255, 256, 264, 275, 281, 282, 288, 296, 311, 331, 352, 367, 369, 370, 409

Canada, 22, 272, 354

Carlyle, Thomas, 255, 256, 408

Carmichael, Robert, 228

Casey, John, 170, 385

cathode rays, 327, 330, 332, 333

Catholic Church, 63, 81, 313, 318, 319, 321–324, 326

Catholic Emancipation, 19, 144, 157, 162, 165, 204, 213, 216

Cavendish Laboratory, 326, 327, 331

Celtic Revival, 317, 326

Celtic Tiger, 25, 26, 361, 366, 367, 370, 374, 379, 382, 388, 398, 399, 407

CERN, 386, 388, 400

Church of Ireland, 8, 17, 23, 64, 72, 107, 115, 133–136, 157, 158, 162, 309, 315, 318, 355

Clarke, Samuel, 111, 251

Coey, Michael, 383

Colburn, Zerah, 217–219

Coleridge, Samuel Taylor, 176, 177, 180, 185, 187–189, 206, 208, 211, 212, 214–216, 235

Collins, Steven, 365

computer, 25, 219, 220, 364–368, 371, 385, 386, 393–395, 398–400, 407, 408

Congreve, William, 85

conical refraction, 19, 189, 192–194, 201, 202, 206, 209, 210, 215, 221, 242

Conway, Arthur W., 180, 182, 347–349, 358, 359

Copernicus, Nicolaus, 14, 29, 169

Cork, 21, 34, 36, 82, 134, 162, 220, 228, 250, 268, 282, 285, 312, 318, 319, 322, 334, 355, 373, 394

corpuscular theory of light, 194, 203

Crick, Francis, 353

Cromwell, Henry, 41, 48

Cromwell, Oliver, 8, 14, 15, 27–30, 32, 41, 47, 48, 64, 66, 74, 137

Crookes, William, 330

cryptography, 368

Curie, Marie, 330, 331, 341

d'Albe, E.E. Fournier, 326
d'Alembert, Jean, 122, 181
D'Arcy, Patrick, 72
Darwin, Charles, 114, 266
Davis, Thomas, 194, 317, 327, 332, 382
Davitt, Michael, 304
de Broglie, Louis, 359
De Morgan, Augustus, 228, 230, 242, 246, 252, 253
De Valera, Éamon, 24, 311, 313, 345, 349–351
De Vere, Aubrey, 175, 205, 206, 210, 233
deism, 63
Deloitte, 378
Descartes, Rene, 57, 75, 83, 87, 92
Dillon, John, 346, 359
Dirac, Paul A.M., 341, 349, 351
Ditchburn, R.W., 325
Dixon, Robert Vickers, 276
Dixon, Stephen, 276, 326
DNA, 353
Down Survey, 14, 30, 42, 45, 47, 48, 54, 57
Drennan, William, 137, 138
Dublin Institute for Advanced Study, 345, 359
Dublin Institute of Technology, 301, 379

Dublin Philosophical Society, 66, 67, 70–73, 85, 86
Dunsink Observatory, 5, 17, 105, 110, 124, 173, 266, 325
dynamics, 20, 123, 182, 183, 188, 243, 275, 276, 296, 297, 300, 314, 385, 397, 398

Easter Rising, 24, 308–312, 314, 315, 321, 355, 376
Eddington, Arthur Stanley, 291, 295–297, 349
Edgeworth, Francis Beaufort, 176, 266
Edgeworth, Maria, 114, 175, 266
Edgeworth, Richard Lovell, 114, 115, 175
Einstein, Albert, 339, 342, 345, 351, 353, 354, 358, 402
electromagnetic theory, 280, 284, 389
electron, 292–295, 297, 310, 328, 329, 332, 339, 342, 360, 389, 392
Emmet, John Patten, 138
Emmet, Robert, 138
engineering, 14, 21, 22, 25, 30, 130, 263, 267–274, 277, 282, 285, 301, 305, 326, 332, 337, 359, 365, 366, 400

ether, 19, 23, 76–78, 82, 132, 194–198, 203, 284, 286–289, 291–295, 297, 310, 311, 328, 342, 391

Euclid, 56, 223, 300

Euler, Leonhard, 122, 123, 181, 223

European Union, 370, 376, 381, 385

evolution, 302, 324

examinations, 10, 18, 142, 144, 145, 147–150, 153, 166–169, 171, 191, 273, 293, 303, 329, 334, 355

Famine, 25, 28, 210, 211

Faraday, Michael, 200, 277, 278, 300

Fermat, Pierre de, 69, 181

Fermi, Enrico, 341

Fermilab, 393, 394

Field, Cyrus, 278

Fields Medal, 356, 404

financial services, 370, 371, 373, 374

Fitzgerald, George Francis, 16, 22, 64, 100, 262, 263, 273, 274, 279, 282–292, 294–296, 299–305, 315, 325–328, 342, 360, 367, 382, 383, 391

Florides, Petros, 354–357

fluxions, 95, 123, 160, 227

Formalism, 253

freethinking, 323

Fresnel, Augustin, 191

Galbraith, Joseph Allen, 317

Galileo, 14, 29, 57, 169, 300, 302

Gamow, George, 336

Gauss, Carl Friedrich, 265, 281

Geiger, Hans, 333

Gell-Mann, Murray, 390

genetics, 351

geometry, 20, 37, 38, 87, 92, 97, 123, 124, 146, 152, 155, 160, 179–183, 188, 192, 219–222, 224–226, 231, 232, 236, 242, 257, 277, 357, 385, 404

Gibbs, Josiah Willard, 245

Gladstone, William E., 315

Goldsmith, Oliver, 308, 382

Google, 371, 399

Grattan, Henry, 102

Graunt, John, 48

Graves, Charles, 228

Graves, John, 241

Gregory, D.F., 228, 317

Guinness Brewery, 282, 363

Gulliver's Travels, 96

Halberstam, H., 224, 359

Hales, William, 132

Halley, Edmond, 68

Hamilton, Eliza, 174, 210

Hamilton, Sir William, 159
Hamilton, Sir William Rowan, 5,
    18, 24, 220, 235, 296,
    313, 348
Hamiltonian function, 183, 275,
    350, 408
Hartley, Sir Walter, 328, 329
Hartlib, Samuel, 32–35, 41, 65
Harvard University, 41, 76, 79,
    122, 264
Haughton, Samuel, 168, 192, 317
Havok, 365, 393
Heaviside, Oliver, 245, 287, 292,
    294, 296, 298
Hegarty, John, 386
Heisenberg, Werner, 341
Heitler, Walter, 353
Helmholtz, Hermann von, 288,
    289
Helsham, Richard, 65, 73–76, 78
Hely-Hutchinson, John, 116
Hennessy, Henry, 170
Herschel, John, 156, 183, 184,
    198, 227, 253
Hertz, Heinrich, 260, 287, 300
Hincks, Thomas Dix, 162
Hobbes, Thomas, 55–57, 64, 83,
    91
Home Rule, 304, 309, 315–317
Horn, Chris, 388
Humboldt, Alexander von, 265
Hutcheson, Francis, 82

Hutton, Charles, 122, 123, 155
Hyde, Douglas, 317, 349

IBM, 371, 394, 395, 399
IDA, 364, 371, 372, 379, 388
idealism, 16, 64, 176, 184, 205,
    206, 208, 209, 214, 289,
    290, 324
imaginary numbers, 8, 11, 219,
    223, 224, 226, 233, 236,
    238
imagination, 175, 185, 186, 198,
    221, 236, 242, 257
imperialism, 271, 280, 315
Indian Civil Service, 167, 273
infinitesimal, 10, 16, 84, 92, 93,
    219, 224
Ingram, John Kells, 224, 359
internet, 25, 366, 369, 385
Iona Technologies, 388

Jameson, Annie, 22, 261
Jellett, J.H., 192, 228, 285, 367
Jesuits, 37, 38, 333
Joly, Charles Jasper, 314
Joly, John, 307, 308, 314, 315,
    321, 322, 326, 330, 331
Joyce, James, 390

Kane, Sir Robert, 303
Kant, Immanuel, 185, 231, 232
Kemp, George, 262
King, William, 53, 82, 85
Kinsella, Raymond, 380

Kirchhoff, Gustav, 328
Kirwan, Richard, 114, 115, 138

Lagrange, Joseph Louis, 106, 122, 126, 129, 156, 181, 227
Lanczos, Cornelius, 314, 353
Laplace, Pierre-Simon, 106, 122, 123, 126–129, 156, 181
Laputa, 13, 96–98, 100, 131
Larcom, Thomas, 42, 45
Lardner, Dionysius, 149, 156, 326
Large Hadron Collider, 388, 389
Larmor, Joseph, 274, 290–293, 295–297, 310, 311, 320, 326, 328, 332, 403
lattice quantum chromodynamics, 385, 389
least action, 181, 291, 295–298
Leibniz, Gottfried Wilhelm, 79, 92, 246, 300
Lemass, Seán, 370
light, 19, 20, 22, 31, 69, 73, 75, 77, 87, 127, 152, 181–183, 188–191, 193, 194, 196, 197, 200–204, 215, 222, 260, 263, 280, 284, 287, 292, 294, 295, 328–330, 332, 342, 383, 389
Lloyd, Bartholomew, 145, 146, 148, 150–153, 156, 163–166, 170, 178, 192, 199, 277

Lloyd, Humphrey, 18, 19, 21, 152, 156, 189, 190, 192, 193, 199, 201–203, 263, 265–271, 274, 277, 282, 283, 305, 315, 384, 385
Locke, John, 63, 69, 79, 81, 85, 86
Lodge, Oliver, 287, 288, 294, 295, 298, 299, 304
logic, 21, 33, 117, 142, 150, 153, 155, 220, 245–249, 251–253, 269
Lorentz, Hendrik, 295, 342, 344
Lyle, Thomas Ranken, 326

MacCullagh, James, 18, 152, 191, 192, 291, 315, 384
MacDonnell, Richard, 149, 150
MacNeven, William James, 138
Magnetic Crusade, 265–267
Mahon, John, 364
Marconi, Guglielmo, 22, 259–263, 299
Maskelyne, Nevil, 114, 124
materials science, 383, 399
Mather, Samuel, 41
Maupertuis, Pierre de, 181
Maxwell, James Clerk, 109, 243, 260, 280, 284, 287, 288, 295, 300, 402
Maxwellians, 264, 284, 292, 298–301, 328

Maynooth, 43, 139, 319, 334, 347, 406

McBrierty, Vincent, 335, 341, 360, 380

McClelland, John Alexander, 333, 334

McConnell, A.J., 152, 182, 314, 348, 349, 357, 359

McHenry, John, 334

McNulty, Eneas, 364

mechanics, 14, 23, 29, 56, 66, 73, 99, 122, 129, 152, 177, 181–183, 247, 250, 272, 274, 276, 310, 327, 333, 342–344, 348, 350, 357, 392

meritocracy, 9, 166, 169

Microsoft, 371

Mill, John Stuart, 206–209, 215, 246, 311

Millikan, Robert, 332

Molloy, Gerald, 262

Molyneux, Samuel, 85

Molyneux, William, 66, 67, 89, 181, 183

Moore, Michael, 72

Morawetz, Constance, 355

Murphy, Robert, 228

Murphy, Tim, 386

Nash, John, 357

National University of Ireland, 322, 333

natural theology, 117, 127, 144, 205, 320

Navier-Stokes equation, 397

navigation, 14, 30, 35, 37, 38, 59, 69, 123, 258, 266

neutron, 338, 340, 389, 391

Nevin, T.E., 349

Newman, John Henry, 318

Newton, Isaac, 14, 23, 29, 57, 63, 73, 75–77, 82, 83, 88, 90, 92, 127, 132, 184, 190, 194–196, 214, 300, 310, 311

Newtonianism, 17, 63, 73, 83, 84, 101, 106

Nicolas de Caritat, marquis de Condorcet, 128, 131

Nobel Prize, 24, 332, 334, 335, 341, 356, 357, 389, 390, 397

Nolan, P.J., 334

O'Ceallaigh, Cormac, 349

O'Connell, Daniel, 157, 163, 213

O'Neill, Eoin, 232, 384, 387, 388, 392

O'Toole, Annrai, 373, 388

optics, 20, 68, 69, 86, 87, 91, 123, 152, 177, 180–183, 188, 194, 201, 202, 204, 210, 247, 277, 284, 286, 299, 328, 357, 358, 378

Oxford Philosophical Society, 68

Oxford University, 27, 28, 95, 134, 147, 152, 160, 177, 185, 206, 279, 341

Parnell, Charles Stewart, 304, 317
Parsons, Charles, 384
Parsons, Sir William, 35, 293
Pauli, Wolfgang, 341, 349
Peacock, George, 227–231, 236, 245
Peardon, Mike, 395, 399
Petty, William, 14, 15, 27–30, 33, 34, 37, 38, 42–60, 64, 65, 67, 70, 71, 366
Phelan, William, 154, 155
Plato, 13, 176, 398
Playfair, John, 123, 124, 127, 132, 155, 158, 224, 228
Plummer, H.C., 325
Power, C.J., 50, 190, 334
Preece, William Henry, 262, 298–300
Presbyterians, 133, 317, 322
Preston, Thomas, 327, 329, 333, 334
Priestley, Joseph, 115, 129–131, 137, 162
Protestant Ascendancy, 18, 107, 133–135, 139, 144, 158, 165, 171, 213, 316
proton, 327, 335–340, 389–391, 393

Puritanism, 28, 30–32, 35, 40, 41, 320
Purser, Michael, 273, 367–369

quantum chromodynamics (QCD), 385, 389–396, 399–401, 405, 408
quantum electrodynamics (QED), 349, 389, 392
quark, 390–393, 396, 402–404
quaternions, 6–8, 11, 20, 21, 220, 221, 225, 236, 239, 240, 242–246, 257, 314, 344, 404
Queen's University Ireland, 18, 144, 168, 250, 268, 273, 285, 290, 293, 318, 319, 322, 328, 332, 333, 394

radiation, 329–331
radio, 22, 263, 298, 327, 361
radium, 330, 331, 336
Reid, Thomas, 86, 194–196
Reilly, Tony, 115, 175
relativity, 8, 23, 295, 314, 333, 342–344, 348, 349, 353, 354, 357, 385, 402
Reynolds, Hugh, 175, 365
Robinson, Bryan, 73, 76–79, 82, 196, 287, 297
Robinson, Thomas Romney, 146, 154, 199

Robison, John, 131, 132, 195, 196
Romanticism, 185, 206
Rowe, C.H., 356
Royal College of Science for Ireland, 283, 303, 318
Royal Irish Academy, 16, 110, 111, 113–115, 126, 134, 138, 145, 155, 170, 180, 192, 233, 317, 324, 327, 332, 358
Royal Society of London, 14, 29, 67, 68, 114, 264, 293
Royal University of Ireland, 311, 319, 322, 327, 329, 334
Russell, Thomas, 138, 162, 250
Rutherford, Ernest, 332, 333, 335, 336, 338, 339, 341
Ryan, Sinéad, 206–208, 392, 394, 395

Sabine, Edward, 265, 266
Salmon, George, 276, 303
Scaife, B.K.P., 152, 191, 264, 359
Schuster, Arthur, 340, 357
Science Foundation Ireland, 377, 406
Second Life, 364, 365, 393
Sexton, Jim, 393–397, 399
Shelley, Percy, 124, 254, 255
simulation, 392, 396, 398, 399, 401

Smith, Erasmus, 74, 137, 162, 185, 190, 191, 227, 243, 275–277, 317, 325, 344, 355, 356, 390
Smolin, Lee, 403
Smurfitt, Michael, 384
software, 26, 365, 368–371, 386, 388, 400
Solvay Conference, 341–343
Sommerfeld, Arnold, 353
spectroscopy, 327, 348, 360
Spinoza, Baruch, 83, 251
Stack, John, 105, 138
statistics, 28, 53, 371
Stevenson, Walter, 331
Stokes, George Gabriel, 291, 295, 367, 397
Stokes, Whitley, 138
Stoney, George Johnstone, 285, 289, 292, 293, 328
string theory, 385, 402–406, 408
Sullivan, William K., 79, 169
surveying, 14, 30, 35–37, 41, 44, 45, 48, 57, 59, 266
Susskind, Leonard, 402
Swift, Jonathan, 12, 13, 16, 54, 62, 70, 74, 84, 96–102, 106, 119, 120, 163, 317, 338
Symner, Miles, 35, 41, 65, 70
Synge, Edward, 82, 355

Synge, Edward Hutchinson, 355, 357, 358
Synge, John Lighton, 354–359, 367, 404
Synge, John Millington, 317, 355

Tait, Peter Guthrie, 5, 6, 183, 243, 275, 276
telegraph, 22, 114, 264, 272, 274, 277, 278, 280, 287, 298
theology, 15, 17, 40, 64, 90, 91, 93, 100, 103, 106, 107, 117, 127, 130, 144, 153, 154, 205, 303, 320
Thomson, J.J., 291, 332
Thomson, William (Lord Kelvin), 162, 183, 228, 243, 275, 277–280, 282, 287, 288, 291, 298, 300, 331, 383
Thrift, Billy, 325
tithe, 133, 134, 137, 158, 315
Toland, John, 15, 16, 63, 79–84, 86, 93, 133, 137
Tone, Wolfe, 102
Townsend, J.S.E., 326, 332, 333
Traill, Anthony, 282
Trinity College Dublin, 14, 16, 17, 19, 23, 35, 38, 41, 61, 62, 65, 66, 68, 74, 75, 84, 95, 101, 105, 112, 116, 117, 120, 124, 130, 136, 143, 152, 160, 163, 178, 189, 191, 204, 221, 225, 231, 243, 262, 263, 265, 268, 269, 276, 279, 282, 285, 293, 307, 310, 311, 319, 320, 323, 326, 332, 333, 335, 340, 343, 348, 351, 354, 355, 361, 365, 367, 368, 381–383, 386, 387, 393, 396, 405
triplets, 6, 235, 236, 238, 239, 242
Trouton, Frederick, 326

U2, 363, 364
Ulster, 27, 29, 42, 110, 116, 135, 137, 317, 324, 333
Unionism, 304
University College Dublin, 24, 152, 262, 318, 329, 333
University College London, 162, 273, 279, 326, 353
useful knowledge, 162, 175
Ussher, James, 40, 41
Utilitarianism, 129, 165, 205–208, 213

vector, 11, 243, 245
vision, 8, 9, 23, 33, 34, 37, 64, 65, 68, 71, 86, 88, 102, 183, 252, 264, 270, 305, 308, 315, 407

Wallis, John, 36, 91
Walton, E.T.S., 335–342, 349, 357, 360, 361, 389, 409

Watson, James, 353, 395
Weaire, Denis, 383
Weber, Wilhelm, 265, 281
Weingarten, Don, 394
Whewell, William, 154, 156, 161, 185, 197–199, 201, 206, 208, 211, 215
Whitehouse, E.O. Wildman, 278, 298
Whittaker, E.T., 190, 220, 236, 343, 344, 347, 350, 357
Wilczek, Frank, 390, 391, 393
Winthrop, John, 40
wireless telegraphy, 22, 260, 298, 299, 333
Witten, Edward, 404
Woodhouse, Robert, 228, 229, 245
Woodward, Richard, 134–136
Wordsworth, William, 20, 173–177, 186, 206, 208, 211, 215
Worsley, Benjamin, 34, 35

x-ray, 263, 327, 330, 333, 337

Yeats, William Butler, 16, 64, 102, 134, 317
Young, Thomas, 191, 196

Zeeman, Pieter, 327, 329, 334

Made in United States
Orlando, FL
07 June 2022